PEROVSKITE SOLAR CELLS
Technology and Practices

PEROVSKITE SOLAR CELLS
Technology and Practices

Kunwu Fu, PhD
Anita Wing Yi Ho-Baillie, PhD
Hemant Kumar Mulmudi, PhD
Pham Thi Thu Trang, PhD

AAP | APPLE ACADEMIC PRESS

Apple Academic Press Inc.	Apple Academic Press Inc.
3333 Mistwell Crescent	1265 Goldenrod Circle NE
Oakville, ON L6L 0A2	Palm Bay, Florida 32905
Canada USA	USA

© 2019 by Apple Academic Press, Inc.

First issued in paperback 2021

Exclusive worldwide distribution by CRC Press, a member of Taylor & Francis Group

No claim to original U.S. Government works

ISBN 13: 978-1-77463-411-0 (pbk)
ISBN 13: 978-1-77188-728-1 (hbk)

Library and Archives Canada Cataloguing in Publication

Title: Perovskite solar cells : technology and practices / Kunwu Fu, Anita Wing Yi Ho-Baillie, Hemant Kumar Mulmudi, Pham Thi Thu Trang.

Names: Fu, Kunwu, author | Anita Wing Yi Ho-Baillie, author | Hemant Kumar Mulmudi, author | Pham Thi Thu Trang, author

Description: Includes bibliographical references and index.

Identifiers: Canadiana (print) 20190064471 | Canadiana (ebook) 2019006448X | ISBN 9781771887281 (hardcover) | ISBN 9780429469749 (PDF)

Subjects: LCSH: Solar cells—Materials. | LCSH: Photovoltaic cells—Materials. | LCSH: Perovskite.

Classification: LCC TK2960. F8 2019 | DDC 621.31/244—dc23

CIP data on file with US Library of Congress

Apple Academic Press also publishes its books in a variety of electronic formats. Some content that appears in print may not be available in electronic format. For information about Apple Academic Press products, visit our website at **www.appleacademicpress.com** and the CRC Press website at **www.crcpress.com**

ABOUT THE AUTHORS

Kunwu Fu, PhD, is now working as an Application Scientist at Bruker (Beijing) Scientific Technology Co., Ltd. He received his PhD in 2016 at the School of Materials Science and Engineering, Nanyang Technological University in Singapore in organo-halide perovskite materials in mesoscopic solar cells after earning his bachelor degree at the same school, studying the morphological and electrical properties of methylammonium lead iodide perovskite in the mesoscopic solar cells. His research focused on understanding the varying nanostructures of perovskite materials and its implications to high-efficiency solar cell device. It also extended to exploring novel small molecule organic materials for hole-transporting layer in high-efficiency perovskite solar cells.

Anita Wing Yi Ho-Baillie, PhD, is an Associate Professor at the University of New South Wales (UNSW), Australia. Her research interests in the field of photovoltaics include high-efficiency silicon solar cells, tandem solar cells, perovskite solar cells, integration of photovoltaics for a wide range of applications, and manufacturing cost analysis. She has been leading the perovskite solar cell research group at UNSW since 2013 and in 2016 announced the energy conversion efficiency records for the largest certified monolithic perovskite solar cells. She obtained her PhD at UNSW in 2005.

Hemant Kumar Mulmudi, PhD, is currently working as a research fellow at the Research School of Engineering, Australian National University, on the processing and characterization of perovskite solar cells. This encompasses investigating the phase segregation and the quantum yield of perovskite films for application in silicon-perovskite tandem solar cells. This involves advanced concepts of particle-wave-matter interaction at the nanoscale regime. His interests span from synthesizing nanomaterials via inexpensive solution-processed methods to integrating them in functional solar photovoltaic devices and fuel cells. He was the first to demonstrate stable and functional inorganic material-based lead-free perovskite solar cell (cesium tin iodide) as photon absorber using solution-processed methods. His inquisitive nature drives him to go beyond to understand the underlying physics and modus operandi of these devices, which has been crucial in expanding this technology to other applications and configurations. Owing to his passion and drive, he led a group of PhD students and postgraduate staff to develop lead-free alternative materials for energy harvesting applications at the Energy Research Institute @NTU, leading to several publications. His contributions to the field of materials science and engineering have been communicated to several high-impact refereed journals. He also believes in scientific outreach and is passionate about expressing his views and scientific findings at various conferences, workshops, and guest lectures at universities. He loves to teach and has taken every opportunity to teach about solar cells and their global imprint to high school and undergraduate students respectively. He loves innovative approaches in device engineering that minimize the processing steps in building complex systems leading to enhancements in photovoltaic parameters of a solar cell. With a PhD in materials science from the Nanyang Technological University, Singapore, an MS in nanotechnology from the Ulm University, Germany, and a bachelor's degree in metallurgical and materials engineering from the Indian Institute of Technology-Roorkee, India, he is highly knowledgeable and proficient in his subject.

Pham Thi Thu Trang, PhD, is currently working as R&D director at IREX Energy Joint stock company, a member of SolarBK, Vietnamese's leading green energy utility company. Previously, she was a lecturer at the University of Engineering and Technology, Vietnam National University, Hanoi. Her research varies from chemical synthesis of nanostructured materials for dye-sensitized solar cells and perovskite solar cells and device fabrication and characterization to performance enhancement for solar modules in solar farm, rooftop, and BIPV applications. She obtained her PhD degree from Nanyang Technological University (Singapore) and completed her postdocs at the Centre National de la Recherche Scientifique, France.

CONTENTS

CONTRIBUTORS

Kunwu Fu
NanoSurface Division, Bruker (Beijing) Scientific Technology Co., Ltd, Beijing, China
E-mail: kunwu.fu@gmail.com

Anita Wing Yi Ho-Baillie
School of Photovoltaic and Renewable Energy Engineering, University of New South Wales, New South Wales, 2052 Australia
E-mail: a.ho-baillie@unsw.edu.au

Hemant Kumar Mulmudi
College of Engineering and Computer Science, Australian National University, Australian Capital Territory, 2600 Australia
E-mail: hemant.mulmudi@anu.edu.au

Pham Thi Thu Trang
Bach Khoa Investment & Development of Solar energy corporation, No. 47, Le Van Thinh Street, Quarter 5, BrinhTrung Dong ward, District 2, HCMC, Vietnam
E-mail: sandy.phamtrang@gmail.com

ABBREVIATIONS

2D	two-dimensional
3D	three-dimensional
AFM	atomic force microscopy
ALD	atomic layer deposition
APCE	absorbed photon-to-current conversion efficiency
AZO	aluminum-doped zinc oxide
BAFB	benzoic acid fullerene bis-adducts
BCP	bathocuproine
BDT	benzodithiophene
CB	conduction band
CCE	carbon counter electrode
CE	counter electrode
CID	conventional interdiffusion
CL	cathodoluminescence
CoO	cost of ownership
DMF	dimethyl formamide
DMSO	dimethyl sulfoxide
DSSCs	dye-sensitized solar cells
EIS	electrochemical impedance spectroscopy
EPBT	energy payback time
EQE	external quantum efficiency
ETA	extremely thin absorber
ETL	electron transporting layer
ETM	electron transporting material
EVA	ethylene vinyl acetate
FA	formamidinium
FESEM	field emission scanning electron microscope
FF	fill factor
FTO	fluorine-doped tin oxide
GO	graphene oxide
GWP	global warming potential
HI (Chemical name)	hydroiodic acid

Abbreviations

HI	hysteresis index
HJ	heterojunction
HOMO	highest occupied molecular orbital
HTL	hole-transporting layer
HTM	hole transport materials
IEC	International Electrochemical Commission
IPCE	incident photon-to-current conversion efficiency
IQE	internal quantum efficiency
iQY	internal photoluminescence quantum yield
IR	infrared
ITO	tin doped indium oxide
JSC	short-circuit photo-current density
LCA	life cycle assessment/life cycle analysis
LCOE	levelizedcost of electricity
LED	light-emitting diode
LHE	light-harvesting efficiency
Li-TFSI	lithium bis (trifluoromethanesulfonyl) imide
LUMO	lowest unoccupied molecular orbital
MA	methylammonium
mp	mesoporous
MPP	maximum power point
MSP	minimum sustainable price
MWNT	multi-walled carbon nanotube
NF	nanofibers
NIMS	National Institute For Materials Science, Tsukuba, Japan
NP	nanoparticle
NREL	National Renewable Energy Laboratory, Colorado, U.S.A
NRs	nanorods
NS	nanosheet
OMC	ordered mesoporous carbon
P3HT	poly (3-hexylthiophene)
PCE	power-conversion efficiency
PDS	photothermal deflection spectroscopy
PEI	polyethylenimine
PET	polyethylene terephthalate
PIB	polyisobutylene
PL	photoluminescence
POZ	phenoxazine
PSC	perovskite solar cells

PTAA	bis (4-phenyl)(2,4,6-trimethylphenyl)-amine
PT	polythiophene
PVT	physical vapor transport
RCP	random copolymer
rGO	reduced graphene oxide
RH	relative humidity
SAED	selective area electron diffraction
S-AMI	solvent-assisted molecule inserting
SEM	scanning electron microscopy
SJTU	Shanghai Jiangtong University
SWCNTs	single-walled carbon nanotubes
TCSPC	time-correlated single photon counting
TEM	transmission electron microscopy
TFs	thin films
TPA	triphenylamine
TTIP	titanium tetraisopropoxide
UNSW	University of New South Wales
UV	ultraviolet
UV-vis	ultraviolet–visible
VBM	valence band maximum
VBs	valence bands
VOC	open-circuit voltage
WF	work function
XPS	x-ray photoelectron spectroscopy
XRD	x-ray diffraction

PREFACE

The global demand for low-cost, environmentally benign and efficient energy is increasing rapidly. As responsible inhabitants of the earth and as scientists, it is pertinent for us to think beyond the conventional energy resources. Renewable energy is promising, and with meticulous and conscientious efforts, the world's energy demands can be fulfilled without affecting the socioeconomic status or the environment. Solar energy is the most powerful source of renewable energy, and researchers have been trying to harness it for ages, giving rise to many innovative approaches and new solar technologies. Recently, a new class of solar cells based on perovskite crystal structure materials has shown remarkable results in a very short time, owing to very desirable photovoltaic-enabling properties, ease of fabrication resulting in intensive research, which benefitted from the earlier work and expertise in dye-sensitized solar cells. Since the first demonstration, perovskite solar cells have emerged to be the front-runners for low-cost, high-performance solar cells, generating fundamental research interests, and even commercial prospects. This book aims to provide comprehensive knowledge on the properties, syntheses, and strategies to acquire high efficiencies, use of alternative materials, tandem applications, life cycle assessments, and commercial aspects of perovskite solar cells. This book will be highly beneficial to a broad spectrum of engineers, and scientists pertaining to the field of material science, physics, and chemistry. The chapters in this book were compiled carefully to meet the requirements of readers from diverse backgrounds. This book is inspired by various works of different researchers in the field of photovoltaics and materials science.

The authors would like to thank Apple Academic Press for providing the opportunity and time required to publish this book. The resources provided by the Australian Centre for Advanced Photovoltaics (ACAP) through the Australian Renewable Energy Agency, supported by the Australian Government, are highly appreciated. Hemant would like to thank the support of his colleagues (Klaus Weber, Sachin Surve, Mark Lockrey, Fiacre Rogeruix, Mark Saunders, Chog Berahujin, The Duong, Marco Ernst, Heping Shen) at the Australian National University, Canberra, and his wife, Dharani for their constant support and encouragement to complete this book. Anita Wing Yi

Ho-Baillie would like to acknowledge the support provided by the School of Photovoltaic and Renewable Energy Engineering at the University of New South Wales and by her family.

PART I

INTRODUCTION TO PEROVSKITE SOLAR CELLS

ANITA WING YI HO-BAILLIE, HEMANT KUMAR MULMUDI, and PHAM THI THU TRANG

CHAPTER 1

PEROVSKITES THIN FILMS FOR PHOTOVOLTAIC APPLICATIONS

1.1 INTRODUCTION

Organic metal halide perovskite solar cells rose to prominence since the first demonstration in solid-state form with energy conversion efficiency of 9.7% in mid-2012.[1] The history of metal halide perovskite can be tracked back to 1839 when the distinctive crystal structure was named.[2] The first reports of organic–inorganic halide perovskites appeared in 1884 and 1892 followed by reports of methyl-ammonium lead halide ($CH_3NH_3PbI_3$) and formamidinium lead halide ($HC(NH_2)_2PbI_3$) in 1978 and in 1995 which are now commonly used in state of the perovskite solar cells.[3,4] Studies of the materials were focused on investigating their prospects for transistors and light-emitting diodes in the 1990s until 2006 when the connection between perovskite and solar cells was publicized in a conference presentation in Japan.[5–11] They have gained tremendous interest over the past few years due to their high efficiency, ease of fabrication, and cost-effectiveness.[12–17] Since then very rapid improvements have led to the many demonstrations of independently certified cell results making this cell technology the fastest solar cell technology to date, with the certified efficiency of 23.7% (NREL Chart, Figure 1.1).[18]

The key attribute that enables such rapid progress is their strong optical absorption,[19–20] which is comparable with those of direct bandgap group III–V semiconductors. Although their carrier mobilities range from 10^1 $cm^2/V \cdot s$ (for thin film[21–23]) to 10^2 $cm^2/V \cdot s$ (for single crystal[24]), lower than those of Si (480 $cm^2/V \cdot s$ for hole and 1430 $cm^2/V \cdot s$ for electrons[25]) and GaAs (≤ 400 $cm^2/V \cdot s$ for hole and ≤ 8500 $cm^2/V \cdot s$ for electrons[26]), they are still three or more orders of magnitude higher than many organic electronic materials.[27] Together with sufficient carrier mobilities, long carrier lifetime (>200 ns[28] or even 1 μs[29]) and long diffusion lengths (up to ~ 1 μm[29]) make these organic metal halide perovskites favorable for photovoltaics.

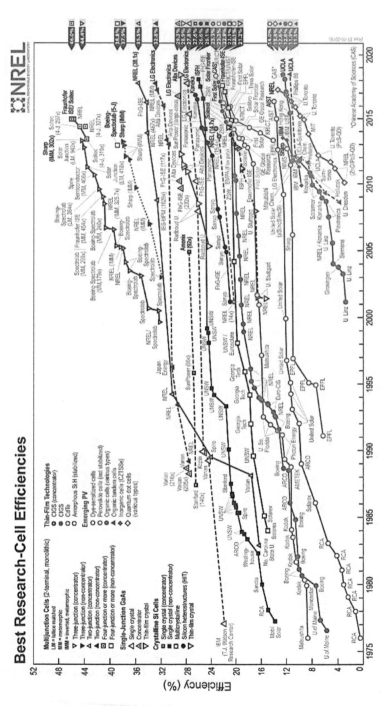

FIGURE 1.1 National Renewable Energy Laboratory (NREL) certified solar efficiency chart.
Source: NREL website, adapted from ref 18.

Figure 1.2 shows a rapid increase in the number of documents in the field of perovskite solar cells. These include an article, conference paper, and review, article in press and book chapters. The data were collected from Scopus which is a large database of peer-reviewed literature such as scientific journals, books, and conference proceedings. The steep growing bar graph from a broad spectrum of scientific community emphasizes the interest and importance of this topic.

Given the numerous advantages of perovskite, a clear understanding of the crystal structure will shed more light on the optical and electrical properties of these systems. An in-depth look is taken at the $CH_3NH_3PbI_3$ ($MAPbI_3$) system in this section. The physical parameters and transitions of phases of bulk $MAPbI_3$ can be found in the following references.[21,30] In this chapter, we focus on the tetragonal and cubic phases. In fact, there are no critical differences between the two phases, except a slight rotation of PbI_6 octahedra along the c axis. The atomic structures of $MAPbI_3$ of the two phases are shown in Figure 1.3. Thus, the tetragonal phase can be treated as a pseudo cubic phase with a*=a/, c*=c/2. Below 54°C, the cubic phase transforms into the tetragonal phase and by annealing at 100°C, the tetragonal phase transforms to the cubic phase.[5,31]

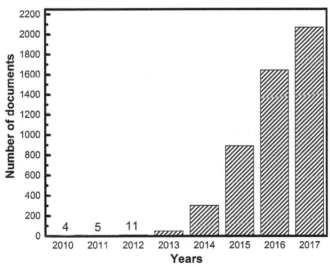

FIGURE 1.2 Bar graph showing number of documents per year with search key word "Perovskite solar cell" from Scopus.

Another phase which should be noted is the amorphous phase. Pair distribution function analysis of X-ray scattering showed that after annealing at

100°C for 30 min, the MAPbI$_3$ in mesoporous TiO$_2$ has about 30 atom% in medium range crystalline order and the other 70 atom% in a disordered state with a coherence length of 1.4 nm.[33] The poor crystallization of the MAPbI$_3$ in mesoporous TiO$_2$ was studied by high-resolution transmission electron microscopy.[34] Quartz crystal microbalance measurements suggest that during sequential method only half of PbI$_2$ is converted to MAPbI$_3$ instantly, while the other half is involved in reversible transformation with MAPbI$_3$. Additionally, the amorphous character with a very small average crystallite size may be present after the transformation.[35] MAPbI$_3$ is a direct bandgap (1.55 eV) material (Figure 1.3c) with an absorption coefficient in the order of $1.5 \times 10^4 cm^{-1}$ at 550 nm.[36]

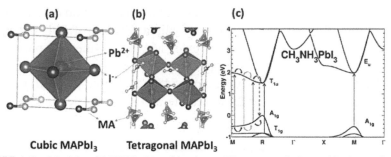

FIGURE 1.3   Models of MAPbI$_3$ (a) cubic phase, (b) tetragonal phase, (c) electronic band structure for the high-temperature cubic phase of MAPbI$_3$ with spin–orbit coupling at the local density approximation level of theory.
Source: Reprinted with permission from ref 32. © 2015 American Chemical Society.

There are several ways of depositing the perovskite layer which include single step and sequential deposition. In single step deposition, all the precursors are mixed together in an organic solvent such as dimethyl formamide (DMF) and are cast onto the substrate or chemical vapor deposition methods can be used to deposit the films. Single/dual source evaporation methods have also been employed to get perovskite films.[37,38] In a sequential deposition, PbI$_2$ is first deposited on the substrate and is later converted into perovskite phase. There are several ways in which perovskite films can be obtained from lead halide films which are described in the following sections.

1.2   TWO-STEP/SEQUENTIAL DEPOSITION

The first demonstration of a sequential method to form perovskite film on a metal oxide scaffold was reported by Burschka et al.[39] The PbI$_2$ was first introduced into the TiO_2 scaffold via spin coating from an organic solvent. These films are annealed at 70°C and then dipped in a CH_3NH_3I solution in isopropanol.[39] This method provides a uniform coverage on the TiO_2 scaffold and also high reproducibility in the power conversion efficiencies (>15%) was achieved. This section focuses on some of the topics related to sequential deposition and its modified versions. Also, the physical and optical properties of perovskite films produced via these methods are discussed.

A study by Pei et al. has investigated the effect of substrate heating before depositing PbI_2.[40] The group investigated the influence of the different preheating substrate temperatures (unheated, 30, 50, 70, and 100°C) on the perovskite film layers before spin coating PbI_2 as shown in Figure 1.4.

As the preheating substrate temperature increased, there is a severe change in the appearance of the films (Figure 1.4). By comparison, the PbI_2 film without preheating substrate does not completely cover the TiO_2 substrate (Figure 1.4a), while the PbI_2 film of preheating substrate has better coverage (Figure 1.4b). At the same time, the morphology of PbI_2 formed at 50°C (Figure 1.4c) is extremely uniform with complete coverage, which is relatively similar in appearance to that formed at 70°C (Figure 1.4d). Furthermore, the gaps between the PbI_2 crystallites are slightly small in the case of 70°C. Increasing the preheating substrate temperature up to 100°C, the morphology of PbI_2 is relatively homogeneous, but there are still areas of exposed TiO_2 nanoparticle on the surface (Figure 1.4e). The reasons for the different PbI_2 morphology are that the crystallization rate of PbI_2 becomes higher than the solvent evaporation rate as the preheating substrate temperature increases. Therefore, the preheating substrate is beneficial to promote the crystallization rate of PbI_2 and obtaining more uniform temperature (Figure 1.4f–j), and denser film morphology. In addition, after forming the $MAPbI_3$ films, the cuboids of $MAPbI_3$ crystals are closely packed as the preheating substrate temperature increases. It is noted that the $MAPbI_3$ grain size increases with increase in the preheating which is related to the conversion rate of PbI_2 to $MAPbI_3$. In other words, the coverage rate of the PbI_2 film and the perovskite film become better as the preheating temperature increases. Therefore, it avoids the direct contact between the electron transport layer and the hole transport layer, reducing the chances of recombination of photo-generated electrons and holes.

To further study the perovskite crystallinity dependence on the preheating temperature, X-ray diffraction (XRD) analysis of PbI_2 films was performed (Figure 1.5). From the XRD spectrum, the PbI_2 peak at 12.67° corresponds to the (001) lattice plane. As a comparison, the full width at half maximum of PbI_2 films becomes narrow and the PbI_2 peak intensity increases with increase in preheating substrate temperature. The results indicate that the peak intensity is higher, and the crystallinity of PbI_2 film is better, as the preheating temperature increases. The characteristics of the $MAPbI_3$ films formed were also measured by XRD (Figure 1.5). As shown in Figure 1.5, the $MAPbI_3$ peak at 14.19° is assigned to the (110) lattice plane. By comparison, as the preheating substrate temperature increases, the intensity of the $MAPbI_3$ (110) peak increases, which is attributed to increasing the conversion rate of perovskite. Moreover, the PbI_2 peak appears, and the peak intensity increases due to the enhanced crystallinity of PbI_2. The results lead to the residual PbI_2 conversion fully into the $MAPbI_3$ for a long time. But when the preheating substrate temperature is up to 70°C, the peak for the PbI_2 is absent and only the peak for the perovskite material remains.

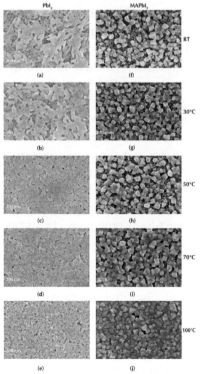

FIGURE 1.4   Top-view scanning electron microscope (SEM) images of the PbI_2 layers deposited on the mp-TiO_2 coated fluorine-doped tin oxide (FTO) substrate, where the substrate was preheated at (a) RT (without preheating), (b) 30°C, (c) 50°C, (d) 70°C, and (e) 100°C. The $MAPbI_3$ thin film layers (f–j) correspond to the PbI_2 layers deposited on the different preheating substrate.

Source: Reprinted with permission from ref 40. https://creativecommons.org/licenses/by/4.0/

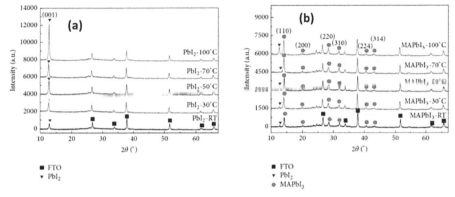

FIGURE 1.5 (a) X-ray diffraction (XRD) patterns of PbI$_2$ films deposited on the different preheating substrate temperate of room temperature, 30, 50, 70, and 100°C, respectively. (b) XRD patterns of MAPbI$_3$ films, where MAI was deposited on the top of the PbI$_2$ layers. The MAPbI$_3$ films depended on the different preheating substrate temperature of room temperature, 30, 50, 70, and 100°C, respectively.
Source: Reprinted with permission from ref 40. https://creativecommons.org/licenses/by/4.0/.

Patel et al. have also studied the formation dynamics of conventional CH$_3$NH$_3$PbI$_3$ perovskite by two-step thermal evaporation method.[41] In this study, the room temperature transformation as methyl ammonium iodide (MAI) enters into PbI$_2$ was studied in both ambient air and vacuum. Figure 1.6 shows the dynamic evolution of absorption features of MAI/PbI$_2$ bilayers following the commencement of interdiffusion in a vacuum and subsequent exposure to ambient air. When the bilayers are in a vacuum, the most dramatic changes occur in the visible part of the absorption region (Figure 1.6a). Within the first 2 h, the PbI$_2$ absorbance band edge (520 nm) disappears, and the MAPbI$_3$ band edge begins to establish itself near 750 nm. These changes indicate that the MAI slowly diffuses across the PbI$_2$ layer and completely disrupts the layered arrays of I−Pb−I. As a result, the formation of the three-dimensional (3D) MAPbI$_3$ perovskite structure is initiated. Even after the disappearance of PbI$_2$, however, the MAPbI$_3$ band edge continues to sharpen and redshift and overall absorption increases above the band edge. After 24 h, the MAPbI$_3$ band edge is clearly defined, and no further spectral changes are observed.

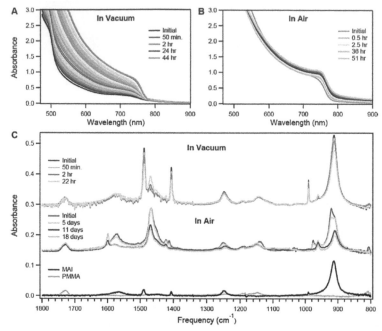

FIGURE 1.6   Visible and IR (infrared) transmission spectra of an initially deposited bilayer MAI/PbI$_2$ as a function of time. Spectra were taken immediately following heating to room temperature (21 ± 1°C), which initiates the formation of MAPbI$_3$ from MAI and PbI$_2$. (a) Changes in absorbance in the visible region for films under vacuum, taken for the initial bilayer MAI/PbI$_2$ (darkest red line), through the progression to a nascent MAPbI$_3$ perovskite structure over a period of 44 h (lightest blue line). (b) Changes in absorbance in the visible region following the subsequent exposure of the film to ambient air. (c) IR transmission spectra for the bilayer MAI/PbI$_2$ (red line), the nascent MAPbI$_3$ perovskite structure formed in a vacuum (light green line), and the more crystalline MAPbI$_3$ following exposure of the film to ambient air (blue line).

Source: Reprinted with permission from ref 41. © 2015 American Chemical Society. https:// creativecommons.org/licenses/by/4.0/.

While changes in the content of PbI$_2$ and MAPbI$_3$ leave clear features in the ultraviolet–visible (UV-vis) absorption spectra, the presence of MAI cannot be as easily inferred because it does not have absorption features in this spectral region. Therefore, Fourier resonance infrared (IR) spectroscopy was employed (Figure 1.6c), which provides valuable insight into the different chemical environments MAI experiences inside the 3D perovskite cage through the vibrational modes associated with the CH$_3$NH$_3^+$ (MA$^+$) bonds. Initially, the spectrum closely resembles that of the evaporated film of MAI, and little change is observed in the first 50 min. After 50 min, however, the incorporation of MAI into the MAPbI$_3$ structure becomes evident from changes in the IR signatures.

From visible and IR transmission spectra it is clear that, in vacuum, MAI and PbI$_2$ interdiffuse to form a nascent MAPbI$_3$ perovskite structure. However, ambient air is required to fully crystallize MAPbI$_3$ and remove excess MAI. The hygroscopic nature of MAI allows it to absorb moisture, making it more mobile and allowing for better intermixing of PbI$_2$ and MAI. It was, however, found to be essential that no excess MAI remains within the film, as the unreacted MAI absorbs moisture to form the dihydrate $(CH_3NH_3)_4PbI_6 \cdot 2H_2O$, which is a relatively stable crystalline material in itself.

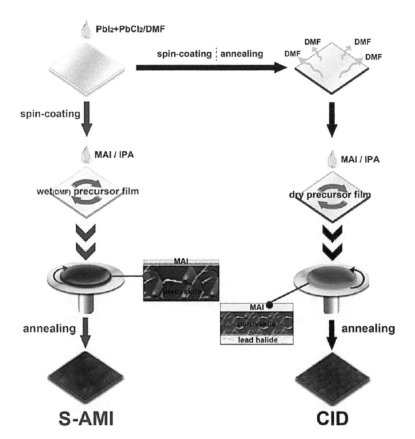

FIGURE 1.7 The scheme showing the steps involved in making the perovskite films prepared by Yuan et al. CID: conventional interdiffusion and S-AMI: solvent assisted molecule inserting strategy.

Source: Reprinted with permission from ref 42. © 2016 American Chemical Society.

Yuan et al. used a solvent-assisted molecule inserting (S-AMI) technique to grow high-quality perovskite.[42] The scheme used to make the films is shown in Figure 1.7. It was found that the retained DMF aids in inserting the MAI molecules in the lead iodide films and provides an environment for grain growth of perovskite films. The starting precursor was a mixture of PbI_2 and $PbCl_2$ mixed in various ratios in DMF. The highest power conversion efficiency obtained by the S-AMI method was 18.3% with fill factor over 80%.

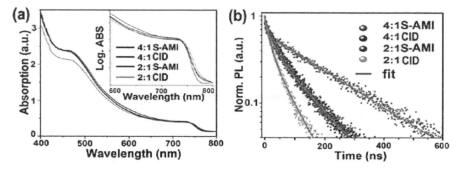

FIGURE 1.8 (a) Ultraviolet–visible (UV–vis) absorption spectra of different perovskite films synthesized by S-AMI or CID with varied precursor composition; the inset shows the absorption curves with absorption intensity in logarithm scale. (b) time-resolved photoluminescence (PL) spectra of the corresponding perovskite films, where the Y-axis symbolizes the normalized PL intensity. A 509.6 nm pulsed diode laser (pulse width: 148.8 ps) with a fluence of ~ 30 nJ/cm^2 was used to excite the samples from the glass substrate side. *Source:* Reprinted with permission from ref 42. © 2015 American Chemical Society.

The optical properties of the films prepared by both conventional inter-diffusion (CID) and S-AMI were tested and it was found that the absorption onset stays same for all the films prepared at 770 nm (Figure 1.8a). The absorption intensity though is higher for the films prepared by the S-AMI technique as compared to the CID method. The time-resolved photoluminescence (PL) spectroscopy also performed on these films (Figure 1.8b). The perovskite films deposited by S-AMI have longer average lifetimes (110 ns for the S-AMI 2:1 device and 212 ns for the S-AMI 4:1 device) than those (61 ns for the CID 2:1 device and 90 ns for the CID 4:1 device) of CID-grown samples.

Another two-step modified method was demonstrated by Mao et al. where PbI_2 films were cast from a solution of DMF into which dimethyl sulfoxide

(DMSO) was introduced as an additive.[43] It was then reacted with MAI which was dissolved in a mixture of ethanol and isopropanol. The highest power conversion efficiency obtained by this method was 15.76% with a fill factor of 77%. It was found that having 8% DMSO in the PbI_2–DMF solution enhanced the photovoltaic parameters of devices. The reaction rate of the formation of $MAPbI_3$ is retarded due to the presence of DMSO. One of the merits of this reaction rate-retarded method is that it hinders the uncontrolled morphology of perovskite without a metal oxide scaffold. But this could lead to unreacted PbI_2 which are in the underneath layers. It was found that traces of unreacted PbI_2 was found when 8% DMSO was used in the precursor solution (Figure 1.9a and b). When DMSO was not used in the precursor solution, there was no unconverted PbI_2 left as seen in Figure 1.9a.

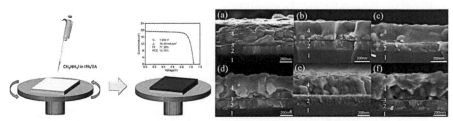

FIGURE 1.9 Cross-sectional images o of $CH_3NH_3PbI_3$ perovskite deposited on glass/ITO/PEDOT: PSS substrates. (a) PbI_2 (in dimethyl formamide [DMF]) and MAI (in IPA) as precursor solutions. PbI_2 dissolved in DMF with 8% DMSO and MAI dissolved in (b) IPA, (c) 25% EA, (d) 50% EA, (e) 75% EA, and (f) pure EA as precursor solutions. Layer 1, 2, 3, and 4 refer to glass, ITO, PEDOT: PSS, and perovskite layers, respectively. Layer 5 in panel b refers to unconverted PbI_2.
Source: Reprinted with permission from ref 43. © 2016 American Chemical Society.

To study the impact of ethanol/IPA mixtures on perovskite morphology, scanning electron microscope (SEM) was performed on these samples. It was found that only a mixed solvent containing 25% of ethanol could give a flat, homogeneous, and pin hole-free films (Figure 1.9c). Using pure IPA as a solvent yielded nonhomogeneous films with grain boundaries. Uneven films with pin holes were observed when ethanol was more than 50% (Figure 1.9d–f). Such mixed solvent methods can be handy to produce smooth, pin hole-free, and homogeneous films.

Cui et al. demonstrated preparation of four different color-tuned perovskites by direct contact of the lead halide films with a heated methylamine halide powder which works by intercalation.[44] To deposit the lead halide films, the group employed PbI_2, PbI_2/$PbBr_2$ mixtures, and $PbBr_2$

dissolved in DMF. The schematic for the preparation of the perovskite films is shown in Figure 1.10a. $MAPbI_3$ has a tetragonal crystal structure while $MAPbBr_3$ has a cubic structure. The XRD pattern in the window between 28° and 31° for the four different perovskites films ($MAPbBr_xI_{3-x}$ ($x=0$, 1, 2, 3)) prepared in this study are shown in Figure 1.10b. As the bromine content increases, the peak shifts to higher angles. If the tetragonal structure is assumed to be a pseudocubic structure, the (220) plane of the tetragonal $MAPbI_3$ can be treated as a (200) plane of pseudocubic $MAPbI_3$ structure. The extracted lattice parameters decrease linearly with an increase in bromine content which is in good agreement with the Vegard's law. Figure 1.10c shows the photographs of different halide perovskite films prepared in this study. It is clearly evident that as the bromine content increases the color of the perovskite changes from black to yellow. This shows that the method employed via intercalation can result in perovskite films with varying colors. Optical properties of these films were also measured to get an estimate of the bandgap. The onset of the absorption for the pure $MAPbI_3$ films was found to be at 786 nm, while that of the pure bromide film was found to be at 542 nm. A linear dependence was observed as the amount of bromine increases in the film as shown in the figure and a mathematical fit was also used to get the fitting parameters (inset). The power conversion efficiencies of champion photovoltaic devices for the pure iodide films was found to be 12.76% while that for the pure bromide films was found to be 3.53%. Such color tunability can be employed for building facades and useful for building integrated photovoltaics.

Sutter-Fella et al. took a two-step low-pressure vapor-assisted solution process to grow high-quality homogeneous mixed-halide perovskites.[45] The intrinsic optoelectronic property of a semiconductor is judged by its internal PL quantum yield (iQY), which is defined by the number of photons which are emitted to the number of photons which are absorbed (taking into account for the correction required for refractive index). The perovskite films with varying amounts of bromine were prepared using the schematic shown in Figure 1.11a. The mixed-halide solutions were spun cast on the glass substrates and annealed at 110°C and were later treated with the vapors of methylamine halides to get the desired composition under low pressure (~0.4 Torr at 120°C for 2 h). With the incorporation of bromine the optical bandgap of the perovskites widened as discussed previously. To iden- tify the recombination regimes in the perovskites, pump-probe-dependent steady state PL was performed on the samples as shown in Figure 1.11b. The generation rate on the x-axis was obtained by dividing the excitation intensity with the film thickness. For the mixed-halide samples, the power law fits has two regimes for the mixed-halide samples. The curves follow a dependence $\propto G^{1.7}$ up to a generation rate of 3×10^{22} $s^{-1} \cdot cm^3$, and for generation rates above 3×10^{22} $s^{-1} \cdot cm^3$ the luminescence intensity is $\propto G^{0.9}$. Power law dependences of more than one are indicative of Shockley-Read-Hall like recombination, that is intra-gap trap-assisted recombination instead of direct electron−hole recombination (bimolecular recombination). The pure bromide film shows a linear relationship with $G^{1.8}$ which is indicative of monomolecular recombination. The iQY for the samples increases with the increase in pump power (Figure 1.11b). This is an indication of trap states leading to low iQY at low optical injection levels. At higher optical injection levels, Auger recombination kicks in. All the samples investigated in this study have remarkably high iQY of more than 30% as seen in Figure 1.11c and the peak iQY varies with the halide addition ()bromine with phase segregation becoming prominent as the bromine content increases (x>1.5, Figure 1.11d).

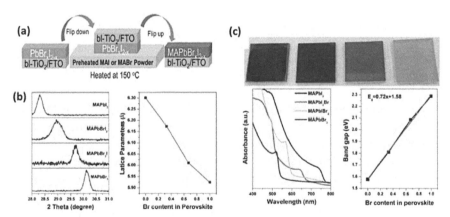

FIGURE 1.10   (a) Schematic illustration of the method employed by Cui et al. to prepare the perovskite films, (b) XRD patterns of mixed lead halide perovskite films and the extracted lattice parameters with respect to the bromine concentration, (c) photos of the films prepared with increasing bromine content and the optical properties measured by absorbance and the linear relationship between the optical bandgap and the bromine content in the films. *Source:* Reprinted with permission from ref 44. © 2016 American Chemical Society.

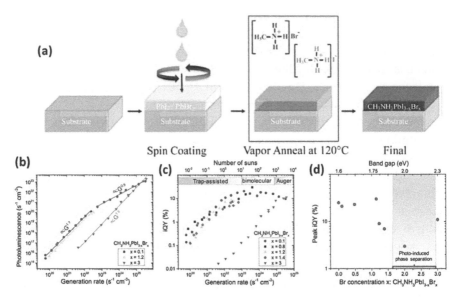

FIGURE 1.11   (a) Schematic illustration of the method employed by Sutter-Fella et al. to prepare the perovskite films, (b) pump-power dependence of steady-state integrated PL, (c) power dependence of iQY for various samples, (d) iQY versus the sample composition with the bandgap.

Source: Reprinted with permission from ref 45. © 2016 American Chemical Society.

1.3   SINGLE-STEP DEPOSITION

There are several ways by which perovskite films have been deposited by the single-step method. These include solution processed, chemical vapor deposition, and dual source evaporation. This section describes some of these methods employed and their physical and optical properties. Snaith et al. used a precursor consisting of $PbCl_2$ and MAI to get the perovskite ($CH_3NH_3PbI_{3-x}Cl_x$) films and demonstrated a 10.9% single junction solar cell.[17]

The role of chlorine has been tentatively attributed to the doping effect. Yantara et al. studied the procedures and compositions commonly utilized in the formation of the single step $CH_3NH_3PbI_3$ and $CH_3NH_3PbI_{3-x}Cl_x$.[46] It was found that $CH_3NH_3PbI_3$ films turn black immediately upon drying at 70°C, while the $CH_3NH_3PbI_{3-x}Cl_x$ films take 30 mins of drying at 100°C. In fact, these films have a color evolution from brown to yellow to attain a final black color. From, the top-view SEM image (Figure 1.12a) it is clear that the $CH_3NH_3PbI_{3-x}Cl_x$ film forms a pin hole-free, smooth film over mesoporous TiO_2, whereas the $CH_3NH_3PbI_3$ film has fibrous structure with plenty of pin holes. From the XRD pattern (Figure 1.12b), it is clearly evident that the $CH_3NH_3PbI_3$ films form polycrystalline tetragonal film with no preferred orientation, while the $CH_3NH_3PbI_{3-x}Cl_x$ film has a preferred orientation along the (110) direction. Both the films exhibit an optical bandgap of 1.55 eV (Figure 1.12c) with identical features at 480 and 760 nm which are signatures of the transitions from the valence band to the conduction band. Figure 1.12d shows the compositional evolution of $CH_3NH_3PbI_{3-x}Cl_x$ films monitored by XRD. The spin-coated sample is brown in color as shown in the inset and was found to have the following phases: $CH_3NH_3PbI_3/CH_3NH_3PbI_{3-x}Cl_x$, PbI_2, $CH_3NH_3PbCl_3$, and traces of CH_3NH_3Cl. After 10 min of annealing at 100°C, the color of the film changes to yellow. It was found to consist of CH_3NH_3Cl and $CH_3NH_3PbCl_3$. Further annealing, the color of the film transforms into black. The phases identified were $CH_3NH_3PbI_3/CH_3NH_3PbI_{3-x}Cl_x$ with a preferred growth direction of the (110) plane. It was concluded that due to better film morphology with minimum effect on the series resistance of the device, $CH_3NH_3PbI_3$-based photovoltaic devices made with excess CH_3NH_3Cl or with $PbCl_2$ precursors showed better performance than the PbI_2 counterpart when deposited by a single-step procedure.

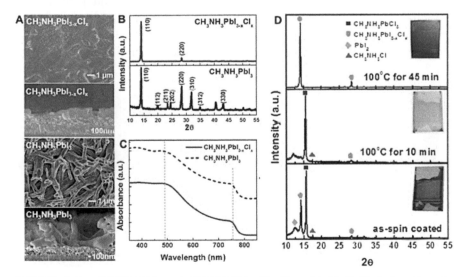

FIGURE 1.12 $CH_3NH_3PbI_3$ and $CH_3NH_3PbI_{3-x}Cl_x$ films (a) Field Emission Scanning Electron Microscope (FESEM) images both top view and cross-section images, (b) XRD patterns, (c) absorption spectrum of the films, (d) evolution of film composition as gleaned through XRD measurements of the $CH_3NH_3PbI_{3-x}Cl_x$ film. The insets show the optical images of the film.

Source: Reprinted with permission from ref 46. © 2016 American Chemical Society

In another study by Ma et al. pure inorganic mixed-halide perovskite were investigated by dual source evaporation.[37] Figure 1.13a shows the XRD patterns of the CsPbIBr$_2$ films deposited via dual source thermal evaporation on c-TiO2/ fluorine-doped tin oxide (FTO) glass substrates at 20 or 75°C followed by post-annealing at 100 or 250°C for 10 min. The reflections match very well with the orthorhombic phase of CsPbIBr$_2$. Figure 1.13b show morphological evolution of perovskite films deposited on the blocking layer before and after annealing. The deposition temperature of 75°C gives larger grains than deposition at 20°C. The grain size of films deposited at 75°C increase to 500–1000 nm when annealed at 250°C. The optical bandgap (Figure 1.13c) of the CsPbIBr$_2$ sample varies linearly with the addition of bromine and was found to be 2.05 eV. The PL peak was found close to the bandgap at 2.0 eV and in contrast to the mixed-halide organic–inorganic perovskite there was no observation of photo-induced phase segregation in pure inorganic perovskites. The PL image (Figure 1.13d) was obtained by exciting the sample with 470 nm laser and recording the signal at 620 nm. The brighter grains imply higher PL efficiency while the darker grains have lower PL efficiency. The PL decay curves were fitted using bifunctional exponential curves. The PL decay curves were fitted with bifunctional curves for the brighter regions and with single exponential curve for the darker regions. Typical lifetimes around 9.35 and 17.7 ns were found for these films which are in the same order as that for the pure CH$_3$NH$_3$PbBr$_3$ films (Figure 1.13e).

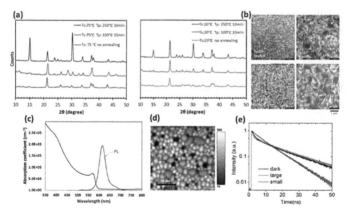

FIGURE 1.13   (a) XRD patterns of CsPbIBr$_2$ films deposited at various conditions, Ts represents the substrate temperature during deposition and Tp represents the temperature of the post-anneal (b) top view SEM images of the films under various conditions, (c) optical properties of CsPbIBr$_2$ sample deposited at 75°C and annealed at 250°C for 10 min (d) PL image of the film, (e) representative PL decay curves of a small-bright grain (blue line), a large-dim grain (red line), and a dark grain (black line).

Source: Reprinted with permission from ref 37. © 2016 John Wiley & Sons.

Lau et al. took a spray-assisted solution processed method to prepare the CsPbIBr$_2$ films.[47] As shown in Figure 1.14a, CsI was sprayed in air on the PbBr$_2$ films which were annealed in air to get the desired perovskite phase. These films were found to be thermally stable at 300°C. To study the effect of annealing temperature on the film formation, CsI was sprayed at room temperature and the films were annealed at 275–350°C for 10 min. Figure 1.14b represents the XRD pattern for the samples prepared at these temperatures. The crystallinity of these films was found to be good for all the annealing temperatures between 275 and 350°C.

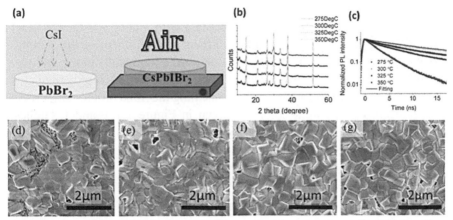

FIGURE 1.14   (a) Schematic illustration of the methodology to prepare the CsPbIBr$_2$ films by Lau et al. (b) XRD patterns for films annealed at various temperatures, (c) PL decay curves for the films annealed at various conditions, (d)–(g) top view SEM images of samples annealed at 275, 300, 325 and 350°C, respectively.
Source: Reprinted with permission from ref 47. © 2016 American Chemical Society.

The optical bandgap was found to be at 2.05 eV with an absorption coefficient of 5×10^4 cm^{-1} at 575 nm. Figure 1.14c shows PL decay curves of the samples annealed at different temperatures. For the films annealed at 275–325°C, a typical lifetime of 11 ns was measured. The lifetime dramatically dropped for the film annealed at 350°C and is highest for the film annealed at 300°C. The morphological evolution of the films annealed at various processing conditions is presented in Figure 1.14d–g. It was found that the grain size increased with an increase in the annealing temperature and also the size of the grains was found to be in the order of 500–1000 nm. The highest power conversion efficiency of 6.7% was achieved with negligible hysteresis for samples annealed at 300°C.

Another approach to get highly reproducible perovskite films was reported by Jeon et al. who employed the solvent engineering.[48] The group used DMSO as a co-solvent in γ-butyrolactone consisting of MAI and PbI_2. The films were washed with toluene to get rid of the solvent and form a MAI-DMSO-PbI_2 phase and a highest power conversion efficiency of 16.46% was reported.

Ahn et al. employed a Lewis base adduct of lead iodide.[49] Figure 1.15 shows a schematic of the process used to fabricate the films. DMF solutions (50 wt.%) containing MAI and PbI_2 (1:1 mol.%) and MAI, PbI_2, and DMSO (1:1:1 mol.%) were prepared. One step approach used to make the films via (a) and (b) lead to nonhomogeneous films with rod-like features. However, a nice dense film is formed if the films were washed with a nonpolar diethyl ether while spin coating. Figure 1.15c and d show a smooth finish of the samples after they were spin coated. DMSO-contained solution results in the pinhole-free film (Figure 1.15d), which indicates that crystal growth is highly regulated possibly due to adduct formation. After annealing the films turn to dark brown in color from a transparent color as shown in Figure 1.15e and f. Highest power conversion efficiency of 19.71% was reported using this method.

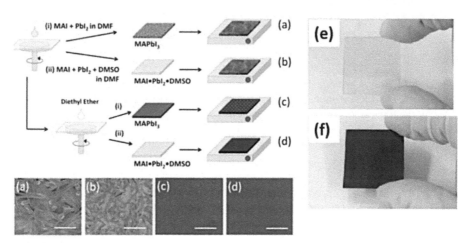

FIGURE 1.15 Schematic illustration of the methodology to prepare the perovskite films (a) one step spin coating of films from DMF, (b) one step spin coating from DMSO in DMF, (c) and (d) were prepared by the same solution as that in (a) and (b), but diethyl ether was dripped during the film spinning, (e) photos of $MAPbI_3$ films before annealing and (f) after annealing at 65°C for 1 min and at 100°C for 2 min.
Source: Reprinted with permission from ref 49. © 2015 American Chemical Society.

Nejand et al. introduced a new cold-rolled process to enhance the micro-structure of the perovskite layer and to get pinhole-free films.[50] A schematic illustration of the process used is shown in Figure 1.16. Columnar well-distributed perovskite films were obtained by spraying a DMF solution containing $PbCl_2$ and $CH_3NH_3PbI_3$ in the ratio 1:3 on a hot compact layer of TiO_2/FTO (Figure 1.16, stage 1 and 2). In stage 3, these structures were exposed to DMF vapor. These slightly wetted perovskite films were passed through the rollers (step 4), the exact process was repeated again (step 5 and 6). This eventually led to a smooth and uniform film (step 7). To investigate the evolution of microstructure at each stage, Field Emission Scanning Electron Microscope (FESEM) was performed on the samples. As seen from Figure 1.16a and b the as sprayed films have a columnar structure with a grain size of 2 µm. After the first pass in cold pressing, the microstructure of the films smoothens as seen in Figure 1.16c and d. By repeating the process, smooth films without pinholes are obtained as seen in Figure 1.16e–g. They called the intermediate perovskite film (stage 5) obtained as "compressing perovskite" and the one obtained at the end as "continuous perovskite."

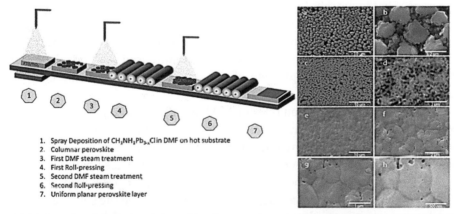

1. Spray Deposition of $CH_3NH_3Pb_{3-x}Cl$ in DMF on hot substrate
2. Columnar perovskite
3. First DMF steam treatment
4. First Roll-pressing
5. Second DMF steam treatment
6. Second Roll-pressing
7. Uniform planar perovskite layer

FIGURE 1.16 Schematic of the cold roll processing employed by Nejand et al. FESEM images of (a, b) as-deposited columnar perovskite layer by spray coating; (c, d) compressed perovskite layer by first step of pre-wetting and cold-roll pressing; (e, f) continuous perovskite layer prepared by second pre-wetting and cold-roll pressing; (g, h) and higher magnification of continuous perovskite layer.
Source: Reprinted with permission from ref 50. © 2016 American Chemical Society.

Power conversion efficiencies obtained from the continuous perovskites were found to be more than the compressing and the columnar perovskite. Transformation of the perovskite layer from columnar microstructure into a continuous and almost pinhole-free structure resulted in a 62% increase in the PCE of the cells. Reducing the recombination regions and increase in the absorbance were proposed as the most effective factors on the cell performance enhancement. Maximum power conversion efficiency of 13.24% was achieved.

Miyadera et al. employed a laser evaporation system to deposit the perovskite films.[51] The laser evaporation system used here is completely different from the pulsed laser deposition system. In the system used in this work, a continuous-wave laser is used to heat the materials locally and the locally heated materials evaporate. This leads to a controlled deposition of perovskite films from the lead halide and amine halide with reduced gas generation. The schematic illustration of the technique employed is shown in Figure 1.17. The evaporation rate was controlled precisely and quickly by tuning the laser parameters. Proportional integral differential control was integrated and a stable deposition rate was demonstrated for several hours. The surface morphology of the perovskite films formed by the laser deposition method is shown in Figure 1.17b. Smooth films of perovskite can be visualized with an average grain size of 50 nm. The root mean square roughness of the films was found to be 5–10 nm. There are some obvious differences in the morphology that is obtained from the vacuum process and the solution-based method. While the morphology in solution processed method depends on the vaporization of the solvent and is accelerated by the anti-solvent used. The morphology in vacuum-based methods depends on the availability and supply rate of the source material and the growth temperature.

The ratio of the deposition rates of MAI and PbI_2 was adjusted to get the perovskite stoichiometry. As MAI is volatile and can re-evaporate, hence co-evaporation was performed at different evaporation rates and the XRD patterns obtained are presented in Figure 1.17c and d. These patterns were obtained for the samples deposited on NiO_x and NiO_x/PCDTBT, respectively. It was found that the optimized deposition rate depends on the substrate used and is very sensitive to the substrate surface conditions. When excess PbI_2 was supplied, the diffraction peak of PbI_2 was observed. When excess MAI was supplied on the NiOx/PCDTBT, unidentified peak at 11.3° was observed (possibility of delta phase). Wide range XRD pattern of the perovskite film is shown in Figure 1.17e. Power conversion efficiency of 16% was achieved using the method described above.

Koh et al. reported the first formamidinium (HC $(NH_2)_2^+$) cation as a replacement to the traditionally used methyl-ammonium ($CH_3NH_3^+$) cation in inorganic–organic metal halide perovskites.[52] The advantage of using formamidinium was found to be in its optical properties. It has an absorption onset in the near-IR region (1.47 eV), hence can be used to extract higher short circuit current from the solar cells.

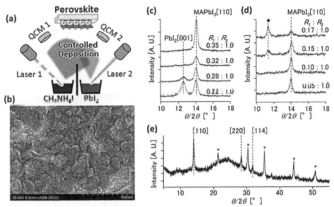

FIGURE 1.17 (a) Schematic diagram of the laser evaporation system employed by Miyadera et al. (b) SEM image of the CH₃NH₃PbI₃ film fabricated using laser deposition method, (c) MAPbI₃ (40 nm) film fabricated on NiOₓ film, (d) MAPbI₃ film fabricated on NiOₓ/PCDTBT with R1 and R2 being the deposition rates of MAI (sensor 1) and PbI₂ + MAI (sensor 2) respectively, (e) XRD pattern of MAPbI₃ film (* being peaks of ITO).
Source: Reprinted with permission from ref 51. © 2016 American Chemical Society

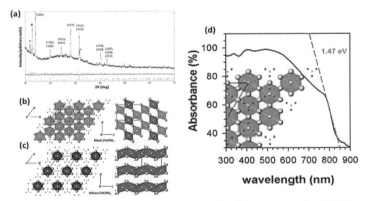

FIGURE 1.18 (a) A Pawley fit of the powder XRD of the pattern for FAPbI₃ perovskite thin film prepared via a sequential deposition process. The reflections indicated by * and # represent the Bragg reflections associated with the yellow polymorph of FAPbI₃ and PbI₂, respectively. The indexed reflections associated with the black perovskite polymorph are indicated in the plot. Polyhedral representations of the black (b) and yellow (c) polymorphs of FAPbI₃. The blue polyhedra represent the PbI₆ octahedra with the Pb and I atoms shown as yellow and orange spheres respectively. Both structures are shown viewed along the crystallographic c (left) and a (right) axes, respectively. In the black polymorph, the inorganic component consists of a three-dimensional network of corner linked PbI₆ octahedra with the yellow polymorph containing linear chains of face-sharing octahedra. The N and C ions of the formamidimium cations are shown as blue and green spheres, respectively, (d) absorption spectrum of the black FAPbI₃ film with absorption onset extended beyond 800 nm.
Source: Reprinted with permission from ref 52. © 2014 American Chemical Society.

To understand the crystal structure formation, these films were cast on glass and Pawley fit was performed on the XRD patterns obtained (Figure 1.18a). The XRD pattern showed the coexistence of yellow polymorph as well as the 2H-polymorph of PbI_2. The $FAPbI_3$ exists as two polymorphs: the black (alpha) phase and the yellow (delta) phase. The black phase has a trigonal symmetry while the yellow phase with a hexagonal symmetry (Figure 1.18b and c). Optical bandgap of 1.47 eV was extracted as seen in Figure 1.18d. Power conversion efficiency of 4.3% was reported.

Wang et al. used low-volatile additives in the precursor solution to assist the formation of the black (alpha) phase of $FAPbI_3$. Insights into the formation mechanisms of black $FAPbI_3$ phase using various additives such as NH_4Cl, MACl, FACl with different volatilities are compared. Figure 1.19a shows a schematic diagram of the various additives used and the process involved.[53] It was found that the lower volatility additives aid in the formation of the black perovskite phase.

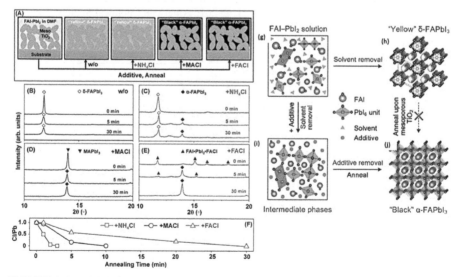

FIGURE 1.19   (a) Schematic diagram of the $FAPbI_3$ phase formed on mesoporous TiO_2 by Wang et al. XRD patterns showing the phase evolution starting from spin-coated precursor films: (b) without additive; (c) with NH_4Cl additive; (d) with MACl additive; and (e) with FACl additive, (f) Cl content (Cl/Pb atomic ratio) in the films at different annealing durations (140°C) measured using EDS, (g) schematic illustration of the mechanisms for the evolution of $FAPbI_3$ phase.

Source: Reprinted with permission from ref 53. © 2015 American Chemical Society.

The representative picture in Figure 1.19a shows that without the additive and with NH_4Cl as an additive result in the formation of yellow (delta) phase of perovskite. MACl and FACl result in the formation of the black alpha phase of $FAPbI_3$. To understand the evolution of different phases, XRD was performed on these samples which were thermally annealed at 140°C for various time durations (Figure 1.19b–e).

When FACl was used as an additive, XRD pattern of the freshly coated film does not show any traces of yellow (delta) phase of $FAPbI_3$, but an unknown phase is present. This intermediate phase was attributed to FAI. $PbI_2.FACl$ adduct formed during reaction of Lewis base halide with Lewis acid PbI_2. Increase in annealing time leads to the formation of the black (alpha) phase of perovskite. This is also reflected in Figure 1.19f. Figure 1.19b shows the XRD pattern for the sample where no additive was added and it was found to have prominent peaks of the yellow polymorph. Even after 30 min of annealing, the alpha phase is not visible. A similar effect is found when NH_4Cl was added. A trace amount of the alpha phase is formed after annealing for 5 min, but the dominant phase still is the yellow phase. When MACl was added as an additive, a black film appeared as soon as spin-coating was completed. There is a slight mismatch in the peaks observed for annealed and the unannealed films. It was found out that the films initially formed are $MAPbI_3$ which then gradually evolve to $FAPbI_3$ upon annealing. When FACl was added, the initial film does not show any traces of the delta yellow phase but has peaks of an intermediate phase which was attributed to the adduct formation of $FAI.PbI_2.FACl$. Upon annealing to 30 min, the amount of Cl in the film decreases and the black alpha phase becomes the only dominant phase present. On the basis of experimental results, the mechanism involved in the evolution of the black alpha phase is shown in Figure 1.19g. The existence of the intermediate phase is shown when additives are added to the DMF solution containing $FAI+PbI_2$. The highest efficiency (13.37%) in this study was obtained for the perovskite sample where FACl was added as an additive.

1.4 LOWER DIMENSIONAL PEROVSKITES

Long-term stability of 3D perovskite solar cells is plagued by their vulnerability to decompose under humid conditions. To overcome this issue several groups have proposed multidimensional/lower dimensional perovskites as an alternative.[54–57] Yao et al. studied a series of two-dimensional (2D) layered perovskites fabricated by incorporating polyethylenimine (PEI) in the layered structures.[58] The use of multidimensional perovskites is manifold including moisture stability and better charge transport compared to their 3D counterparts. Intermediate multilayered perovskite $(PEI)_2(MA)_{n-1}Pb_nI_{3n+1}$ (n=3, 5, 7) by replacing partially MA in MAPbI$_3$ with PEI cations were investigated in this study. Tight stacking of the inorganic layers was achieved by intercalating the polymeric ammonium cation which improved the charge transfer of the perovskite stack (Figure 1.20a).

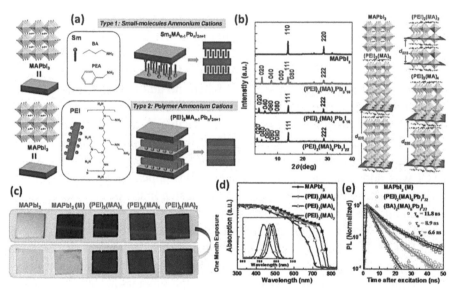

FIGURE 1.20 (a) Schematic illustration of multi-layered perovskites by different ammonium cations (b) XRDs of bulk powder of MAPbI$_3$ and $(PEI)_2(MA)_{n-1}Pb_nI_{3n+1}$ perovskites, with the illustration (right) of their respective diffraction planes, (c) images of perovskite films before and after exposure to 50% humidity, (d) UV−visible absorption spectra of spin-coated MAPbI$_3$ and $(PEI)_2(MA)_{n-1}Pb_nI_{3n+1}$ films and the corresponding PL spectra (inset), (e) PL decay curves for various samples studied.
Source: Reprinted with permission from ref 58. © 2016 American Chemical Society.

XRD patterns of the various powder perovskites prepared in this study are presented in Figure 1.20b. For the $(PEI)_2(MA)_2Pb_3I_{10}$ perovskite, the reflections represent the vertical orientation of MAPbI$_3$ at (111) and (202) and the also reflections of the layered structure $(PEI)_2PbI_4$. As n increases to 5,7, the competition between PEI cations to grow in layered structures increases with the MA cations which tend to grow out of the plane. Therefore, multilayered perovskites have growth in the perpendicular direction rather than in the layered plane. All the films were cast from a DMF solution in a single-step process. With the multilayered perovskites, smooth continuous pinhole-free films were attained without annealing. Annealing was mandatory for the pure MAPbI$_3$ films. The moisture stability test was also performed for the films (50% RH for 30 days). It is clearly evident from the optical images of Figure 1.20c, that the MAPbI$_3$ films immediately decompose when exposed to humidity and PbI$_2$ films are obtained with the loss of MA$^+$ cation. However, for the multilayered perovskite (PEI)$_2$(MA)$_{n-1}$Pb$_n$I$_{3n+1}$, there was no obvious color change even after a month of exposure to moisture. The optical properties of the films were studied by measuring the absorption and PL (Figure 1.20d). It was found that by introducing PEI in the films, the bandgap of the perovskite increases with decrease in n values. The inset presents the steady-state PL of the samples. The PL emission shifted from 783 nm of MAPbI$_3$ to 720 nm of (PEI)$_2$(MA)$_2$Pb$_3$I$_{10}$. This distinct feature is consistent with the absorption spectra. Time-resolved PL was performed to understand the kinetics of charge carriers in perovskite films (Figure 1.20e). Bimolecular recombination was observed for both MAPbI$_3$ and (PEI)$_2$(MA)$_6$Pb$_7$I$_{22}$ films. This indicates improved charge transfer due to the PEI intercalation in the films.

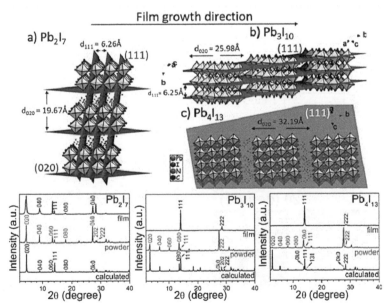

FIGURE 1.21 XRDs of thin films versus bulk materials of (a) (BA)$_2$(MA)Pb$_2$I$_7$, (b) (BA)$_2$(MA)$_2$Pb$_3$I$_{10}$, and (c) (BA)$_2$(MA)$_3$Pb$_4$I$_{13}$ perovskites, with the illustration of their respective diffraction planes. Note that the Miller indices are different from those of (BA)$_2$PbI$_4$ and MAPbI$_3$ because of the different assignment of the orthogonal unit cell axes.
Source: Reprinted with permission from ref 54. © 2016 American Chemical Society.

Cao et al. studied the effect of aliphatic alkylammonium cation on the surface coverage and self-assembly of perovskites.[54] The perovskite films of $(BA)_2(MA)_{n-1}Pb_nI_{3n+1}$ were spun cast from a solution of DMF by single-step deposition. Generally, the orientation of 2D perovskites is in such a way that the layers grow parallel to the substrate. In the case of $(BA)_2(MA)_{n-1}Pb_nI_{3n+1}$, this trend appears to be true only for the n=1 compound, where preferential growth along the (110) direction occurs. As n>1, a competition between the BA and MA cation starts and clearly mixed reflections are observed. From Figure 1.21, for the n=2 structure, the reflections are a mix of (111) and (202) which reveal the vertical growth of the compound to the substrate. The n=3, 4 compounds have dominant (111) and (202) reflections while lack the (0k0) reflection, revealing the vertical growth of the compound.

As highlighted before, one of the major advantages of working with 2D homologs of perovskite is their enhanced moisture stability. (Figure) It is well known that $MAPbI_3$ turn to yellow in color upon exposure to moisture due to the loss of MA^+ cation (Figure 1.22b). However, films of $(BA)_2(MA)_2Pb_3I_{10}$ remain unchanged upon exposure to 40% humidity over a period of 2 months.

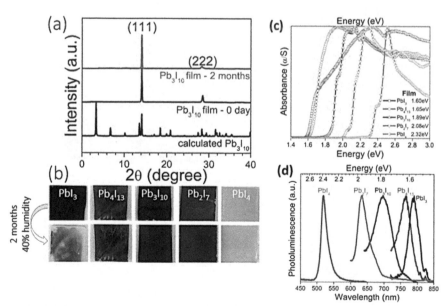

FIGURE 1.22   (a) XRDs of fresh and aged (2 months) $(BA)_2(MA)_2Pb_3I_{10}$ film. (b) images of perovskite films before and after exposure to humidity, (c) absorption spectra of the films prepared in this study, (d) PL spectra of the films.
Source: Reprinted with permission from ref 54. © 2016 American Chemical Society.

The optical properties of these films were studied using absorption/ luminescence spectroscopy and the results are presented in Figure 1.22c and d. The optical bandgap increases with a decrease in the n values from n=∞ to 1. The bandgap was found to be 1.52 eV for the 3D MAPbI$_3$. In addition to the primary absorption edge, another peak above the absorption edge of 2D perovskite was observed. This peak is attributed to the long-lived excitonic state trapped due to the electric field of BA$^+$ cations and negatively charged $(MA)_{n-1}Pb_nI_{3n+1}$. PL measurements on the $(BA)_2(MA)_{n-1}Pb_nI_{3n+1}$ perovskite films cast on glass substrates was performed. The PL spectrum of all the 2D perovskites and 3D MAPbI$_3$ films are consistent with the optical bandgaps measured. For n=1 compound, the PL emission is consistent with the high-energy optical absorption peak. With increasing n values, the PL emission is red shifted and is also in agreement with the optical bandgaps measured.

Saidaminov et al. synthesized Cs$_4$PbBr$_6$ via a solution processed method with a remarkably high 45% PL quantum yield. CsPbBr$_3$ has a cubic crystal structure with corner shared PbBr$_6^{4-}$ octahedra (Figure 1.23a). The crystal structure of Cs$_4$PbBr$_6$ is shown in Figure 1.23b. The Cs$^+$ ion minimizes the electronic overlap between the adjacent octahedra, resulting in confined individual octahedra. The luminescence properties of the Cs-based compounds were studied under a UV light, seen in Figure 1.23c. The crushed powder of CsPbBr$_3$ has no luminescence, while the unwashed Cs$_4$PbBr$_6$ emits green light. The washed (purified) Cs$_4$PbBr$_6$ has a very bright green emission. Using such 0D structures may have very promising applications in light emission devices.

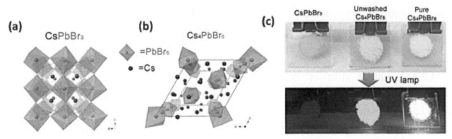

FIGURE 1.23 Crystal structure of (a) CsPbBr$_3$, (b) Cs$_4$PbBr$_6$ with corner shared and isolated PbBr$_6^{4-}$ octahedral, (c) different samples prepared in the study under UV lamp (365 nm). *Source:* Reprinted with permission from ref 59. © 2016 American Chemical Society.

1.5   STRUCTURE–PROPERTY RELATIONSHIP IN LEAD-FREE PEROVSKITES

There have been several studies to replace the toxic Pb^{2+} ion in the perovskite structure.[60,61] The major replacement has been the Sn^{2+} ion. This replacement has led to several structural changes. In Figure 1.24, these changes in the crystal structure are shown. The A site in the ABX_3 crystal structure has been replaced by Cs and the B site with Sn in the case of $CsSnI_3$. The X site is also changed by varying it from I to Br.

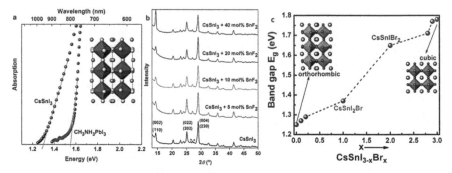

FIGURE 1.24   (a) Optical absorption spectrum of $CsSnI_3$ and $CH_3NH_3PbI_3$ illustrating the onset of $CsSnI_3$ absorption is extended into the infrared region. The inset shows the typical perovskite structure adopted by ABX_3 compounds (b) XRD patterns of pristine $CsSnI_3$ and samples with 5, 10, 20 and 40 mol% added SnF_2, with all the patterns matching the black orthorhombic phase of $CsSnI_3$. The pristine sample has two peaks which represent the yellow polymorph of $CsSnI_3$ marked by asterisk, (c) bandgap variation with respect to Br concentration. The crystal structure also changes from orthorhombic ($CsSnI_3$) to cubic ($CsSnBr_3$).

Source: A and C: Reprinted with permission from ref 61. © 2015 American Chemical Society.
B: Reprinted with permission from ref 60. © 2014 John Wiley & Sons.

$CsSnI_3$ has a smaller bandgap (1.3 eV) when compared to $MAPbI_3$ (1.55 eV) and hence a higher theoretical short-circuit photocurrent density of 34.3 mA/cm^2 can be achieved compared to 25.9/mA · cm^2 for $CH_3NH_3PbI_3$-based systems (Figure 1.24a). $CsSnI_3$ is prone to form intrinsic defects associated with Sn-cation vacancies that result in metallic conductivity.[60] In order to utilize $CsSnI_3$ as a semiconducting photoabsorber, SnF_2 was added to control the metallic conductivity. Samples were prepared by stoichiometric mixing of CsI, SnI_2, and SnF_2 in appropriate solvents and cast by spin-coating followed by 70°C drying. Figure 1.24b shows XRD patterns for films of pristine $CsSnI_3$ and those with 5, 10, 20, and 40 mol.% SnF_2, cast from DMSO solution onto mesoporous TiO_2 films. All the reflections observed in the XRD patterns of Figure 1.24b are in good agreement with the reported orthorhombic (Pnam) structure of $CsSnI_3$.[62] The structure-property relationship of anionic bromine substitution in $CsSnI_{3-x}Br_x$ films is shown in Figure 1.24c. As the amount of Br increases, the structure of the perovskite transforms from orthorhombic to a cubic structure. The optical bandgap can be tuned from 1.23 eV for the pure iodide to 1.78 for the pure bromide. The highest V_{oc} obtained for pure bromide film was 410 mV and the notable decrease in charge carrier densities were reported due to bromine incorporation.

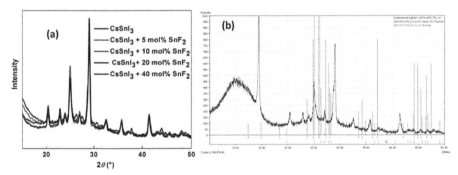

FIGURE 1.25 (a) XRD patterns of all the samples superimposed on each other showing that addition of SnF_2 promotes the formation of a black orthorhombic polymorph of $CsSnI_3$. All the films were cast from DMSO solution. The cell parameters extracted via a Pawley fit, yielded lattice parameters a=8.708 (9) Å, b=8.622 (9) Å and c=12.390 (12) Å. (b) XRD pattern showing the absence of SnF_2 and $CsSnF_3$ from the $CsSnI_3$+40 mol.% SnF_2 sample. The red lines represent $CsSnF_3$ while the green lines represent SnF_2. The main reflection for SnF_2 overlaps with the (202)/(022) reflection of the $CsSnI_3$ perovskite but the absence of other expected reflections of SnF_2 suggest that it is not a crystalline component of the thin film sample. Similarly, no reflections associated with $CsSnF_3$ can be assigned indicating its absence in the sample.

Source: Reprinted with permission from ref 60. © 2014 John Wiley & Sons.

The addition of SnF_2 does not result in a significant variation in the lattice parameters for the films and unwanted phases (Figure 25a,b), which suggests that the F^- does not substitute the I^- in the perovskite structure due to the much smaller ionic radii of F (1.33 Å) compared to I (2.20 Å). However, the addition of SnF_2 does eliminate the formation of the weak unidentified reflections attributed to the yellow non-perovskite polymorph of $CsSnI_3$.[63]

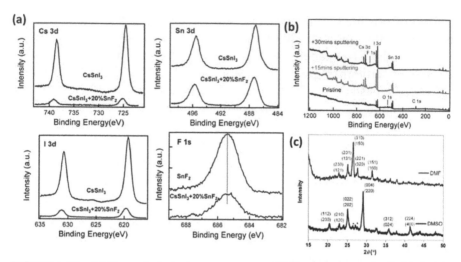

FIGURE 1.26 (a) X-ray photoelectron spectroscopy (XPS) narrow scans for Cs 3d, Sn 3d, I 3d and F 1 s. The F 1 s peak observed in the 20 mol.% SnF_2-$CsSnI_3$ sample displays no shift in the binding energy with respect to the SnF_2 reference sample. The symmetric F 1 s peak shape implies all fluorine atoms in the 20 mol.% SnF_2-$CsSnI_3$ samples are surrounded by a uniform chemical environment, demonstrating that there is no chemical bonding of SnF_2 with $CsSnI_3$, (b) wide scan of $CsSnI_3$ + 20%SnF_2 sample before and after sputtering, (c) yellow polymorph ($Cs_2Sn_2I_6$) is obtained when DMF was used as solvent compared to DMSO which leads to the more prominent orthorhombic phase of $CsSnI_3$.
Source: Reprinted with permission from ref 60. © 2014 John Wiley & Sons.

X-ray photoelectron spectroscopy measurements on pristine $CsSnI_3$ and 20 mol.%SnF_2-$CsSnI_3$ (Figure 1.26a,b) confirm the presence of F within the film. However, the F 1 s peak observed in the 20 mol.%SnF_2-$CsSnI_3$ sample does not display any shift in the binding energy with respect to the SnF_2 reference sample. The symmetric F 1s peak shape implies all fluorine atoms in the 20 mol.% SnF_2-$CsSnI_3$ sample are surrounded by a uniform chemical environment. The Cs 3d, I 3d, and Sn 3d spectra of both the samples can always be fitted with one dominant peak, indicating that SnF_2 is uniformly mixed in $CsSnI_3$ sample, but does not get integrated into the lattice to distort the local environment. Interestingly, this was the first demonstration of the formation of orthorhombic $CsSnI_3$ from a solution processed method at low temperatures (70°C). A significant dependence on the solvent is noted with other solvents such as DMF not resulting in the formation of the orthorhombic phase under identical conditions (Figure 1.26c). This change is also reflected in the color of the solution (not shown here). The yellow color solution is obtained when $CsSnI_3$ is dissolved in DMSO while the solution turns reddish almost immediately when DMF is used as a solvent.

$A_3Sb_2I_9$ compounds are an interesting class of materials as they can exist in both dimer or layered phases. The $Cs_3Sb_2I_9$ compound exists in the dimer phase while the Rb counterpart exists in the layered phase. Solution-processed method was employed to form these films as shown in Figure 1.27a. The stoichiometric ratio of the precursor was dissolved in DMF and spin-coated after filtering the solution. The samples were annealed at 120°C for 10 min. Figure 1.27b shows XRD pattern of the $Rb_3Sb_2I_9$ films which have a single phase with no traces of unreacted precursor. Iodine vacancies have been shown to alter the electrical properties of such perovskites. To study this effect, SbI_3 was dissolved in toluene and dripped while spin-coating. It was also found that using such a dripping method to cast the films yield homogenous, continuous smooth films.

X-ray photoelectron spectroscopy (XPS) measurements in Figure 1.27c show that the films treated with SbI_3 yield films with near ideal stoichiometry. The optical absorption coefficient of the material was found to be 105/cm which indicate that very thin layers would suffice for adequate absorption (Figure 1.27d). Such high values for absorption coefficient are a result of p-to-p direct transitions. The cathodoluminescence (CL) intensity of SbI_3 treated samples was found to be twice as high as compared to the pristine samples. The CL spectrum from the SbI_3 treated sample can be deconvoluted to three different peaks centered at 1.98, 1.63, and 1.41 eV as shown in Figure 1.27e. To understand further the origin of these peaks, the variation of the intensities of the peaks with excitation power was plotted and the saturation rate of the emissions was found. From the power resolved measurements it was found that the 1.98 eV emission (close to the absorption edge) has as a near linear dependence on excitation power ($I_{CL} \propto I_b^{0.82}$) indicating that it could be from band-to-band transition as reported in the case of other perovskites. The slight shift of the emission from the bandgap could be due to excitonic effects as observed in other layered materials. The 1.41 and 1.63 eV peaks saturated faster with increasing excitation powers ($I_{CL} \propto I_b^{0.61}$ and $I_{CL} \propto I_b^{0.71}$, respectively), indicating that these peaks may be associated with some recombination center with a long relaxation time such as deep-level defects. A power conversion of 0.66% was achieved using layered $Rb_3Sb_2I_9$ as an absorber.

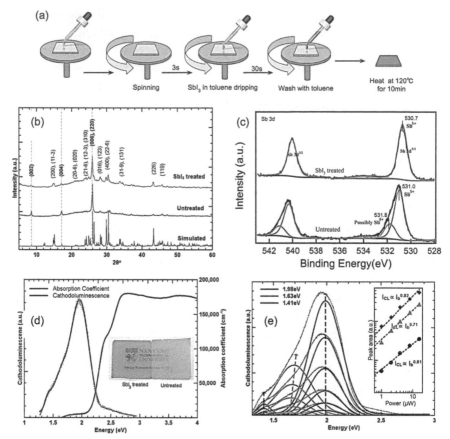

FIGURE 1.27 (a) Procedure adopted to form $Rb_3Sb_2I_9$ films, (b) comparison of XRD of untreated and SbI_3 treated films and (c) XPS comparison of both films showing the Sb $3d^{3/2}$ and Sb $3d^{5/2}$ peaks. (d) absorption coefficient and cathodoluminescence (at accelerating voltage of 5 kV and beam current of 0.2 nA) of SbI_3 treated $Rb_3Sb_2I_9$ and (e) variation of cathodoluminescence with fixed accelerating voltage (5 kV) and varying excitation power (in log scale). I_b is the beam current and I_{CL} is the CL intensity.

Source: Reprinted with permission from ref 64. © 2016 American Chemical Society.

KEYWORDS

- **perovskite solar cells**
- **photovoltaics**
- **mesoporous TiO_2**
- **metal halide perovskite**
- **perovskite fabrication methods**

REFERENCES

1. Kim, H.-S.; et al. Lead Lodide Perovskite Sensitized All-Solid-State Submicron Thin Film Mesoscopic Solar Cell with Efficiency Exceeding 9%. *Sci. Rep.* **2012,** *2,* 591.

2. Rose, G. Description of Some New Minerals from the Urals. *Pogendorff Annalen der Physik und Chemie (in german)* **1839,** *48,* 551–572.

3. Wang, S.; et al. Synthesis and Characterization of $[NH_2C(I):NH_2]_3MI_5$ (M = Sn, Pb): Stereochemical Activity in Divalent Tin and Lead Halides Containing Single .ltbbrac.110.rtbbrac. Perovskite Sheets. *J. Am. Chem. Soc.* **1995,** *117*(19), 5297–5302.

4. Topsöe, H. Crystallographic-Chemical Investigations of Homologous Compounds. *Z. Kristallogr.* (in German) **1884,** *8,* 246–296.

5. Poglitsch, A.; Weber, D. Dynamic Disorder in Methyl Ammonium Trihalogeno Plumbates (II) Observed by Millimeter-Wave Spectroscopy. *J. Chem. Phys.* **1987,** *87*(11), 6373–6378.

6. Ishihara, T.; Takahashi, J.; Goto, T. Exciton State in Two-Dimensional Perovskite Semiconductor $(C_{10}H_{21}NH_3)_2PbI_4$. *Solid State Commun.* **1989,** *69*(9), 933–936.

7. Onoda-Yamamuro, N.; Matsuo, T.; Suga, H. Dielectric Study of $CH_3NH_3PbX_3$ (X = Cl, Br, I). *J. Phys. Chem. Solids* **1992,** *53*(7), 935–939.

8. Mitzi, D. B.; et al. Conducting Tin Halides with a Layered Organic-Based Perovskite Structure. *Nature* **1994,** *369*(6480), 467–469.

9. Mitzi, D. B.; et al. Conducting Layered Organic-Inorganic Halides Containing <110>-Oriented Perovskite Sheets. *Science* **1995,** *267*(5203), 1473–1476.

10. Kagan, C. R.; Mitzi, D. B.; Dimitrakopoulos, C. D. Organic-Inorganic Hybrid Materials as Semiconducting Channels in Thin-Film Field-Effect Transistors. *Science* **1999,** *286*(5441), 945–947.

11. Kojima, A.; et al. Novel Photovoltaic Solar Cell Sensitized by Lead-Halide Compounds (II). *J. Soc. Photogr. Imaging Jpn. 69,* 28–29 (in Japanese).

12. McMeekin, D. P.; et al. A Mixed-Cation Lead Mixed-Halide Perovskite Absorber for Tandem Solar Cells. *Science* **2016,** *351*(6269), 151–155.

13. Quarti, C.; et al Structural and Optical Properties of Methylammonium Lead Iodide Across the Tetragonal to Cubic Phase Transition: Implications for Perovskite Solar Cells. *Energy Environ. Sci.* **2016,** *9*(1), 155–163.

14. You, J.; et al. Improved Air Stability of Perovskite Solar Cells via Solution-Processed Metal Oxide Transport Layers. *Nature Nanotechnol.* **2016,** *11*(1), 75–81.

15. Burschka, J.; et al. Sequential Deposition as a Route to High-Performance Perovskite-Sensitized Solar Cells. *Nature* **2013,** *499*(7458), 316–319.

16. Kojima, A.; et al. Organometal Halide Perovskites as Visible-Light Sensitizers for Photovoltaic Cells. *J. Am. Chem. Soc.* **2009,** *131*(17), 6050–6051.

17. Lee, M. M.; et al. Efficient Hybrid Solar Cells Based on Meso-Superstructured Organometal Halide Perovskites. *Science* **2012,** *338*(6107), 643–647.

18. Best Research-Cell Efficiencies, NREL. http://www.nrel.gov/pv/assets/images/efficiency_chart.jpg.

19. Green, M. A.; Ho-Baillie, A.; Snaith, H. J. The Emergence of Perovskite Solar Cells. *Nat. Photonics* **2014,** *8*(7), 506–514.

20. Green, M. A.; et al. Optical Properties of Photovoltaic Organic–Inorganic Lead Halide Perovskites. *J. Phys. Chem. Lett.* **2015,** *6*(23), 4774–4785.

21. Stoumpos, C. C.; Malliakas, C. D.; Kanatzidis, M. G. Semiconducting Tin and Lead Iodide Perovskites with Organic Cations: Phase Transitions, High Mobilities, and Near-Infrared Photoluminescent Properties. *Inorg. Chem.* **2013,** *52*(15), 9019–9038.

22. Wehrenfennig, C.; et al. High Charge Carrier Mobilities and Lifetimes in Organolead Trihalide Perovskites. *Adv. Mater.* **2014,** *26*(10), 1584–1589.

23. Motta, C.; El-Mellouhi, F.; Sanvito, S. Charge Carrier Mobility in Hybrid Halide Perovskites. *Sci. Rep.* **2015,** *5,* 12746.

24. Lian, Z.; et al. Perovskite $CH_3NH_3PbI_3(Cl)$ Single Crystals: Rapid Solution Growth, Unparalleled Crystalline Quality, and Low Trap Density Toward 108 cm^{-3}. *J. Am. Chem. Soc.* **2016,** *138*(30), 9409–9412.

25. Green, M. A. Intrinsic Concentration, Effective Densities of States, and Effective Mass in Silicon. *J. Appl. Phys.* **1990,** *67*(6), 2944–2954.

26. Blakemore, J. S. Semiconducting and Other Major Properties of Gallium Arsenide. *J. Appl. Phys.* **1982,** *53*(10), R123–R181.

27. Bartelt, J. A.; et al. Charge-Carrier Mobility Requirements for Bulk Heterojunction Solar Cells with High Fill Factor and External Quantum Efficiency >90%. *Adv. Energy Mater.* **2015,** *5*(15), 1500577–n/a.

28. Xing, G.; et al. Long-Range Balanced Electron- and Hole-Transport Lengths in Organic-Inorganic $CH_3NH_3PbI_3$. *Science* **2013,** *342*(6156), 344–347.

29. Stranks, S. D.; et al. Electron-Hole Diffusion Lengths Exceeding 1 Micrometer in an Organometal Trihalide Perovskite Absorber. *Science* **2013,** *342*(6156), 341–344.

30. Kawamura, Y.; Mashiyama, H.; Hasebe, K. Structural Study on Cubic–Tetragonal Transition of $CH_3NH_3PbI_3$. *J. Phys. Soc. Jpn.* **2002,** *71*(7), 1694–1697.

31. Oku, T.; et al. Microstructures and Photovoltaic Properties of Perovskite-Type $CH_3NH_3PbI_3$ Compounds. *Appl. Phys. Express* **2014,** *7*(12), 121601.

32. Even, J.; et al. Solid-State Physics Perspective on Hybrid Perovskite Semiconductors. *J. Phys. Chem. C* **2015,** *119*(19), 10161–10177.

33. Choi, J. J.; et al. Structure of Methylammonium Lead Iodide Within Mesoporous Titanium Dioxide: Active Material in High-Performance Perovskite Solar Cells. *Nano Lett.* **2014,** *14*(1), 127–133.

34. Zhou, Y.; et al. Crystal Morphologies of Organolead Trihalide in Mesoscopic/Planar Perovskite Solar Cells. *J. Phys. Chem. Lett.* **2015,** *6*(12), 2292–2297.

35. Harms, H. A.; et al. Mesoscopic Photosystems for Solar Light Harvesting and Conversion: Facile and Reversible Transformation of Metal-Halide Perovskites. *Faraday Discuss.* **2014,** *176,* 251–269.

36. Im, J.-H.; et al. 6.5% Efficient Perovskite Quantum-Dot-Sensitized Solar Cell. *Nanoscale* **2011,** *3*(10), 4088–4093.

37. Ma, Q.; et al. Hole Transport Layer Free Inorganic $CsPbIBr_2$ Perovskite Solar Cell by Dual Source Thermal Evaporation. *Adv. Energy Mater.* **2016,** *6*(7), 1502202-n/a.

38. Fan, P.; et al. High-Performance Perovskite $CH_3NH_3PbI_3$ Thin Films for Solar Cells Prepared by Single-Source Physical Vapour Deposition. *Sci. Rep.* **2016,** *6,* 29910.

39. Burschka, J.; et al. Sequential Deposition as a Route to High-Performance Perovskite-Sensitized Solar Cells. *Nature* **2013,** *499*(7458), 316–319.

40. Pei, Y.; et al. Effect of Perovskite Film Preparation on Performance of Solar Cells. *J. Chem.* **2016,** *2016,* 10.

41. Patel, J. B.; et al. Formation Dynamics of $CH_3NH_3PbI_3$ Perovskite Following Two-Step Layer Deposition. *J. Phys. Chem. Lett.* **2016,** *7*(1), 96–102.

42. Yuan, S.; et al. High-Quality Perovskite Films Grown with a Fast Solvent-Assisted Molecule Inserting Strategy for Highly Efficient and Stable Solar Cells. *ACS Appl. Mater. Interfaces* **2016,** *8*(34), 22238–22245.

43. Mao, P.; et al. Efficiency-Enhanced Planar Perovskite Solar Cells via an Isopropanol/ Ethanol Mixed Solvent Process. *ACS Appl. Mater. Interfaces* **2016,** *8*(36), 23837–23843.
44. Cui, D.; et al. Color-Tuned Perovskite Films Prepared for Efficient Solar Cell Applications. *J. Phys. Chem. C* **2016,** *120*(1), 42–47.
45. Sutter-Fella, C. M.; et al. High Photoluminescence Quantum Yield in Band Gap Tunable Bromide Containing Mixed Halide Perovskites. *Nano Lett.* **2016,** *16*(1) 800–806.
46. Yantara, N.; et al. Unravelling the Effects of Cl Addition in Single Step $CH_3NH_3PbI_3$ Perovskite Solar Cells. *Chem. Mater.* **2015,** *27*(7), 2309–2314.
47. Lau, C. F. J.; et al. $CsPbIBr_2$ Perovskite Solar Cell by Spray-Assisted Deposition. *ACS Energy Lett.* **2016,** *1*(3), 573–577.
48. Jeon, N. J.; et al. Solvent Engineering for High-Performance Inorganic–Organic Hybrid Perovskite Solar Cells. *Nat. Mater.* **2014,** *13*(9), 897–903.
49. Ahn, N.; et al. Highly Reproducible Perovskite Solar Cells with Average Efficiency of 18.3% and Best Efficiency of 19.7% Fabricated via Lewis Base Adduct of Lead(II) Iodide. *J. Am. Chem. Soc.* **2015,** *137*(27), 8696–8699.
50. Abdollahi Nejand, B.; et al. New Scalable Cold-Roll Pressing for Post-Treatment of Perovskite Microstructure in Perovskite Solar Cells. *J. Phys. Chem. C* **2016,** *120*(5), 2520–2528.
51. Miyadera, T.; et al. Highly Controlled Codeposition Rate of Organolead Halide Perovskite by Laser Evaporation Method. *ACS Appl. Mater. Interfaces* **2016,** *8*(39), 26013–26018.
52. Koh, T. M.; et al. Formamidinium-Containing Metal-Halide: An Alternative Material for Near-IR Absorption Perovskite Solar Cells. *J. Phys. Chem. C* **2014,** *118*(30), 16458–16462.
53. Wang, Z.; et al. Additive-Modulated Evolution of $HC(NH_2)_2PbI_3$ Black Polymorph for Mesoscopic Perovskite Solar Cells. *Chem. Mater.* **2015,** *27*(20), 7149–7155.
54. Cao, D. H.; et al. 2D Homologous Perovskites as Light-Absorbing Materials for Solar Cell Applications. *J. Am. Chem. Soc.* **2015,** *137*(24), 7843–7850.
55. Quan, L. N.; et al. Ligand-Stabilized Reduced-Dimensionality Perovskites. *J. Am. Chem. Soc.* **2016,** *138*(8), 2649–2655.
56. Tsai, H.; et al. High-Efficiency Two-Dimensional Ruddlesden–Popper Perovskite Solar Cells. *Nature* **2016,** *536*(7616), 312–316.
57. Liu, J.; et al. Two-Dimensional $CH_3NH_3PbI_3$ Perovskite: Synthesis and Optoelectronic Application. *ACS Nano* **2016,** *10*(3), 3536–3542.
58. Yao, K.; et al. Multilayered Perovskite Materials Based on Polymeric-Ammonium Cations for Stable Large-Area Solar Cell. *Chem. Mater.* **2016,** *28*(9), 3131–3138.
59. Saidaminov, M. I.; et al. Pure Cs_4PbBr_6: Highly Luminescent Zero-Dimensional Perovskite Solids. *ACS Energy Lett.* **2016,** *1*(4), 840–845.
60. Kumar, M. H.; et al. Lead-Free Halide Perovskite Solar Cells with High Photocurrents Realized Through Vacancy Modulation. *Adv. Mater.* **2014,** *26*(41), 7122–7127.
61. Sabba, D.; et al. Impact of Anionic Br⁻ Substitution on Open Circuit Voltage in Lead Free Perovskite $(CsSnI_{3-x}Br_x)$ Solar Cells. *J. Phys. Chem. C* **2015,** *119*(4), 1763–1767.
62. Koji, Y.; et al. Structural Phase Transitions of the Polymorphs of $CsSnI_3$ by Means of Rietveld Analysis of the X-Ray Diffraction. *Chem. Lett.* **1991,** *20*(5), 801–804.
63. Zhou, Y.; et al. Room Temperature "One-Pot" Solution Synthesis of Nanoscale $CsSnI_3$ Orthorhombic Perovskite Thin Films and Particles. *Mater. Lett.* **2013,** *110,* 127–129.
64. Harikesh, P. C.; et al. Rb as an Alternative Cation for Templating Inorganic Lead-Free Perovskites for Solution Processed Photovoltaics. *Chem. Mater.* **2016,** *28*(20), 7496–7504.

CHAPTER 2

TYPICAL CONFIGURATIONS OF PEROVSKITE SOLAR CELLS

2.1 EARLY STAGE OF PEROVSKITE SOLAR CELLS (PSCS): FROM LIQUID TO SOLID-STATE PEROVSKITE-SENSITIZED SOLAR CELLS

Numerous configurations with a wide range of choice of materials have been investigated and developed for perovskite solar cells (PSCs) with success and promising appearance in the mass production line shortly. To obtain a better understanding of the evolution and development of this device, it is worth looking back to the start of this interesting and amazing era of organic–inorganic perovskite materials and its application in the solar cells (Figure 2.1).

In 2009, Professor Miyasaka's group from Tokyo Polytechnique University, Japan published an article in the Journal of American Chemical Society on organometal halide perovskites solar cells.[1] In this pioneering work, the authors studied two materials $CH_3NH_3PbI_3$ and $CH_3NH_3PbBr_3$ and were the very first group to apply them in photovoltaics. In this work, perovskite was regarded as a visible light sensitizer thanks to previous investigations by several groups that confirm their broad and high-absorption spectra.[2] Miyasaka et al. employed lead halide perovskite in liquid-based dye-sensitized solar cell (DSSC) with the perovskite materials replaced dye molecules as light-absorbing material. Perovskites were deposited on the transparent conducting electrode with a mesoporous film titanium dioxide (TiO_2) nanoparticles in sandwiched with a platinized counter electrode. Between the two electrodes, a liquid electrolyte containing redox couple I^-/I_3^- was filled to transport electron back and forth between the counter electrode to the oxidized sensitizer which has injected its excited electron into the conduction band of TiO_2. The working mechanism of this device was expected to have no major difference compared to the conventional

DSSCs. Nonetheless, the main disadvantage is the stability of perovskite material in the electrolyte—which is often regarded as being highly corrosive[3]—making the device working for a short few minutes and degraded quickly after measurement. Due to this serious instability, despite being a very good light absorber, perovskite was not an interesting material for photovoltaic applications for a long time. Professor Nam Gyu-Park's group in 2011 picked up this topic and found out that small perovskite quantum dot of 2.5 nm can be energetically favorably formed on the TiO_2 surface. Nonetheless, in this paper, Park et al. continued to apply $CH_3NH_3PbI_3$ in I^-/I_3^- liquid electrolyte-based DSSC configuration and the durability remained poor, despite obtaining 6.5% efficiency with a thinner TiO_2 film compared to previous study (3.6–8.6 μm compared to 12 μm in Miyasaka et al.'s work)—depicting the high absorption coefficient of $CH_3NH_3PbI_3$.

Remarkably in 2013, Park's group decided to replace the liquid electrolyte with a solid-state hole-transporting material (HTM) in an attempt to make a solid-state DSSC. In a paper published with Grätzel's group, they have successfully produced all-solid-state thin film solar cell with perovskite still being regarded as a sensitizer and spiro-OMeTAD as the HTM. It is worth noting that this solar cell not only had significantly high efficiency of over 9% compared to other perovskite-based solar cell in mentioned articles but it was also much more stable over 500 h.[4] This was a breakthrough report and opened up the new era for PSC. This new device configuration aligns with the extremely thin absorber (ETA) cell structure—which is often applied for inorganic materials or quantum dots that have high-absorption coefficient and does not require a thick layer of TiO_2.[5,6] The new generation of PSCs can be categorized into two main groups: the mesoscopic PSC which contains a layer of nanostructured materials—active or passive—to act as scaffold supporting the perovskite layer and the planar solar cell configuration which consists of a thin layer of perovskite is sandwiched between two selective contacts. In the following sections, these configurations will be discussed in details.

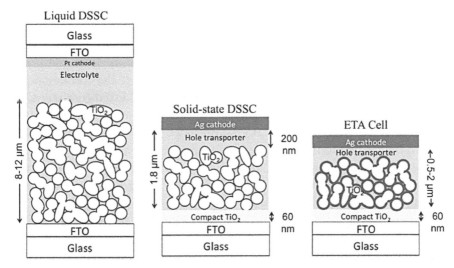

FIGURE 2.1   Historic evolution of device structure: perovskite was first applied in liquid dye-synthesized solar cell (DSSC) with 8–12-µm thick TiO$_2$ film, then in solid-state DSSC with 1.5–3 µm and eventually resembled ETA cell with 0.5–2-µm film. (Note: FTO: fluorine-doped tin oxide)
Source: Adapted from ref 7.

2.2   MESOSCOPIC PSC

The mesoscopic PSC configuration consists of a nanostructured material which acts as a scaffold for the deposition of the perovskite. This nanostructured material is often a semiconducting metal oxide with wide bandgap or it can even be an insulating material. Depending on the function of this metal oxide scaffold, some transport charge carrier and some do not, the mesoscopic PSCs discussed in this chapter can be classified into two categories: the active scaffold and passive scaffold.

2.3   MESOSCOPIC PSC WITH ACTIVE SCAFFOLD

Many metal oxide nanostructures have been studied as the active scaffold for mesoscopic PSCs but the mesoporous TiO$_2$ nanoparticles network similar to the photoanode of DSSC is still by far the most commonly used material. A regular typical configuration of mesoscopic PSC is as follow: FTO/TiO$_2$ compact/TiO$_2$ mesoporous layer/CH$_3$NH$_3$PbI$_3$/spiro-OMeTAD/

Ag. The TiO_2 compact film serves as the hole-blocking layer to reduce recombination at the anode due to perovskite or HTM being in contact with FTO. The TiO_2 mesoporous film's function is still in debate. In principal, it can serve as electron-transporting layer, but its necessity is in question, while on the other hand, it acts as a scaffold to support the formation of the perovskite $CH_3NH_3PbI_3$. Spiro-OMeTAD is the HTM and Ag serves as the hole collector or counter electrode. This device configuration is often regarded as n-i-p mesoscopic due to the presence of the mesoscopic metal oxide layer (mesoporous network of nanoparticles or 2-dimension nano-structures) and the physical bottom layer (closest to the fluorine-doped tin oxide FTO substrate—the front electrode) is the electron transporter and made of n-type semiconductor.

For fabrication, a blocking layer is deposited on the FTO substrate typically done by spray pyrolysis but other chemical and physical deposition methods have been reported including spin-coating or sputtering. The blocking layer should be compact and free of pinhole to prevent the recombination between the perovskite and the FTO substrate. This layer is often made of the same materials with the mesoscopic layer, yet different materials with matching energy levels can also be used. Many efforts have been paid to improve the conductivity through doping. The mesoscopic layer is then deposited on the compact layer by screen printing or hydrothermal or other synthesis methods (Figure 2.2). The perovskite layer can be deposited on the mesoporous layer by single or sequential deposition. In single deposition method, both the precursors (PbI_2 or $PbCl_2$ and CH_3NH_3I) are mixed in the solution in advance and the perovskite is formed during spin-coating and post-annealing. In sequential deposition method introduced by Burschka et al., PbI_2 is first spin-coated on the mesoscopic layer, followed by dipping in CH_3NH_3I solution.[8] During the dipping process, PbI_2 was converted to $CH_3NH_3PbI_3$. The advantage of the mesoscopic layer was shown to have significant effect in the sequential deposition method. In this method, it is believed that only a thin layer of PbI_2 can be converted to perovskite through dip coating, hence it is important to limit the thickness of PbI_2 coated to several tens of nanometer. The small pore in mesoporous TiO_2 film facilitates uniform coating of PbI_2 to be limited to the size of the pore.

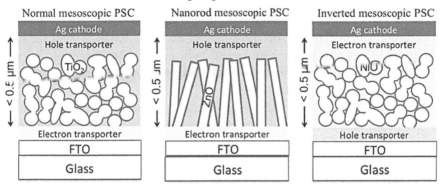

FIGURE 2.2 Schematic diagram of mesoscopic perovskite solar cells (PSCs). In normal structure, the nanostructured film can be n-type semiconductor TiO_2, ZnO, or one-dimensional nanomaterials like nanorod, nanowires. In inverted mesoscopic PSC, the mesoporous film is often made of NiO nanoparticles.
Source: Adapted from ref 7, 14, 20.

In the Miyasaka et al.'s employment of perovskite in liquid-based solar cell, TiO_2 films of 8 and 12 μm were used.[1] In Park et al.'s follow-up work in 2011, the authors have studied the effect of TiO_2 film thickness by varying the 3.6–8.6 μm and observed a significant improvement with decreasing the film thickness thanks to the high adsorption coefficient of perovskite $(1.5 \times 104/cm)$—an order higher than Ruthenium-based N719 dye.[9] When later employed in solid-state DSSCs, the study by Park and Grätzel et al. used a TiO_2 film of only 0.8 μm, a fold lower and two times lower than their liquid-state and solid-state DSSCs counterparts, respectively and the achieve efficiency exceeding 9%.[4] Despite the necessity of this scaffold to obtain high-efficiency PSCs is still in question, the thickness of the TiO_2 layer has been regarded as a crucial parameter influencing the device performance. Conventional solid-state DSSCs require thickness of a TiO_2 mesoporous layer of 2–3 μm, thick enough for sufficient dye-loading responsible for light absorption while thin enough to permit facile pore filling of solid HTM. However, in PSC, sub-micrometer TiO_2 layer thickness has been proven to provide sufficient loading of the light absorbing layer due to perovskite's high absorption coefficient and also because of its pore filling complication in the mesoporous layer.

Leijitens et al. have investigated the pore filling fractions of perovskite and found that it depends strongly on the thickness of the TiO_2 film.[10] The perovskite layer should not only penetrate into the pores between the TiO_2 nanoparticles but also form a thin layer covering the TiO_2 to prevent recombination with the HTM. The quality of the perovskite film is also dependent on the concentration and solvent used in the solution of perovskite precursor and deposition method. In Leijitens et al.'s finding, for perovskite concentration of 40 wt.%, the thickness of the TiO_2 layer should be low (~below 440 nm) to reach 100% pore filling for achieving high device performance. The high efficiency related to high percentage of pore filling is attributed to high electron density in the TiO_2 leading to better charge transport rates and higher charge collection efficiency. The full coverage of TiO_2 by perovskite also reduces the recombination reaction between the injected electron in TiO_2 and hole in spiro-OMeTAD.

The pore-filling efficiency changes not only according to the TiO_2 thickness but also according to its pore size and porosity. Although TiO_2 nanoparticles forming a mesoporous film is the most commonly used nanostructure for mesoscopic PSC, other materials and structures have also been investigated. Various nanocrystals of TiO_2 such as nanorods,[11] nanowires,[12] nanofibers,[13] with different size and shapes have been investigated. Other n-type semiconductors that have similar bandgap and energy level to TiO_2 have also been investigated to construct the scaffold of mesoscopic PSCs, such as ZnO,[14] WO_3. One-dimensional nanomaterials are investigated due to their promising better charge transport compared to the network of nanoparticles (Figure 2.3). The use of TiO2 nanorods in fabrication of perovskite solar cells was realized in this approach.

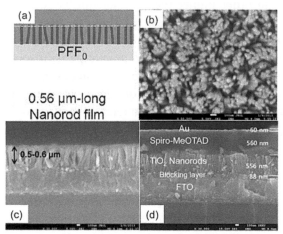

FIGURE 2.3　Illustration (a) and field emission scanning electron microscope (FESEM) images ((b) top view, (c) cross-sectional) of 0.56 μm-long rutile TiO_2 nanorods. The cross-sectional view of corresponding mesoscopic PSC is shown in (d). The authors also fabricated PSCs with 0.92 and 1.58-μm-long nanorods.

Source: Reprinted with permission from ref 15. © 2013 American Chemical Society.

Further enhancement in the performance of PSC can be achieved by improving conductivity for better charge transport and collection, doping of the nanostructured semiconductors, or incorporating them with higher conductivity materials. Han et al.[16] demonstrated the incorporation of reduced graphene oxide (RGO) and mesoporous TiO_2 to form a composite that reduced internal resistance and enhanced charge collection with increasing concentration of RGO (Figure 2.4). The optimized condition of 1% loaded RGO (w/w ratio) has resulted in 14.5% efficiency with higher J_{sc}, V_{oc}, and FF as compared to the 13.5% mesoporous TiO_2 conventional device.

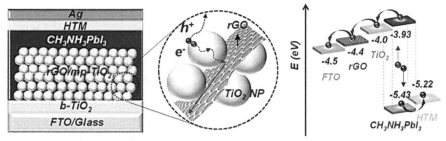

FIGURE 2.4   Diagram of reduced graphene oxide/mesoporous-TiO_2 nanocomposite-based PSCs.
Source: Reprinted with permission from ref 16. © 2015 American Chemical Society.

Conventional perovskite mesoscopic solar cells has n-i-p structure with n-type semiconductor as the mesoscopic nanostructure which accepts and transport electron. However, active scaffold materials can also be p-type semiconductors to create inverted mesoscopic architecture of p-i-n. NiO is the most common p-type metal oxide used in DSSCs now again applied to PSCs. Photoluminescence study performed by Docampo et al. showed photoluminescence (PL) quenching of perovskite in the presence of NiO indicating efficient charge transfer from perovskite to NiO with estimated 95%.[17] The preliminary study by Tian et al.[18] with structure FTO/NiO$_x$/ mp-NiO/$CH_3NH_3PbI_3$/$PC_{61}BM$/BCP/Al achieved 1.5% efficiency (where phenyl-C_{61}-butyric acid methyl ester [$PC_{61}BM$] and bathocuproine [BCP]). Further investigation by Taiwanese groups published in several articles has suggested an optimal thickness of NiO mesoporous layer deposited by spin-coating to be around 250 nm and the compact NiO or electron blocking layer is influenced by not only its thickness but also the doping quality (Figure 2.5).[19,20] By employing low temperature sputtering of NiO in flow of O_2 and Ar (1:9 ratio) to deposit NiO film of 10 ± 0.5-nm thick, an efficiency of

11.6% was obtained. Physical deposition method for NiO electron-blocking layer and mesoporous layer has exhibited more superior characteristics to enhance the performance of inverted mesoscopic PSC. NiO has also been doped with Cu to increase its conductivity and balance the hole-transport diffusion length closer to that of electron.[21]

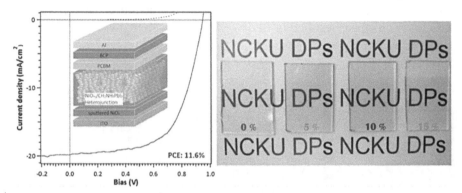

FIGURE 2.5 NiO_x/perovskite heterojunction cell achieving 11.6% power conversion efficiency (PCE) and picture of sputtered NiO films with increasing oxygen flow ratios. *Source:* Reprinted with permission from ref 19. © 2014 American Chemical Society.

Hysteresis of IV characteristic is widely known in PSCs device performance and has been attributed to many factors including ion migration, polarization, trap states, and unbalanced charge transport. With inverted p-i-n configuration TCO/hole-transporting layer (HTL)/perovskite/electron-transporting layer (ETL)/metal electrode, the hysteresis is often suppressed. The lower hole diffusion length of HTM such as NiO compared to the electron diffusion length of its n-type semiconductor counterpart has assisted the smaller hysteresis in inverted solar cell configuration. Nonetheless, the recombination between holes injected in the NiO and free electron in perovskite layer is two orders of magnitude higher than the recombination at $MAPbI_3/TiO_2$ interface.[22]

2.4 MESOSCOPIC PSC WITH PASSIVE SCAFFOLD

Very shortly after the breakthrough finding by Park and Grätzel et al. employing spiro-OMeTAD as HTM in a solid-state PSC instead of the previous liquid-based device,[4] Snaith and his coworkers published another remarkable article in prestigious Science journal that changed scientists' perspective view on perovskite's properties.[23] In this article, Snaith et al. reported new device architecture by replacing the mesoporous TiO_2 layer with a mesoporous Al_2O_3 film. Al_2O_3 is a wide bandgap material (7–9 eV) and is an electrical insulating material. The device with Al_2O_3 surprisingly exhibited better power conversion efficiency compared to that employing TiO_2 scaffold, especially with higher V_{oc} (0.98 V vs. 0.80 V in TiO_2). Since Al_2O_3 is an insulator, electron is not injected into the Al_2O_3 but is transported in the perovskite layer itself, and eliminating the recombination between injected electron in the metal oxide with hole in perovskite in the case of TiO_2 scaffold. It is very important to note that previously perovskite was thought to act mainly as a sensitizer and only absorbs light, but from this finding, it was realized that perovskite can transport electron effectively and the necessity of TiO_2's role in PSC is put under question.

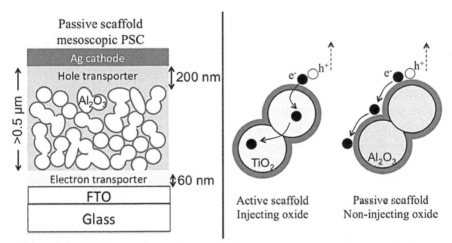

FIGURE 2.6 (Left) Schematic diagram of passive scaffold mesoscopic PSCs, the nanostructured film can be Al_2O_3, SiO_2, ZrO_2. (Right) Illustration of charge transfer and transport in an active scaffold (TiO_2)-based PSC versus a passive scaffold (Al_2O_3). *Source:* Adapted from ref 23.

The first article introducing Al_2O_3 as passive scaffold presented a deposition technique similar to that of TiO_2 mesoporous layer with Al_2O_3 particles incorporated in a paste with binding agents that require high-temperature sintering to remove (500°C). Nonetheless, since electron is not injected in the passive Al_2O_3, it is realized that its crystallinity or conductivity is not crucial and low-temperature processing is highly possible. Ball et al. later demonstrated an improved approach to deposit Al_2O_3 from a commercially available solution of colloids (<50-nm nanoparticles)[24] in a configuration of FTO/compact TiO_2 blocking layer/Al_2O_3/$CH_3NH_3PbI_{3-x}Cl_x$/spiro OMeTAD/ Ag. This process was completed with 150°C annealing and did not require high temperature since there is no binder to be removed. They have achieved a significantly high performance of 12.3% efficiency (V_{oc} was 1.02 V, J_{sc} of 18 mA/cm^2, and FF 0.67). Shortly after, lower than 150°C processable highly crystalline TiO_2 hole-blocking layer was introduced by Wojciechowski et al.[25] allowing the realization of a fully low-temperature fabricated PSC promising for printing flexible device. The low-temperature blocking layer consisted of very small nanoparticles dispersed in an alcohol-based solution of titanium diisopropoxide bis (acetylacetonate) which decomposes to TiO_x upon drying and bridging the small nanoparticles to form continuous crystalline thin film. The authors have also discovered that the conventional high-temperature approach to deposit compact blocking layer by sintering spin-coated film from titanium isopropoxide (TTIP) solution resulted in significantly high series resistance. With low-temperature process, the resistance is lowered and the solar cell's fill factor is greatly improved. The best solar cell with this approach exhibited 15.9% efficiency.

Large bandgap materials other than Al_2O_3 have also been employed as passive scaffold in PSC such as ZrO_2[26] or SiO_2.[27] The working mechanism of passive scaffold PSC has been suggested to follow that of ETA solar cells from the first Science paper that introduced the use of Al_2O_3.[23] It is later demonstrated more clearly by Kim et al. from electrochemical impedance spectroscopy (EIS) measurements of ZrO_2-based PSC.[28] In the EIS spectra, charge accumulation in different parts of the device can be observed as capacitance in an equivalent electrical circuit. This was the first charge accumulation observed in the perovskite layer—which was thought to be a light-absorbing material—and it has proven more firmly that the perovskite layer functions not only as light absorber but also as ambipolar electron and hole transporters. This finding shed a new light on the understanding of not only perovskite as an emerging material for energy conversion but also the working mechanism and architecture of PSC. With passive scaffold, PSC's working mechanism similar to ETA; in contrast to active scaffold, PSC behaving in between a DSSC and a thin-film solar cell. It is also worth mentioning that photovoltage in passive scaffold devices is often higher compared to its active scaffold counterparts. For comparison, the V_{oc} of some notable articles is reported as follow: in Lee et al.'s work, V_{oc} was 0.8 V for TiO_2 and 0.98 V for Al_2O_3 PSC, respectively;[23] in Hwant et al.'s work, V_{oc} was 0.94 V for TiO_2 and 1.05 V for SiO_2 PSC, respectively;[27] in Bi et al.'s work, V_{oc} was 0.89 V for TiO_2 1.07 V for ZrO_2 PSC, respectively.[26] From these work, it is realized that the V_{oc} is often 10–20% higher in passive scaffold compared to active scaffold. An explanation is due to the lower recombination at the interface of HTM with perovskite and the passive scaffold (Figure 2.6). Since there is no electron injected in the scaffold, the recombination reaction happens between injected electron and HTM is eliminated. Moreover, investigation by EIS has shown a reduction of chemical capacitance in the insulating oxide. Chemical capacitance represents the accumulation of charges and is closely related to the material's sub-bandgap or density of states. Unlike semiconductors like TiO_2, insulators such as Al_2O_3, SiO_2, and ZrO_2 do not have sub-bandgap states and the chemical capacitance of the cell is significantly decreased.[23]

The performance of devices with passive scaffold is also affected by the film thickness, particle size, porosity, and pore size. With Al_2O_3 mesoscopic PSC, Ball et al. have varied the thickness of Al_2O_3 layer from 0 to 1.4 μm and investigated the corresponding performance. Both V_{oc} and FF increased slowly with increasing Al_2O_3 thickness until they reached the optimized point at 0.4–0.8 μm, while J_{sc} reached optimum 16.9±1.9 mA/cm² with a thin film of only 80 nm. Investigation of the optical property through internal quantum efficiency (IQE) shows that IQE also peaked with film thickness of 80 nm and reduced with a thicker film. This result justified the predominant spontaneous electron-hole pair dissociation and the transport and collection of electron and hole in the range of over hundreds of nm is poor. Since the best device of 80 nm Al_2O_3 was composed mainly of solid perovskite films had IQE approaching almost 100%, this indicated that a highly crystalline perovskite is capable of ambipolar charge transport. The current drop was attributed to grain boundaries charge trapping and recombination since the grain size of perovskite crystals incorporated in the Al_2O_3 porous structure is smaller than in the solid film over layer.[24]

The effect of porosity and particle size on the infiltration and eventually the performance of passive scaffold-based PSC were investigated by varying SiO_2 monodispersed particles from 15 to 100 nm. It was observed that pore size of over 50 nm was sufficient for the infiltration of perovskite, while 100 nm SiO_2 could result in scattering of either SiO_2 or perovskite leading to higher absorption in the longer wavelengths. However, this complex scattering reduced the overall absorption, and devices with particles larger than 50 nm had a slight decrease in J_{sc}, while those with particles smaller than 50 nm had sharper decrease of J_{sc} due to lower loading of perovskite. The poor infiltration also led to increasing the resistances and lower FF with <50 nm particles.[27]

2.5 PLANAR ARCHITECTURE

The first PSC structure was based on the configuration of liquid electrolyte-based DSSCs and developed towards solid-state by replacing the electrolyte with HTM in the remarkable articles by Park et al.[4] Shortly after Snaith et al.'s report[23] on passive scaffold mesostructured Al_2O_3 signifying the electron transportation capability of lead halide perovskite, another noteworthy study published by Etgar et al. demonstrated a device architecture in which spiro-OMeTAD—the HTM—was eliminated.[29] The structure represented was FTO/TiO_2 blocking layer/TiO_2 sheets/$CH_3NH_3PbI_3$/Au 5.5% efficiency. Although the efficiency was lower than other reports at the same time, this is still a very important finding because, for the first time, perovskite was proven to be able to transport hole. Several groups have quickly followed up this work and eventually proved the ambipolar property of lead halide perovskite allowing it to transport both electron and hole.[30] Together with the investigation of passive scaffold such and Al_2O_3, ZrO_2, SiO_2, it has been suggested that the role of the mesostructure is unnecessary. A planar configuration was hence proposed without the scaffold and the perovskite layer is a thin film. Nonetheless, due to imbalanced diffusion length,[30] it is essential to separate charges by selective contacts and to block the opposite charges, reducing recombination.[31] The planar PSC configuration resembles a thin film or heterojunction solar cell with a layer of perovskite in sandwich with two selective contacts. This configuration allows more facile fabrication as the mesostructure layer which often requires high-temperature deposition is eliminated. This is a major advantage for the development of low-temperature flexible devices on polyethylene terephthalate (PET) substrates using printable processing methods.[32] It also facilitates the integration of perovskite for top or bottom tandem solar cells. Furthermore, different deposition techniques of organometal halide perovskites can be investigated and applied. The choice of electron or HTMs is also wider in planar configuration because it is no more necessary to synthesize nanostructures.

Planar heterojunction PSC can have a conventional n-i-p or an inverted p-i-n architecture (Figure 2.7). In this section, each structure will be discussed in details on choice of materials and fabrication process, as well as the advantages and disadvantages involved.

Planar perovskite solar cells

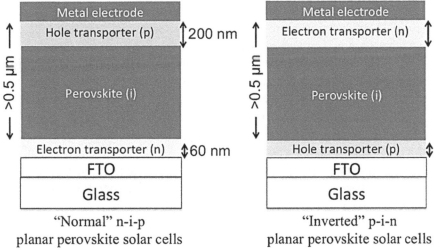

FIGURE 2.7 Schematic diagram of normal n-i-p and inverted p-i-n perovskite solar cells. *Source:* Adapted from ref 62.

2.6 CONVENTIONAL N-I-P STRUCTURE

In an n-i-p architecture, the bottom contact collects negative charges while the top contact collects positive charges. The most common structure of a conventional n-i-p planar PSC is $FTO/TiO_2/CH_3NH_3PbI_3/spiro\text{-}OMeTAD/Ag$. In planar devices, it is critical to optimize the quality of the perovskite layer. Liu et al. were the first to implement this structure using mixed-halide perovskite $CH_3NH_3PbI_xCl_{3-x}$ and comparing two methods of deposition: by dual-source thermal evaporation and by solution process spin-coating.[32] Although both techniques produced high quality of crystallined mixed-halide perovskite, an analysis of the morphology by top-view scanning electron microscopy (SEM) have shown a superior quality of coverage and uniformity of the film deposited by evaporation in contrast to partially coated structure with micron-size voids in between crystal appeared on the solution processed film. The cross-sectional images of these films also revealed much larger crystal size with spin-coated film but had an undulating nature and the film thickness varied widely from 50–410 nm, while the vapor-deposited film had uniform thickness of 330 nm in average. The superior uniformity over a large scale of the vapor-deposited film eventually resulted in better solar cell performance. The optimized planar PSC fabricated from this method with a thickness of the perovskite layer of 330 nm exhibited a J_{sc} of 21.5 mA/cm^2, a V_{oc} of 1.07V, a FF of 0.68, and yielding an efficiency of 15.4% comparing to 8.6% efficiency of the spin-coated device—whose other parameters were all lower. The optimized thickness has not only allowed sufficient absorption but also thin enough to facilitate the diffusion of electron and hole to respective contacts. In addition, the uniformity of the film over a length scales prevent shunting caused by the direct contact of spiro-OMeTAD and TiO_2 compact layer.

Choosing electron-transporting material (ETM) and HTM are crucial in the development of planar PSC as they are the key layers to extract charges, block the opposite charge carriers, and reduce recombination. Since low-temperature processability is a major advantage of planar PSC, synthesis of TiO_2 compact layer with low temperature attracted many research groups. Conings et al. demonstrated a method for deposition of TiO_2 layer at temperature lower than 150°C.[34] This layer was applied as the electron-transporting layer in the architecture of ITO/TiO_2/$CH_3NH_3PbI_xCl_{3-x}$/P3HT/Ag (with P3HT poly (3-hexylthiophene) being the HTM) achieving 13.6% efficiency. This method involves using TTIP precursor with the presence of nitric acid and water as the source for hydrolysis in alcohol solvent to form a stable transparent dispersion of nanoparticles of 6 nm. Prior to spin-coating, a small amount (8%) of titanium diisopropoxide bis (acetylacetonate) (TiAcAc) was added to facilitate the connection between nanoparticles and assist the formation of the thin film. The author also optimized the condition of annealing with the best temperature of 135°C in 45 min, that is sufficient for decomposition of NO_3^-. Annealing longer than 45 min resulted in coarse perovskite morphology formed and too many pinholes in the perovskite layer, leading to degradation of device performance.

Other low-temperature TiO_2 deposition methods have also been developed in the literature. Yella et al. introduced chemical bath deposition of nanocrystalline rutile TiO_2 with $TiCl_4$ precursor. $TiCl_4$ has been widely used in liquid-based DSSC to form a thin blocking layer in a process often known as pretreatment and can easily form a rutile phase crystallined film with low-temperature process (70°C). The thickness and morphology of the film can be controlled by the concentration of the precursor solution. With this simple process, an impressive PCE of 13.7% has been achieved compared to the high-temperature processed anatase TiO_2 film prepared by spin-coating of $TiCl_4$ solution that had only 3.7% efficiency.[35]

Oxides other than TiO_2 have also been employed as the ETL in planar n-i-p PSC such as ZnO. ZnO has high electron mobility and energy band levels align with that of TiO_2. Planar ZnO-based PSC was first introduced by Kumar et al. in 2013[14] using electrodeposition method (Figure 2.8). In this work comparing nanostructured ZnO and planar ZnO, the planar devices exhibited 5.54 and 2.18% efficiency on FTO substrate and ITO/PET flexible substrate, respectively. Although lower than the nanorod counterparts due to lower J_{sc} and FF, this study pioneered and proved the concept of employing ZnO as low-temperature processable electron-transporting layer in PSC.

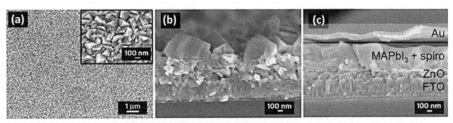

FIGURE 2.8   FESEM images of (a) top view of the ZnO compact layer electrodeposited on FTO, (b) cross-sectional view of perovskite on the ZnO compact layer, and (c) cross-sectional view of the fabricated planar n-i-p device.
Source: Reprinted with permission from ref 33. © 2013 Royal Society of Chemistry.

Liu and Kelly synthesized a thin film of nanocrystallined ZnO by hydrolysis of Zn acetate solution at room temperature[36] with superior electron mobility. The champion cell yielded 15.7% efficiency while the flexible ones had 10.2% on average. Interestingly, a strategy sandwiching TiO_2/ZnO/TiO_2 by spin-coating multiple layers was reported to exhibit maximum incident-photon-to-current efficiency at the range of 400–780 nm. The 12.8% efficiency achieved was attributed to reduced recombination as compared to single layer of TiO_2 or ZnO alone, or TiO_2/ZnO layer. Beside chemical methods, physical deposition of ZnO or TiO_2 have also been investigated such as atom layer deposition[37] or sputtering[38] Semiconductors other than oxides have also been used as electron-transporting layer in PSC such as the demonstration by Wang et al. with CdSe nanoparticles.[39] The device with a structure of FTO/CdSe/$CH_3NH_3PbI_3$/spiro-OMeTAD/Ag exhibited an efficiency of 11.7%.

2.7   INVERTED P-I-N PLANAR STRUCTURE

The first few reports on p-i-n structure were influenced by the design of heterojunction concept of organic photovoltaics. This idea came from the fact that perovskite exhibits ambipolar semiconductor characteristic and can generate and transport both electrons and holes. The first reported p-i-n PSC employed organic hole and electron-transporting layer having the architecture: $FTO/PEDOT:PSS/CH_3NH_3PbI_3/PC_{61}BM$ (or C_{60} fullerene)/Al (with PEDOT:PSS is Poly (3,4-ethylenedioxythiophene)-poly (styrenesulfonate) and $PC_{61}BM$ is fullerene derivative [6,6]-phenyl-C_{61}-butyric acid methyl ester). The initial result was merely 3.9%[40] but device performance enhancement had been developed quickly to over 18% within a short period with $ITO/PEDOT:PSS/CH_3NH_3PbI_3/PC_{61}BM/Au$ configuration.[41] The high efficiency of inverted planar PSC was accounted for the balanced rate of electron and hole transport in the respective layer, in contrast to the imbalance rate in normal cell where hole was transported at a faster rate than electron.

It is worth noting that the differences between p-i-n and n-i-p devices do not only lie on the normal or inverted architecture but also on the choice of materials. Due to organometal halide perovskite's sensitivity to humidity and oxygen, it is very crucial to choose the materials being deposited on top of it while it is more flexible for the choice of the layer underneath. For example, PEDOT:PSS is water-based and cannot be deposited on top of perovskite through solution process, but it is possible to be spin-coated on the FTO (or ITO) substrate before the deposition of perovskite. Hence it is possible to employ PEDOT:PSS in inverted p-i-n solar cell but not in the normal structure. Although its advantage is low-temperature facile process, PEDOT:PSS's work function varies greatly with the ratio of its ionomers (4.9–5.2 eV) and may not perfectly match the valence band of perovskite. Furthermore, PEDOT:PSS has been known for its corrosive nature and hygroscopic affecting PSC's stability, replacing it with graphene oxide has been considered previously for bulk heterojunction and lately for perovskite heterojunction (PHJ). Quenching of photoluminescence of $CH_3NH_3PbI_{3-x}Cl_x$ proves that there is charge transfer between the materials and the perovskite film formed on graphene oxide (GO) had high uniformity and lower roughness compared to PEDOT:PSS, making it a highly potential alternative. Wu et al. fabricated inverted planar PSC with GO as HTL having 12.4% efficiency, higher than 9.3% of the PEDOT:PSS counterpart.

Polythiophene (PT) has a matching highest occupied molecular orbital (HOMO) level (~ 5.2 eV) with the maxima of valence band of perovskite and hence was chosen as hole-transporting layer in a structure of ITO/PT/

$CH_3NH_3PbI_3/C_{60}/BCP/Ag$ with optimized efficiency of 11.8%.[42] Although PT has increasing conductivity with higher thickness, its limitation lies on its light adsorption that when using on the front electrode, it caused loss of photons reaching the perovskite layer. Polander et al. performed investigation on a series of HTMs with varying HOMO level to identify the correlation between the HOMO and V_{oc} of the solar cells.[43] Lowering HOMO lead to increasing V_{oc} with the highest reported of 1.03 V with spiro-TTB (a derivative of spiro-OMeTAD having methyl groups replacing methoxy groups and a HOMO level of -5.3 eV[43]). However, if the HOMO level was too low (lower than -5.3 eV), there would not be sufficient driving force for regeneration and the performance deteriorated.

The choice of p-type materials for hole-transporting layer in inverted PSC not only includes organic semiconductors like PT,[42] spiro-OMeTAD, GO also is extended to inorganic such as NiO_x,[44,45,46] MoO_x,[47] CuI,[48,49] Cu_2O,[50,51] or CuSCN.[52] NiO, in particular, has been long employed for p-type solid-state DSSC and naturally became a suitable choice for PSC since its development from DSSC.[53] The first few groups applying NiO in PSC as a mesoporous layer of nanoparticles, but later on a thin film of NiO could be deposited by electrodeposition or solgel methods with good hole-transporting mobility and giving higher V_{oc} thanks to the suitable energy alignment between NiO_x and perovskite (Figure 2.9).[54] Surface wettability of perovskite on NiO_x film can also be improved by ultraviolet (UV)-ozone treatment that leads to better quality of active area and better charge extraction and transport compared to PEDOT:PSS.

In contrast to planar n-i-p PHJ in which most ETMs are inorganic, in inverted planar PHJ p-i-n most ETM being investigated are organic due to their low-temperature processing. Fullerene C_{60} and its derivatives such as $PC_{61}BM$, indene C60 bisadduct (ICBA), $PC_{71}BM$ are the most widely studied materials. Although $PC_{61}BM$ has shown sufficient charge transfer through photoluminescence quenching from perovskite,[17] it is still a controversial debate that whether the fullerene is only an electron acceptor and transporting layer or it is essential to dissociate charges from the photo-generated excitons in the perovskite. The derivatives of fullerene have high solubility in organic solvents and different electron mobility and energy level and hence for each derivative, it is necessary to optimize the device performance based on the film thickness. Furthermore, the position of the lowest unoccupied molecular orbital (LUMO) level plays an important role to achieve high V_{oc}. To study the trend of V_{oc} with respect to LUMO level, Wang et al. investigated three fullerene derivatives in inverted planar configuration and achieved V_{oc} of 0.53, 0.91, and 1.06 V for C_{60}, $PC_{61}BM$, and ICBA, respectively.[55] PC$_{71}$BM has higher electron mobility and its absorption in the UV-visible range can contribute to the photocurrent when employing in PSC, it yielded a higher efficiency of 16.31%.[56] Other organic ETMs in p-i-n configuration include organoborane compounds (3TPYMB),[57] diperylene diimide.[58] Inorganic n-type semiconductor used in planar p-i-n PSC needs to be processed at low temperature not to damage the perovskite layer, hence ZnO is the most common choice. The choice of solvent is also critical for the stability of the device. Chlorobenzene was chosen by You et al. for deposition of ZnO nanoparticles to fabricate inverted planar PSC with the structure of ITO/PEDOT:PSS/CH$_3$NH$_3$PbI$_{3-x}$Cl$_2$/ZnO/Al achieving 11.5% efficiency.[59]

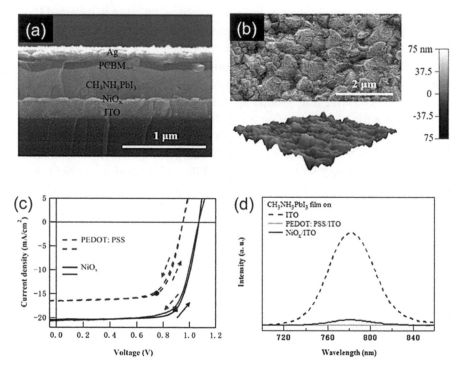

FIGURE 2.9 Flexible planar inverted p-i-n PSCs with NiO$_x$ and PEDOT:PSS as the HTL: (a) cross-sectional scanning electron microscopy (SEM) view of device, (b) top-view SEM and atomic force microscopy images of perovskite on NiO$_x$, (c) I–V characteristics, and (d) photoluminescence spectra showing quenching with the presence of NiO$_x$ and PEDOT:PSS. *Source:* Reprinted with permission from ref 45. © 2016 American Chemical Society.

In planar PSC, it is crucial to couple an interlayer to prevent shunting and reduce recombination. Both p-type and n-type interlayers (coupling with HTL and ETL, respectively) have been investigated, but n-type interlayer is often more critical because of the lower electron-transporting rate in PSC as well as the often poorly coverage of ETL on perovskite. The most common materials demonstrated as n-type interlayers are derived from organic solar cells, such as LiF,[60] ZnO, TiO$_x$,[17] BCP[40] and bathophenanthorline.[61] The roles of the interlayer are widely discussed and studied, including: reduction of energy barrier between the LUMO of the ETL (mainly PCBM) and the Fermi level (Al or Au);[60] acting as an optical spacer to improve absorption in the active layer; block holes; transport electron; and protection of the ETL from possible damage during thermal evaporation of the electrode.[55]

2.8   CONCLUSION

PSC was first developed based on the configuration of DSSC with regard to perovskite's excellent absorption in the visible range. Due to highly unstable devices within electrolyte environment, switching to organic HTM has brought a breakthrough in solar cell research and opened a new era of PSCs. Originally, an active scaffold layer—typically TiO$_2$—was used similar to solid-state DSSC. Upon realization of perovskite's ability to transport both electron and hole, other kinds of scaffolds have been employed: p-type scaffold such as NiO, and passive scaffold which acts as a template to support the deposition of perovskite but does not transport any charges. The ambipolar semiconductor property of perovskite eventually leads researchers to develop planar configuration: the perovskite active layer is in sandwich between an electron-transporting layer and a hole-transporting layer. This resembles well a thin-film organic solar cell and has achieved significantly high efficiency. It opens the debatable question whether the mesoscopic layer is necessary for the performance of PSC or not. The planar heterojunction clearly has more simple architecture and hence fabrication process is promising to be adopted for scale-up production.

There are three major aspects which should be considered when comparing mesoscopic versus planar structure.[62] First, the morphology and crystallization of perovskite layer are greatly dependent on the layer underneath it. The advantage of planar structure is that incomplete pore filling (that can lead to shunting or recombination) is avoided and large crystallites can be achieved. However, it has been shown that grain boundaries are harmless and polycrystalline films have long charge-carrier diffusions—indicating that large grain size is not very desirable. On the other hand, smaller crystallites formed on mesoporous scaffold exhibit broader absorption spectra compared to bigger crystallites deposited on planar substrate.

Second, in terms of diffusion length and charge-carrier generation, it is worth noting that the device's performance is greatly dependent on the film thickness because of exciton and free carrier diffusion lengths constraints. In early reports, it was suggested that PSCs are excitonic and in thick planar solar cell, the diffusion of excitons is not sufficient leading to losses. In contrast, the mesoscopic architecture allows construction of thicker film and efficient charge dissociation at the interface of perovskite/metal oxide semiconductors. The electron-hole diffusion length of $CH_3NH_3PbI_{3-x}Cl_x$ is over 1 μm while that $CH_3NH_3PbI_3$ perovskite is only ~ 100 nm.[63] While a concrete conclusion of exciton diffusion length is still in debate, it is clear that if the diffusion length was short, the preferred device structure would be mesoscopic. In this case, the distance that excitons or free carriers have to travel will depend on the pore sizes of the scaffold and maximum absorption can be achieved without sacrificing exciton and charge recombination.[62]

The third aspect to compare between mesoscopic and planar devices is hysteresis. From the beginning of PSC development, it was observed that the record I–V characteristic is highly dependent on the direction of measurement (forward or reverse bias), the rate of voltage sweep, and light soaking for the device.[64] Although the origin of hysteresis is not yet clear, most devices employing mesoporous TiO_2 scaffold have small hysteresis while planar devices suffer severely. By studying EIS, Kim and Park found that the capacitance at low frequency is higher in a planar device, indicating that the cause of hysteresis is due to the flat substrate inducing dipole polarization in the perovskite film.[65] Snaith et al. demonstrated that in mesoscopic device, charge transfer from perovskite to TiO_2 is efficient due to higher surface area and the impact is less pronounced.[66] Since hysteresis was still observed in Al_2O_3 passive scaffold device, this means that the device architecture is not the key factor, but fast charge dissociation and separation is probably the main cause to lower hysteresis. However, it is remarkably noted that inverted planar device having configuration of PEDOT:PSS/perovskite/$PC_{61}BM$ have been reported without hysteresis. Since the photoluminescence quenching of $CH_3NH_3PbI_{3-x}Cl_x$ to PEDOT:PSS and $PC_{61}BM$ are better than to spiro-OMeTAD and TiO_2, it can be understood that charge transfer of the former combination is better than the latter two materials. Therefore the normal n-i-p device exhibited larger hysteresis compared to the inverted structure.

High-efficiency PSCs can be achieved by the combination of device architecture and a proper choice of materials in each configuration. There remains a variety of options for materials, structures, and doping to enhance the performance of the mesoscopic layer considering its morphology including porosity, surface area, and its electrical properties such as better charge transport and collection attributed to higher conductivity.

KEYWORDS

- dye-sensitized solar cell
- liquid electrolyte
- hole-transporting material
- extremely thin absorber
- fluorine-doped tin oxide

REFERENCES

1. Kojima, A.; Teshima, K.; Shirai, Y.; Miyasaka, T. Organometal Halide Perovskites as Visible-Light Sensitizers for Photovoltaic. *J. Am. Chem. Soc.* **2009,** *131*(17), 6050–6051.
2. Kitazawa, N.; Watanabe, Y.; Nakamura, Y. Optical Properties of $CH_3NH_3PbX_3$ (X=halogen) and their Mixed-Halide Crystals. *J. Mater. Sci.* **2002,** *7*(37), 3585–3587.
3. Hagfeldt, A.; Boschloo, G.; Sun, L.; Kloo, L.; Pettersson, H. Dye-Sensitized Solar Cells. *Chem. Rev.* **2010,** *110,* 6595–6663.
4. Kim, H. S.; Lee, C. R.; Im, J. H.; Lee, K. B.; Moehl, T.; Marchioro, A.; Moon, S. J.; Humphry-Baker, R.; Yum, J. H.; Moser, J. E.; Grätzel, M.; Park, N. G. Lead Iodide Perovskite Sensitized All-Solid-State Submicron Thin Film Mesoscopic Solar Cell with Efficiency Exceeding 9%. *Sci. Rep.* **2012,** *2*(7436), 591.
5. Kamat, P. V.; Christians, J. A.; Radich, J. G. Quantum Dot Solar Cells: Hole Transfer as a Limiting Factor in Boosting the Photoconversion Efficiency. *Langmuir* **2014,** *30*(20), 5716–5725.
6. Emin, S.; Singh, S. P.; Han, L.; Satoh, N.; Islam, A. Colloidal Quantum Dot Solar Cells. *Sol. Energy* **2011,** *85*(6), 1264–1282.
7. Snaith, H. Perovskites: The Emergence of a New Era for Low-Cost, High- Efficiency Solar Cells. *J. Phys. Chem. Lett.* **2013,** *4,* 3623–3630.
8. Burschka, J.; Pellet, N.; Moon, S.-J.; Humphry-Baker, R.; Gao, P.; Nazeeruddin, M. K.; Grätzel, M. Sequential Deposition as a Route to High-Performance Perovskite-Sensitized Solar Cells. *Nature* **2013, 499,** 3–7.
9. Im, J.; Lee, C.; Lee, J.; Park, S.; Park, N. 6.5% Efficient Perovskite Quantum-Dot-Sensitized Solar Cell. *Nanoscale* **2011,** *3*(10), 4088–4093.
10. Leijtens, T.; Lauber, B.; Eperon, G. E.; Stranks, S. D.; Snaith, H. J. The Importance of Perovskite Pore Filling in Organometal Mixed Halide Sensitized TiO_2-Based Solar Cells. *J. Phys. Chem. Lett.* **2014,** *5 (7),* 1096–1102.

11. Kim, H.; Lee, J.; Yantara, N.; Boix, P. P.; Kulkarni, S.; Mhaisalkar, S. G.; Grätzel, M.; Park, N. High Efficiency Solid State Sensitized Solar Cell Based on Submicrometer Rutile TiO_2 Nanorod and $CH_3NH_3PbI_3$ Perovskite Sensitizer. *Nano Lett.* **2013,** *1,* 1–8.

12. Jiang, Q.; Sheng, X.; Li, Y.; Feng, X.; Xu, T. Rutile TiO_2 Nanowire-Based Perovskite Solar Cells. *Chem. Commun.* **2014,** *50*(94), 14720–14723.

13. Dharani, S.; Mulmudi, H. K.; Yantara, N.; Thu Trang, P. T.; Park, N.-G.; Graetzel, M.; Mhaisalkar, S.; Mathews, N.; Boix, P. P. High Efficiency Electrospun TiO_2 Nanofiber Based Hybrid Organic–Inorganic Perovskite Solar Cell. *Nanoscale* **2014,** *6*(3), 1675–1679.

14. Kumar, M. H.; Yantara, N.; Dharani, S.; Graetzel, M.; Mhaisalkar, S.; Boix, P. P.; Mathews, N. Flexible, Low-Temperature, Solution Processed ZnO-Based Perovskite Solid State Solar Cells. *Chem. Commun. (Camb).* **2013,** *49*(94), 11089–11091.

15. Kim, H.; Lee, J.; Yantara, N.; Boix, P. P.; Kulkarni, S.; Mhaisalkar, S. G.; Grätzel, M.; Park, N. High Efficiency Solid-State Sensitized Solar Cell-Based on Submicrometer Rutile TiO_2 Nanorod and $CH_3NH_3PbI_3$ Perovskite Sensitizer. *Nano Lett.* **2013,** *13*(6), 2412–2417.

16. Han, G. S.; Song, Y. H.; Jin, Y. U.; Lee, J. W.; Park, N. G.; Kang, B. K.; Lee, J. K.; Cho, I. S.; Yoon, D. H.; Jung, H. S. Reduced Graphene Oxide/Mesoporous TiO_2 Nanocomposite Based Perovskite Solar Cells. *ACS Appl. Mater. Interfaces* **2015,** *7*(42), 23521–23526.

17. Docampo, P.; Ball, J. M.; Darwich, M.; Eperon, G. E.; Snaith, H. J. Efficient Organometal Trihalide Perovskite Planar-Heterojunction Solar Cells on Flexible Polymer Substrates. *Nat. Commun.* Nov. **2013,** *4,* 2761.

18. Tian, H.; Xu, B.; Chen, H.; Johansson, E. M. J.; Boschloo, G. Solid-State Perovskite-Sensitized p-Type Mesoporous Nickel Oxide Solar Cells. *ChemSusChem* **2014,** *7*(8), 2150–2153.

19. Wang, K.; Shen, P.; Li, M.; Chen, S.; Lin, M.; Chen, P.; Guo, T. Low-Temperature Sputtered Nickel Oxide Compact Thin Film as Effective Electron Blocking Layer for Mesoscopic $NiO/CH_3NH_3PbI_3$ Perovskite Heterojunction Solar Cells. *ACS Appl. Mater. Interfaces* **2014,** *6,* 11851–11858.

20. Wang, K.-C.; Jeng, J.; Shen, P.; Chang, Y.-C.; Diau, E. W.; Tsai, C.; Chao, T.-Y.; Hsu, H.-C.; Lin, P.; Chen, P.; Guo, T.-F.; Wen, T.-C. p-Type Mesoscopic Nickel Oxide/Organometallic Perovskite Heterojunction Solar Cells. *Sci. Rep.* **2014,** *4,* 4756.

21. Kim, J. H.; Liang, P.; Williams, S. T.; Cho, N.; Chueh, C.; Glaz, M. S.; Ginger, D. S.; Jenm A. K. High-Performance and Environmentally Stable Planar Heterojunction Perovskite Solar Cells Based on a Solution-Processed Copper-Doped Nickel Oxide Hole-Transporting Layer. *Adv. Mater.* **2014,** *27,* 695–701.

22. Xu, X.; Liu, Z.; Zuo, Z.; Zhang, M.; Zhao, Z.; Shen, Y.; Zhou, H.; Chen, Q.; Yang, Y.; Wang, M. Hole Selective NiO Contact for Efficient Perovskite Solar Cells with Carbon Electrode. *Nano Lett.* **2015,** *15*(4), 2402–2408.

23. Lee, M. M.; Teuscher, J.; Miyasaka, T.; Murakami, T. N.; Snaith, H. J. Efficient Hybrid Solar Cells Based on Meso-Superstructured Organometal Halide Perovskites. *Science* **2012,** *338*(6107), 1–5.

24. Ball, J. M.; Lee, M. M.; Hey, A.; Snaith, H. J. Low-Temperature Processed Meso-Superstructured to Thin-Film Perovskite Solar Cells. *Energy Environ. Sci.* **2013,** *5*(9), 121–127.

25. Wojciechowski, K.; Saliba, M.; Leijtens, T.; Abate, A.; Snaith, H. J. Sub-150°C Processed Meso-Superstructured Perovskite Solar Cells with Enhanced Efficiency. *Energy Environ. Sci.* **2014,** *7*(3), 1142–1147.

26. Bi, D.; Moon, S.-J.; Häggman, L.; Boschloo, G.; Yang, L.; Johansson, E. M. J.; Nazeeruddin, M. K.; Grätzel, M.; Hagfeldt, A. Using a Two-Step Deposition Technique to Prepare Perovskite ($CH_3NH_3PbI_3$) for Thin Film Solar Cells Based on ZrO_2 and TiO_2 Mesostructures. *RSC Adv.* **2013,** *3*(11), 18762.

27. Hwang, S. H.; Roh, J.; Lee, J.; Ryu, J.; Yun, J.; Jang, J. Size-Controlled SiO_2 Nanoparticles as Scaffold Layers in Thin-Film Perovskite Solar Cells. *J. Mater. Chem. A* **2014,** *2*, 16429–16433.

28. Kim, H.; Mora-Sero, I.; Gonzalez-Pedro, V.; Fabregat-Santiago, F.; Juarez-Perez, E. J.; Park, N.; Bisquert, J. Mechanism of Carrier Accumulation in Perovskite Thin-Absorber Solar Cells. *Nat. Commun.* **2013,** *4*, 1–7.

29. Etgar, L.; Gao, P.; Xue, Z.; Peng, Q.; Chandiran, A. K.; Liu, B.; Nazeeruddin, M. K.; Grätzel, M. Mesoscopic $CH_3NH_3PbI_3/TiO_2$ Heterojunction Solar Cells. *J. Am. Chem. Soc.* **2012,** *134*(42), 17396–17399.

30. Xing, G.; Mathews, N.; Sun, S.; Lim, S. S.; Lam, Y. M.; Gratzel, M.; Mhaisalkar, S.; Sum, T. C. Long-Range Balanced Electron- and Hole-Transport Lengths in Organic-Inorganic $CH_3NH_3PbI_3$. *Science* **2013,** *342*(6156), 344–347.

31. Juárez-Pérez, E. J.; Wussler, M.; Fabregat-Santiago, F.; Lakus-Wollny, K.; Mankel, E.; Mayer, T.; Jaegermann, W.; Mora-Sero, I. The Role of the Selective Contacts in the Performance of Lead Halide Perovskite Solar Cells. *J. Phys. Chem. Lett.* **2014,** *5*, 680–685.

32. Liu, M.; Johnston, M. B.; Snaith, H. J. Efficient Planar Heterojunction Perovskite Solar Cells by Vapour Deposition. *Nature* **2013,** *501*(7467), 395–398.

33. Kumar, M. H.; Yantara, N.; Dharani, S.; Graetzel, M.; Mhaisalkar, S.; Boix, P. P.; Mathews, N. Flexible, Low-Temperature, Solution Processed ZnO-Based Perovskite Solid State Solar Cells. *Chem. Commun. (Camb).* **2013,** *49*(94), 11089–11091.

34. Conings, B.; Baeten, L.; Jacobs, T.; Dera, R.; D'Haen, J.; Manca, J.; Boyen, H. G. An Easy-to-Fabricate Low-Temperature TiO_2 Electron Collection Layer for High Efficiency Planar Heterojunction Perovskite Solar Cells. *APL Mater.* **2014,** *2*(8), 081505-1-8.

35. Yella, A.; Heiniger, L. P.; Gao, P.; Nazeeruddin, M. K.; Grätzel, M. Nanocrystalline Rutile Electron Extraction Layer Enables Low-Temperature Solution Processed Perovskite Photovoltaics with 13.7% Efficiency. *Nano Lett.* **2014,** *14*(5), 2591–2596.

36. Liu, D. Y.; Kelly, T. L. Perovskite Solar Cells with a Planar Heterojunction Structure Prepared Using Room-Temperature Solution Processing Techniques. *Nat. Photonics* **2014,** *8*(2), 133–138.

37. Dong, X.; Hu, H.; Lin, B.; Ding, J. J.; Yuan, N. The Effect of ALD-ZnO Layer on the Formation of $CH_3NH_3PbI_3$ with Different Perovskite Precursors and Sintering Temperatures. *Chem. Commun.* **2014,** *50*(3), 14405–14408.

38. Tseng, Z.-L.; Chiang, C.-H.; Wu, C.-G. Surface Engineering of ZnO Thin Film for High Efficiency Planar Perovskite Solar Cells. *Sci. Rep.* **2015,** *5*, 13211.

39. Wang, L.; Fu, W.; Gu, Z.; Fan, C.; Yang, X.; Li, H.; Chen, H. Low Temperature Solution Processed Planar Heterojunction Perovskite Solar Cells with a CdSe Nanocrystal as an Electron Transport/Extraction Layer. *J. Mater. Chem. C* **2014,** *2*(43), 9087–9090.

40. Jeng, J.; Chiang, Y.-F.; Lee, M.; Peng, S.-R.; Guo, T.-F.; Chen, P.; Wen, T. $CH_3NH_3PbI_3$ Perovskite/Fullerene Planar-Heterojunction Hybrid Solar Cells. *Adv Mater.* **2013,** *25*(27), 3727–3732.

41. Heo, J. H.; Han, H. J.; Kim, D.; Ahn, T. K.; Im, S. H. Hysteresis-Less Inverted $CH_3NH_3PbI_3$ Planar Perovskite Hybrid Solar Cells with 18.1% Power Conversion Efficiency. *Energy Environ. Sci.* **2015,** *8*(5), 1602–1608.

42. Yan, W.; Li, Y.; Li, Y.; Ye, S.; Liu, Z.; Wang, S.; Bian, Z.; Huang, C. Stable High-Performance Hybrid Perovskite Solar Cells with Ultrathin Polythiophene as Hole-Transporting Layer. *Nano Res.* **2015,** *8*(8), 2474–2480.

43. Polander, L. E.; Pahner, P.; Schwarze, M.; Saalfrank, M.; Koerner, C.; Leo, K. Hole-Transport Material Variation in Fully Vacuum Deposited Perovskite Solar Cells. *APL Mater.* **2014,** *2*(8), 1–6.

44. Park, J. H.; Seo, J.; Park, S.; Shin, S. S.; Kim, Y. C.; Jeon, N. J.; Shin, H.-W.; Ahn, T. K.; Noh, J. H.; Yoon, S. C.; Hwang, C. S.; Seok, S. Efficient $CH_3NH_3PbI_3$ Perovskite Solar Cells Employing Nanostructured p-Type NiO Electrode Formed by a Pulsed Laser Deposition. *Adv. Mater.* **2015,** *27*(27), 4013–4019.

45. Yin, X.; Chen, P.; Que, M.; Xing, Y.; Que, W.; Niu, C.; Shao, J. Highly Efficient Flexible Perovskite Solar Cells Using Solution-Derived NiOx Hole Contacts. *ACS Nano* **2016,** *10*(3), 3630–3636.

46. Kwon, U.; Kim, B.-G.; Nguyen, D. C.; Park, J.-H.; Ha, N. Y.; Kim, S.-J.; Ko, S. H.; Lee, S.; Lee, D.; Park, H. J. Solution-Processible Crystalline NiO Nanoparticles for High-Performance Planar Perovskite Photovoltaic Cells. *Sci. Rep.* **2016,** *6*, 30759.

47. Tseng, Z.-L.; Chen, L.-C.; Chiang, C.-H.; Chang, S.-H.; Chen, C.-C.; Wu, C.-G. Efficient Inverted-Type Perovskite Solar Cells Using UV-Ozone Treated MoOx and WOx as Hole Transporting Layers. *Sol. Energy* **2016,** *139*, 484–488.

48. Sepalage, G. A.; Meyer, S.; Pascoe, A.; Scully, A. D.; Huang, F.; Bach, U.; Cheng, Y.-B.; Spiccia, L. Copper(I) Iodide as Hole-Conductor in Planar Perovskite Solar Cells: Probing the Origin of J-V Hysteresis. *Adv. Funct. Mater.* **2015,** *25*(35), 5650–5661.

49. Gharibzadeh, S.; Nejand, B. A.; Moshaii, A.; Mohammadian, N.; Alizadeh, A. H.; Mohammadpour, R.; Ahmadi, V.; Alizadeh, A. Two-Step Physical Deposition of a Compact CuI Hole-Transport Layer and the Formation of an Interfacial Species in Perovskite Solar Cells. *ChemSusChem,* **2016,** *9*(15), 1929–1937.

50. Zuo, C.; Ding, L. Solution-Processed Cu_2O and CuO as Hole Transport Materials for Efficient Perovskite Solar Cells. *Small* **2015,** *11*(41), 5528–5532.

51. Chen, L.-C.; Chen, C.-C.; Liang, K.-C.; Chang, S. H.; Tseng, Z.-L.; Yeh, S.-C.; Chen, C.-T.; Wu, W.-T.; Wu, C.-G. Nano-Structured $CuO-Cu_2O$ Complex Thin Film for Application in $CH_3NH_3PbI_3$ Perovskite Solar Cells. *Nanoscale Res Lett.* **2016,** *11*(1), 402.

52. Ye, S.; Sun, W.; Li, Y.; Yan, W.; Peng, H.; Bian, Z.; Liu, Z.; Huang, C. CuSCN-Based Inverted Planar Perovskite Solar Cell with an Average PCE of 15.6%. *Nano Lett.* **2015,** *15(6),* 3723–3728.

53. Odobel, F.; Pellegrin, Y.; Gibson, E. A.; Hagfeldt, A.; Smeigh, A. L.; Hammarström, L. Recent Advances and Future Directions to Optimize the Performances of p-Type Dye-Sensitized Solar Cells. *Coord. Chem. Rev.* **2012,** *256*(21–22), 2414–2423.

54. Hu, L.; Peng, J.; Wang, W.; Xia, Z.; Yuan, J.; Lu, J.; Huang, X.; Ma, W.; Song, H.; Chen, W.; Cheng, Y. B.; Tang, J. Sequential Deposition of $CH_3NH_3PbI_3$ on Planar NiO Film for Efficient Planar Perovskite Solar Cells. *ACS Photonics* **2014,** *1*(7), 547–553.

55. Wang, Q.; Shao, Y.; Dong, Q.; Xiao, Z.; Yuan, Y.; Huang, J. Large Fill-Factor Bilayer Iodine Perovskite Solar Cells Fabricated by Low-Temperature Solution-Process. *Energy Environ. Sci.* **2014**, *7*(7), 2359–2365.
56. Chiang, C.-H.; Tseng, Z.-L.; Wu, C.-G. Planar Heterojunction Perovskite/PC71BM Solar Cells with Enhanced Open-Circuit Voltage via a (2/1)-Step Spin-Coating Process. *J. Mater. Chem. A* **2014**, *2*, 15897–15903.
57. Malinkiewicz, O.; Roldán-Carmona, C.; Soriano, A.; Bandiello, E.; Camacho, L.; Nazeeruddin, M. K.; Bolink, H. J. Metal-Oxide-Free Methylammonium Lead Iodide Perovskite-Based Solar Cells: The Influence of Organic Charge Transport Layers. *Adv. Energy Mater.* **2014**, *4*(15), 1–9.
58. Kim, S. S.; Bae, S.; Jo, W. H. A Perylene Diimide-Based Non-Fullerene Acceptor as an Electron Transporting Material for Inverted Perovskite Solar Cells. *RSC Adv.* **2016**, *6*(24), 19923–19927.
59. You, J.; Hong, Z.; Yang, Y. M.; Chen, Q.; Cai, M.; Song, T.-B.; Chen, C.-C.; Lu, S.; Liu, Y.; Zhou, H. Low-Temperature Solution-Processed Perovskite Solar Cells with High Efficiency and Flexibility. *ACS Nano* **2014**, *8*(2), 1674–1680.
60. Seo, J.; Park, S.; Chan Kim, Y.; Jeon, N. J.; Noh, J. H.; Yoon, S. C.; Seok, S. Benefits of Very Thin PCBM and LiF Layers for Solution-Processed p–i–n Perovskite Solar Cells. *Energy Environ. Sci.* **2014**, *7*(8), 2642.
61. Hou, F.; Su, Z.; Jin, F.; Yan, X.; Wang, L.; Zhao, H.; Zhu, J.; Chu, B.; Li, W. Efficient and Stable Planar Heterojunction Perovskite Solar Cells with an MoO_3/PEDOT:PSS Hole Transporting Layer. *Nanoscale* **2015**, *7*(21), 9427–9432.
62. Salim, T.; Sun, S.; Abe, Y.; Krishna, A.; Grimsdale, A. C.; Lam, Y. M. Perovskite-Based Solar Cells: Impact of Morphology and Device Architecture on Device Performance. *J. Mater. Chem. A* **2015**, *3*, 8943–8969.
63. Stranks, S. D.; Eperon, G. E.; Grancini, G.; Menelaou, C.; Alcocer, M. J. P.; Leijtens, T.; Herz, L. M.; Petrozza, A.; Snaith, H. J. Electron-Hole Diffusion Lengths Exceeding 1 Micrometer in an Organometal Trihalide Perovskite Absorber. *Science* **2014**, *342*(2013), 341–344.
64. Tress, W.; Marinova, N.; Moehl, T.; Zakeeruddin, S. M. Nazeeruddin, M. K.; Grätzel, M. Understanding the Rate-Dependent J–V Hysteresis, Slow Time Component, and Aging in $CH_3NH_3PbI_3$ Perovskite Solar Cells: The Role of a Compensated Electric Field. *Energy Environ. Sci.* **2015**, *8*(3), 995–1004.
65. Kim, H. S.; Park, N. G. Parameters Affecting I-V Hysteresis of $CH_3NH_3PbI_3$ Perovskite Solar Cells: Effects of Perovskite Crystal Size and Mesoporous TiO_2 Layer. *J. Phys. Chem. Lett.* **2014**, *5*(17), 2927–2934
66. Snaith, H. J.; Abate, A.; Ball, J. M.; Eperon, G. E.; Leijtens, T.; Noel, N. K.; Stranks, S. D.; Wang, J. T. W.; Wojciechowski, K.; Zhang, W. Anomalous Hysteresis in Perovskite Solar Cells. *J. Phys. Chem. Lett.* **2014**, *5*(9), 1511–1515.

CHAPTER 3

CURRENT STATUS OF HIGH-EFFICIENCY SOLAR CELLS[1]

3.1 INTRODUCTION

The performance of the best-certified organic–inorganic halide perovskite solar cells has improved by more than 150% relative and 8% absolute (Figure 3.1) over the last few years making it the fastest advancing photovoltaic technology.[1] There has been a move from the widely researched $CH_3NH_3PbI_3$ $(MAPbI_3)^{1-2}$ toward mixed perovskites by incorporating $HC(NH_2)_2^+$ (FA^+) to take advantage of lower bandgap and therefore higher current but is often compromised by the addition of Br^- ions for better perovskite phase stability.[3] The addition of Cs^4 and most recently $Rb^{5,6}$ also produce cells with excellent efficiencies (although not certified) and stability.

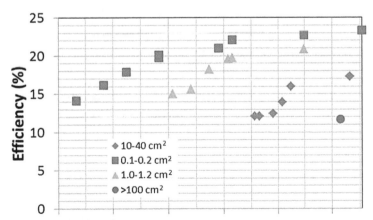

FIGURE 3.1 Evolution of certified perovskite solar cell performance.
Source: Adapted from ref 1.

---

[1]Some of the contents in this section are taken from ref 1.

TABLE 3.1 Independently Certified Efficiencies for Small Perovskite Solar Cells.

Certification time	Publication time	Eff. (%)	Area (cm^2)	V_{OC} (V)	J_{SC} (mA/ cm^2)	Fill Factor (%)	Comments	References
May 13	July 13	14.1	0.209	1.00	21.3	0.66	Sequential solution process for perovskite. Glass/FTO/c-TiO$_2$/ mp-TiO$_2$/ MAPbI$_3$/spiro-OMeTAD/ Au	[2]
November 13	July 14	16.2	0.094	1.11	19.6	0.74	Solvent engineering during solution process of perovskite. Glass/FTO/c-TiO$_2$/mp-TiO$_2$/ MAPbI$_{1-x}$Br$_x$/ PTAA/Au	[7]
April 14	January 15	17.9	0.096	1.11	21.8	0.74	Compositional engineering of perovskite. Glass/FTO/c-TiO$_2$/mp-TiO$_2$/ (FAPbI$_3$)$_{0.85}$ (MABr$_3$)$_{0.15}$/ PTAA/Au	[3]
November 14	November 2015	19.7	0.096	1.13	22.5	0.78	Incorporate excess PbI$_2$ into perovskite. Glass/FTO/c-TiO$_2$/mp-TiO$_2$/ (FAPbI$_3$)$_{0.85}$ (MAPbBr$_3$)$_{0.15}$/ PTAA/Au	[8]
November 14	May 2015	20.1	0.096	1.06	24.7	0.77	Intramolecular exchange for perovskite. Glass/FTO/c-TiO$_2$/ mp-TiO$_2$/ (FAPbI$_3$)$_X$ (MAPbBr$_3$)$_{1-Y}$/ PTAA/Au	[9]

TABLE 3.1   *(Continued)*

Certification time	Publication time	Eff. (%)	Area (cm²)	V_{OC} (V)	J_{SC} (mA/cm²)	Fill Factor (%)	Comments	References
December 15	September 2016	21.0	0.105	1.13	23.8	0.78	Polymer-templated nucleation. Glass/FTO/c-TiO₂/mp-TiO₂/ $(FAI)_{0.81}$ $(PbI_2)_{0.85}$ $(MAPbBr_3)_{0.15}$/ Spiro-OMeTAD/ Au	[10]
March 2016	June 2017	22.1	0.095	1.10	25.0	0.80	Iodide management. Glass/ FTO/c-TiO₂/ mp-TiO₂/ $(FAPbI3)_X$ $(MAPbBr3)_{1-Y}$/ PTAA/Au	[11]
July 2017	–	22.7	0.094	1.14	24.9	0.80	Details not published yet	[12]
July 2018	–	23.7					Details not published yet	[31]

Source: Adapted from ref 1.

The "standard architecture" is still the most popular choice for the state of the art cells (Table 3.1) which consists of fluorine-doped tin oxide (FTO)-coated glass which is then coated with hole-blocking or compact titanium dioxide (c-TiO₂) layer overlaid by mesoporous (mp) TiO₂ to form a "meso-structure" on which perovskite absorber layer is deposited. The most commonly used hole-transport layer (HTM) on these high performing devices are 2,2',7,7'-tetrakis(N,N-di-p-methoxyphenylamine)9,9'-spirobifluorene (spiro-OMeTAD) and Poly[bis(4-phenyl)(2,4,6-trimethylphenyl)amine] (PTAA). Although the exact composition of the 20.1% and 22.1% devices listed in Table 3.1 are unknown (Figure 3.2a), their spectral responses shown in Figure 3.2b show very good long wavelength response. This may be due to the lower Br content (less than 15%) delivering a higher short current density at around 25 mA/cm². The better short wavelength responses may also indicate the use of less absorptive substrates. In terms of deposition process, the use of mixed solvents is common with dimethylformamide and dimethyl sulfoxide being the common ones. There has been an emphasis on the two-step approach[9,11] with the latest result produced by minimizing iodide deficiency, thereby reducing defects and therefore recombinations in the film according to Seok et al.[11] The fill factor of 80% is the highest reported for all cells. As the quality of the perovskite film continues to improve, more effort will be placed on interface engineering where an additional layer between the perovskite and blocking or transport layer has shown to be effective in passivation[13] and morphology control.[10]

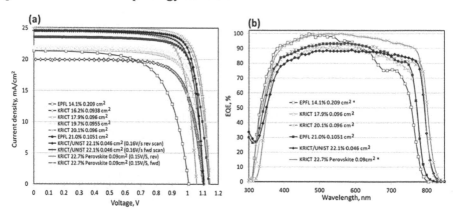

FIGURE 3.2   (a) Current density–voltage; (b) Quantum efficiency of certified small area perovskite solar cells.
Note: * denote normalized data while others are as measured.
Source: Reproduced from ref 1, 12, 14, 15. Ref 12 © John Wiley & Sons. Ref 14; © 2017 John Wiley & Sons; Ref. 15 © 2017 John Wiley & Sons.

Since the year 2015, there are increasing activities on 1 cm² cell demonstrations (Table 3.2 and Figure 3.3) rather than tiny (0.1–0.2 cm²) devices initiated by National Institute for Materials Science, Tsukuba, Japan (NIMS).[16,17] Interesting, many of these cells adopt an "inverted" cell architecture with the first improvement in cell performance attributed to improved carrier extractions after doping of inorganic interlayers. The second improvement in cell performance came from the use of PCBM dissolved in toluene as an "antisolvent" for the perovskite nucleation process which was reported to also form a "perovskite–fullerene graded heterojunction." For the 19.6% cell, instead of using a solution method of reducing the solubility of perovskite precursor in the solvent for rapid nucleation, a physical vacuum-based process is used that effectively removes the perovskite solvent for rapid nucleation.[18] This method is effective on large area allowing a high efficiency 1 cm² cell to be achieved. Another scalable process for improving the perovskite crystallization process over a large area is also developed by a team at the University of New South Wales (UNSW) who demonstrated a slightly larger device on

1.2 cm^2 at 18%. In the second half of the year 2016, cells and modules that are more realistic in size (> 10 cm^2) are starting to be certified (Table 3.3) as fabrication, cell interconnection, and encapsulation techniques mature. The first certified result was on a 12.1% efficient 36 cm^2 minimodule fabricated by Shanghai Jiangtong University (SJTU) in conjunction with NIMS[14] in August 2016. This was followed by the certification of the 12.1% efficient 16-cm^2 single monolithic cell in September 2016 fabricated by UNSW who developed a scalable anti-solvent deposition process.[19] While IMEC's 16-cm^2 eight cell minimodule was certified to be 12.4% efficient in December 2016 and SJTU/NIMS's 36-cm^2 minimodule was certified to be 13.9% efficient in February 2017, these results were surpassed by Microquanta's 16.0% efficient 16-cm^2 minimodules certified in April [14] 2017 and 17.25% efficient 17.3-cm^2 minimodules certified in May 2018. [31] Recently larger sub-modules have also been demonstrated on 703cm^2 with 11.7% certified efficiency and 802cm^2 with 11.6% certified efficiency [31]

TABLE 3.2 Independently Certified Efficiencies for 1 cm^2 Perovskite Solar Cells.

Certification time	Publication time	Eff. (%)	Area (cm^2)	V_{OC} (V)	J_{SC} (mA/ cm^2)	Fill factor (%)	Comments	References
February 15	October 15	15.0	1.10	1.09	20.6	0.67	Doping of inorganic inter-layers. Glass/ FTO/NiMgLiO (p-type)/MAPbI$_3$/ PCBM/Ti(Nb) O$_x$/Ag	[16]
October 15	September 16	18.2	1.02	1.08	21.5	0.78	Perovskite– fullerene graded heterojunction. Glass/FTO/ NiO/(FAPbI$_3$)$_{0.85}$ (MAPbBr$_3$)$_{0.15}$/ PCBM/ Ti(Nb) O$_x$/Ag	[17]
February 16	May 16	19.6	1.00	1.14	22.6	0.76	Vacuum flash-assisted solution process. Glass/ FTO/c-TiO$_2$/ mp TiO$_2$/ (FAPbI$_3$)$_{0.85}$ (MAPbBr$_3$)$_{0.15}$/ spiro-OMeTAD/ Au	[18]

TABLE 3.2 *(Continued)*

Certification time	Publication time	Eff. (%)	Area (cm²)	V_{OC} (V)	J_{SC} (mA/ cm²)	Fill factor (%)	Comments	References
March 16	June 2017	19.7	0.99	1.10	24.7	0.72	Iodide management. Glass/ FTO/c-TiO$_2$/ mp-TiO$_2$/ (FAPbI$_3$)$_X$ (MAPbBr$_3$)$_{1-Y}$/ PTAA/Au	[11]
September 16	August 17	18.0	1.19	1.13	21.4	0.75	Scalable solution process for improving perovskite crystallization. Glass/FTO/c-TiO$_2$/mp-TiO$_2$/ (FAPbI$_3$)$_{0.85}$ (MAPbBr$_3$)$_{0.15}$/ spiro-OMeTAD/ Au	[19]
July 17	–	20.9	0.99	1.13	20.9	0.75	Details not published yet	[12]

Source: Adapted from ref 1.

TABLE 3.3 Independently Certified Efficiencies for > 10 cm² Perovskite Solar Cells and Module.

Certification time	Description	Eff. (%)	Area (cm²)	References
August 16	10 cells in series	12.1	36	[14]
September 16	Single cell	12.1	16	[19]
December 16	Eight cells in series	12.4	16	[14]
February 17	10 cells in series	13.9	36	[14]
April 17	Six cells in series	16.0	16	[14]
May 18	7 cells in series	17.25	17.3	[32]
March 18	44 cells in series	11.7	703	[32]
April 18	22 cells in series	11.6	802	[32]

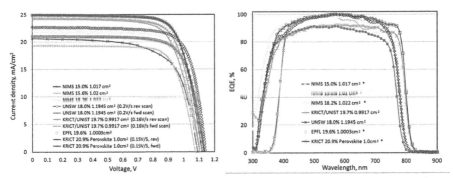

FIGURE 3.3 (a) Current density—voltage; (b) quantum efficiency of certified 1 cm² and slight larger 1.2 cm² perovskite solar cells.

Note: * denote normalized data while others are as measured.

Source: Reproduced from ref 1, 12, 14, 15. Ref 12 © John Wiley & Sons. Ref 14; © 2017 John Wiley & Sons; Ref. 15 © 2017 John Wiley & Sons.

TABLE 3.4 Independently Certified or Independently Verified (for the Second Entry) Efficiencies for Planar Perovskite Solar Cells.

Certification time	Publication time	Eff. (%)	Area (cm²)	V_{OC} (V)	J_{SC} (mA/ cm²)	Fill factor (%)	Comments	References
July 15	February 16	15.3	0.050	1.10	18.97	0.74	Quasi-two-dimensional perovskite. Glass/ FTO/c-TiO_2/PEA$_2$(MA)$_{n-1}$Pb$_n$I$_{3n+1}$/ spiro-OMeTAD/ Au	[20]
N. A.	August 16	19.4*	0.113	1.10	22.44	0.79	Mortification of TiO_2 using [BMIM]BF4. Glass/ FTO/c-TiO_2/MAPbI$_3$/ PTAA/Au	[21]
April 16	November 16	19.9	0.074	1.07	24.31	0.77	Low-temperature solution processed SnO_2 nanoparticle. Glass/ FTO/ SnO$_2$/ (FAPbI$_3$)$_{0.97}$ (MABr$_3$)$_{0.03}$/ Spiro-OMeTAD/ Au	[22]

TABLE 3.4 *(Continued)*

Certification time	Publication time	Eff. (%)	Area (cm²)	V_{oc} (V)	J_{sc} (mA/cm²)	Fill factor (%)	Comments	References
November 16	February 17	20.1	0.049	1.17	21.70	0.79	Chlorine-capped TiO_2 colloidal nano-crystal. Glass/ITO/c-TiO_2-Cl/ $Cs_{0.05}FA_{0.81}$ $MA_{0.14}$ $PbI_{2.55}B_{r0.45}$/ Spiro-OMeTAD /Au	[23]
February 17	June 17	20.6	0.072	1.13	22.99	0.79	Defect passivation by quaternary ammonium halides. Glass/ ITO/PTAA/ $(FAPbI_3)_{0.85}$ $(MAPbBr_3)_{0.15}$/ Choline chloride/ C60/ BCP/ Cu	[24]
January 17	October 17	20.9	0.074	1.12	24.02	0.78	PbI_2 control. Glass/ ITO/ SnO_2/ $(FAPbI_3)$ $_x(MAPbBr_3)_{3-x}$/ spiro-OMeTAD/ Au	[25]

Note: *Verified by National Center of Supervision and Inspection on Solar Photovoltaic Product Quality (CPVT).
Source: Adapted from ref 1.

Simpler planar devices (Table 3.4) with "standard" polarity that eliminate the use of the mp-TiO_2 are slowly catching up in conversion efficiencies, as new strategies are developed to overcome conspicuous hysteresis found in these devices in general. This probably made the certifying of these devices difficult at the early stages of development. The first of such planar device is developed by Liu et al.[26] with the $CH_3NH_3PbI_3$ perovskite layer deposited by dual-source thermal evaporation. This 15.4% efficiency device is first of its kind as a physical deposition is used instead of the typically used solution process such as spin-coating to deposit the perovskite layer. An 18% planar cell that uses SnO_2 via atomic layer deposition (ALD) as the blocking layer and $(FAPbI_3)_{0.85}(MAPbBr_3)_{0.15}$ as the absorber layer for the first time was reported by Baena et al.[27] A more efficient (19.3%) planar $CH_3NH_3PbI_3$ device was reported by Zhou et al.[28] replacing the FTO glass with ITO glass coated with polyethyleneimine ethoxylated, which is then overlaid by a Yttrium-doped c-TiO_2. Since then, more creditable certified efficiencies have been reported. The first of which, interestingly, is on a device that uses a less conventional perovskite material that intercalates phenylethylammonium ($C_8H_9NH_3$ or PEA) between the three-dimensional perovskite layers results in quasi-two-dimensional structures with layered perovskites through van der Waals interactions. The glass/FTO/c-TiO_2/$PEA_2(MA)_{n-1}Pb_nI_{3n+1}$/spiro-OMeTAD/Au cell where $n = 60$[20] is certified to be 15.3% efficient. The increased stability can be due to the higher energy required to remove the PEAI from the perovskite to initiate decomposition. The devices also overcome hysteresis that is commonly seen in current–voltage curve of planar devices. It was suggested that the introduction of these multilayered structures might have altered, for example, suppress electronic and ionic motion across the film thereby reducing the hysteresis. New treatments of TiO_2 have led to great improvements in efficiencies to 19.4%[21] and 20.1%.[23] The report of a solution process for SnO_2 allowing 20% cell efficiency to be achieved[24] is most useful as SnO_2 by ALD is not always easily accessible. This will likely generate more work using SnO_2 to eliminate the UV-instability associated with TiO_2 previously reported.[29]

In terms of HTM, Spiro-OMeTAD, and PTAA are still heavily used in the state of the art devices despite their limitations in terms of stability (especially when additives are incorporated) and being the possible cause of device degradation when in contact with perovskite.[30] Opportunities using other types of charge transport layers are starting to emerge especially in "inverted" devices. Recent work has also reported the importance of passivating interface layers. These results are important for improving the voltage outputs of larger bandgap cells for tandem applications. Demonstrations of highly efficient devices that eliminate gold are also important to reduce the cost of state of the art devices.

KEYWORDS

- world record perovskite solar cells
- mesoscopic cell architecture
- planar perovskite cell
- large area perovskite cell
- perovskite module

REFERENCES

1. Ho-Baillie, A. W. Y. Current Status of High-Efficiency Perovsksite Solar Cells Photovoltaics International 36th Edition, May 2017.
2. Burschka, J.; Pellet, N.; Moon, S.-J.; Humphry-Baker, R.; Gao, P.; Nazeeruddin, M. K.; Grätzel, M. *Nature* **2013,** *499,* 316–319.
3. Jeon, N. J.; Noh, J. H.; Yang, W. S.; Kim, Y. C.; Ryu, S.; Seo J.; Seok, S. I. Compositional Engineering of Perovskite Materials for High Performance Solar Cells. *Nature* **2015,** *517,* 476–480.
4. Saliba, M.; et al. Cesium-Containing Triple Cation Perovskite Solar Cells: Improved Stability, Reproducibility and High Efficiency. *Energy Environ. Sci.* **2016,** *9,* 1989.
5. Saliba, M.; et al. Incorporation of Rubidium Cations into Perovskite Solar Cells Improves Photovoltaic Performance. *Science* **2016,** DOI: 10.1126/science.aah5557.
6. Zhang, M.; Yun, J. S.; Ma, Q.; Zheng, J.; Lau C. F. J.; Deng, X.; Kim, J.; Kim, D.; Seidel, J.; Green, M. A.; Huang. S.; Ho-Baillie. A. W. Y. High-Efficiency Rubidium-Incorporated Perovskite Solar Cells by Gas Quenching. *ACS Energy Lett.* **2017,** *2*(2), 438–444.
7. Jeon, N. J.; Noh, J. H.; Kim, Y. C.; Yang, W. S.; Ryu, S.; Seok, S. I. Solvent-Engineering for High Performance Inorganic-Organic Hybrid Perovskite Solar Cells. *Nat. Mater.* **2014,** *13,* 897.
8. Kim, Y. C.; Jeon, N. J.; Noh, J. H.; Yang, W. S.; Seo, J.; Yun, J. S.; Ho-Baillie, A.; Huang, S.; Green, M. A.; Seidel, J.; Ahn, T. K.; Seok, S. I. Beneficial Effects of PbI$_2$ Incorporated in Organo-Lead Halide Perovskite Solar Cells. *Adv. Energy Mater.* **2016,** *6,* 1502104.DOI: https://doi.org/10.1002/aenm.201502104.
9. Yang, W. S.; Noh, J. H.; Jeon, N. J.; Kim, Y. C.; Ryu, S.; Seo, J.; Seok, S. I. High-Performance Photovoltaic Perovskite Layers Fabricated Through Intramolecular Exchange. *Science* **2015,** *348,* 1234–1237.
10. Dongqin, B.; Chenyi, Y.; Luo, J., Décoppet, J.-D.; Zhang, F.; Zakeeruddin, S. K.; Li, X.; Hagfeldt, A.; Grätzel, M. DOI: 10.1038/NENERGY.2016.142.
11. Yang, W. S.; Park, B.-W.; Jung, E. H.; Jeon, N. J.; Kim, Y. C.; Lee, D. U.; Shin, S. S.; Seo, J.; Kim, E. K.; Noh, J. H.; Seok, S. Iodide Management in Formamidinium-Lead-Halide–Based Perovskite Layers for Efficient Solar Cells. *Science* **2017,** *356,* 1376–1379.

12. Green, M. A.; Hishikawa, Y.; Dunlop, E. D.; Levi, D. H.; Hohl-Ebinger, J.; Ho-Baillie, A. W. Y. Solar Cell Efficiency Tables (Version 51). *Prog. Photovoltaics: Res. Appl.* **2018**, *26*, 3–12.

13. Peng, J.; Wu, Y.; Ye, W.; Jacobs, D. A.; Shen, H.; Fu, X.; Wan, Y.; Duong, T.; Wu, N.; Barugkin, C.; Nguyen, H. T.; Zhong, D.; Li, J.; Lu, T.; Liu, Y.; Lockrey, M. N.; Weber, K. J.; Catchpole, K. R.; White, T. P. Interface Passivation Using Ultrathin Polymer–Fullerene Films for High-Efficiency Perovskite Solar Cells with Negligible Hysteresis. *Energy Environ. Sci.* **2017**, *10*, 1792–1800.

14. Green, M. A.; Hishikawa, Y.; Warta, W.; Dunlop, E. D.; Levi, D. H.; Hohl-Ebinger, J.; Ho-Baillie, A. W. Y. Solar Cell Efficiency Tables (Version 50). *Prog. Photovoltaics: Res. Appl.* **2017**, *25*, 668–676.

15. Green, M. A.; Emery, K.; Hishikawa, Y.; Warta, W.; Levi, D. H.; Dunlop, E. D.; Ho-Baillie, A. W. Y. Solar Cell Efficiency Tables (Version 49). *Prog. Photovolt: Res. Appl.* **2017**, *25*, 3–13.

16. Chen, W.; et al. Efficient and Stable Large-Area Perovskite Solar Cells with Inorganic Charge Extraction Layers. *Science* **2015**, *350*, 944–948. DOI: 10.1126/science.aad1015.

17. Wu, Y.; et al. Perovskite Solar Cells with 18.21% Efficiency and Area Over 1 cm^2 Fabricated by Heterojunction Engineering. DOI: 10.1038/NENERGY.2016.148.

18. Li, X.; et al. A Vacuum Flash–Assisted Solution Process for High-Efficiency Large-Area Perovskite Solar Cells. DOI: 10.1126/science.aaf8060.

19. Kim, J.; Yun, J. S.; Cho, Y.; Lee, D. S.; Wilkinson, B.; Mahboubi Soufiani, A.; Deng, X.; Zheng, J.; Shi, A.; Lim, S.; Chen, S.; Hameiri, Z.; Zhang, M.; Lau, C. F. J.; Huang, S.; Green, M. A.; Ho-Baillie, A. W. Y. Overcoming the Challenges of Large-Area High-Efficiency Perovskite Solar Cells. *ACS Energy Lett.* **2017**, *2*, 1978–1984.

20. Quan, L.; et al. Ligand-Stabilized Reduced-Dimensionality Perovskites. *J. Am. Chem. Soc.* **2016**, *138*, 2649–2655.

21. Yang, D.; et al. Surface Optimization to Eliminate Hysteresis for Record Efficiency Planar Perovskite Solar Cells. *Energy Environ. Sci.* **2016**, *9*, 3071.

22. Jiang, Q.; et al. Enhanced Electron Extraction Using SnO$_2$ for High-Efficiency Planar-Structure HC(NH$_2$)$_2$PbI$_3$-Based Perovskite Solar Cells. DOI: 10.1038/NENERGY.2016.177.

23. Tan, H.; et al. Efficient and Stable Solution-Processed Planar Perovskite Solar Cells via Contact Passivation. *Science* **2017**, *355*, 722–726.

24. Zheng, X.; Chen, B.; Dai, J.; Fang, Y.; Bai, Y.; Lin, Y.; Wei, H.; Zeng, X. C.; Huang, J. Defect Passivation in Hybrid Perovskite Solar Cells Using Quaternary Ammonium Halide Anions and Cations. *Nat. Energy* **2017**, *2*, 17102.

25. Jiang, Q.; Chu, Z.; Wang, P.; Yang, X.; Liu, H.; Wang, Y.; Yin, Z.; Wu, J.; Zhang, X.; You, J. Planar-Structure Perovskite Solar Cells with Efficiency Beyond 21%. *Adv. Mater.* **2017**, *29*, 1703852.

26. Liu, M.; Johnston, M.; Snaith, H. J. Efficient Planar Heterojunction Perovskite Solar Cells by Vapour Deposition. *Nature* **2013**, *501*, 395–398.

27. Baena, J. P. C.; et al. Highly Efficient Planar Perovskite Solar Cells Through Band Alignment Engineering. *Energy Environ. Sci.* **2015**, *8*, 2928–2934.

28. Zhou, H.; et al. Interface Engineering of Highly Efficient Perovskite Solar Cells. *Science* **2014**, *345*, 542.

29. Leijtens, T.; et al. Overcoming Ultraviolet Light Instability of Sensitized TiO_2 with Meso-Superstructured Organometal Tri-Halide Perovskite Solar Cells. *Nat. Commun.* **2013,** *4,* Article Number: 2885.
30. Kim, J.; Park, N.; Yun, J. S.; Huang, S.; Green, M. A.; Ho-Baillie, A. W. Y. Long-Term Stability and Thermal Degradation of Planar $CH_3NH_3PbI_3$ and $HC(NH_2)_2PbI_3$ Perovskite Solar Cells with the Hole Transfer Materials of Spiro-OMeTAD and PTAA. *Sol. Energy Mater. Sol. Cells* **2017,** *162,* 41–46.
31. Best Research-Cell Efficiencies, NREL. http://www.nrel.gov/pv/assets/images/efficiency_chart.jpg.
32. Green, M. A.; Hishikawa, Y.; Dunlop, E. D.; Levi, D. H.; Hohl-Ebinger, J.; Ho-Baillie, A. W. Y. Solar Cell Efficiency Tables (Version 52). ProgPhotovolt Res Appl. 2018;26:427–436.

CHAPTER 4

DURABILITY AND STABILITY OF PEROVSKITE SOLAR CELLS

Early generation perovskite solar cells have a lifetime of 1000–2000 h. Under constant illumination, an encapsulated two-dimensional Ruddlesden–Popper perovskite solar cell has been reported to last for 2250 h.[1] The biggest challenge of perovskite cell is its instability when exposed to moisture[2-4] under high temperature,[5-8] under light (especially ultraviolet (UV)[9] or concentrated sunlight),[10] and under electrical bias, especially in the presence of moisture.[11,12]

Although a plethora of literature on perovskite stability including methods of improving cell durability can be found, a consistent reporting method or a standardized testing regime is still lacking. This is due to the early stage of the technology development, the many facets of instability exhibited by perovskite solar cells not only in device performance but also in device operation itself, and different behaviors and failure modes in cells of different architecture.

4.1 HISTORY OF REPORTING

The first report of a stability study is on the first efficient solid-state perovskite cell at 9.7% in mid-2012.[13] The unencapsulated fluorine-doped tin oxide (FTO)/c-TiO$_2$/mp-TiO$_2$/CH$_3$NH$_3$PbI$_3$/spiro-MeOTAD/Au device was stored at room temperature although illumination conditions or humidity levels were not specified. Electrical characteristics were monitored irregularly over the reporting period of 500 h showing a drop of short circuit current by ~9%, relative fluctuations in open-circuit voltage and improvement in fill factor and, therefore, improvement in energy conversion efficiency by ~14% relative. However, longer-term stability tests were not carried out in that work.

The effect of humidity during air exposure and the effect of mixed halide content on the durability of FTO/c-TiO$_2$/mp-TiO$_2$/CH$_3$NH$_3$PbI$_{3-x}$Br$_x$/

PTAA/Au solar cell were reported in March 2013.[2] An increase in humidity from 35% to 55% reduces the conversion efficiency of $CH_3NH_3PbI_3$ and $CH_3NH_3PbI_{2.94}Br_{0.06}$ dramatically. $CH_3NH_3PbI_{2.8}Br_{0.2}$ and $CH_3NH_3PbI_{2.71}Br_{0.29}$ cells with higher bromide content retain the conversion efficiency over the reporting period of 20 days (~ 480 h) at room temperature with occasional current-voltage measurements. Illumination conditions during storage were not specified and longer-term stabilities were not reported in this work.

In terms of encapsulated devices, the first report was published in July 2013. Burschka et al. encapsulated an FTO/c-TiO$_2$/mp-TiO$_2$/CH$_3$NH$_3$PbI$_3$/spiro-MeOTAD/Au cell in argon using a 50-mm-thick hot-melting polymer and a microscope coverslip.[14] The sealed device was then subjected to ~ 100mWcm^{-2} light-emitting diode (LED) soaking at a temperature of 45°C. The device lost 20% of its initial conversion efficiency after a period of 500 h. No longer-term stabilities were reported in this work.

In their work, Leijtens et al.[9] encapsulated FTO/c-TiO$_2$/mp-TiO$_2$/CH$_3$NH$_3$PbI$_{3-x}$Cl$_x$/spiro-MeOTAD/Au and FTO/c-TiO$_2$/mp-Al$_2$O$_3$/CH$_3$NH$_3$PbI$_{3-X}$Cl$_X$/spiro-MeOTAD/Au cells using an epoxy resin and a glass coverslip in a nitrogen-filled glove box and the UV stability.[9] The devices were then subjected to a 100 mWcm^{-2} light soaking (Xenon arc lamp) at 40°C with and without UV filtering. Although unexpected, efficiency of the encapsulated TiO$_2$-based device dropped by 80% in 4 h while the unencapsulated device lost only 50% of its initial efficiency in the same period. The device performed better when a UV filter was used. It was suggested that UV-instability of such a device was due to light-induced desorption of surface-adsorbed oxygen in the mp-TiO$_2$ layer leaving unoccupied, deep surface trap sites and a free electron per site which recombines readily with holes on the spiro-OMeTAD hole transporter. When mp-Al$_2$O$_3$ is used instead, devices were more stable under UV. A lifetime of 1000 h was claimed for these encapsulated devices under continuous illumination at 40°C although a 50% drop in efficiency is evident in the first 200 h of the illumination test. No longer-term stabilities were reported in this work.

In mid-2014, a perovskite cell of very different architecture FTO/c-TiO$_2$/mp-TiO$_2$/mp-ZrO$_2$/(5-HOOC (CH$_2$)$_4$NH$_3$)$_X$(CH$_3$NH$_3$)$_{1-X}$PbI$_3$/Carbon using different perovskite material deposited by drop-casting through the carbon electrode is reported to have better stability.[15] Under constant (AM 1.5 simulated) illumination at room temperature, the unencapsulated device retains its initial conversion efficiency over a reporting period of 1008 h. It is unclear how the temperature of the cells is measured and humidity levels during the reporting period were not specified. Again, longer-term stabilities were not reported in this work. A simpler version of such device reported by Zhou et al. which consists of FTO/c-TiO$_2$/ mp TiO$_2$/CH$_3$NH$_3$PbI$_3$/Carbon displays stability over the reporting period of 2446 h ($\sim$480 h) when the unencapsulated device is stored in the dark in the air with electrical characteristics monitored regularly over the reporting period.[16] Humidity levels were not specified and no longer-term stabilities were reported in this work.

4.2 DEGRADATION MECHANISM IN MOISTURE

Since then, more systematic investigations into the degradation of the CH$_3$NH$_3$PbI$_3$ solar cell have appeared around the same time in January 2015.[3,4] Yang et al. carried out in situ absorption spectroscopy and in situ grazing incidence X-ray diffraction (XRD) under controlled relative humidity at temperature $=22.9\pm0.5°C$.[3] It was found that the time ($\tau_{1/2}$) it takes for the absorbance of the perovskite film to deteriorate half of its initial is 10,000 h in relative humidity (RH) of 20%, or 1,000 h in RH=50%, or 34 h in RH=80%, or 4 h in RH=98%. The absorbance of the film remains unchanged for the reporting period of 2 weeks under 0% RH dry air that contains oxygen. No longer-term stabilities under dry air were reported. The effect of hole transport materials (HTM) with and without lithium bis (trifluoromethanesulfonyl) imide (Li-TFSI) additives were also investigated. When the perovskite film is in contact with HTLs, 2,2',7,7'-tetrakis (N, Ndi-4-methoxyphenylamino)-9, 90-spirobifluo-rene (Spiro-OMeTAD), the $\tau_{1/2}$ lifetime deteriorates or remains the same depending on the presence of Li-TFSI. When a polymer-based HTM is used either poly [bis (4-phenyl)(2,4,6-trimethylphenyl)-amine] (PTAA) or poly (3-hexylthiophene) (P3HT), $\tau_{1/2}$ improves. The authors suggested that the polymer-based HTMs act as barriers to moisture ingress by absorbing and trapping moisture into the HTM slowing down the water percolation to the perovskite layer. It was suggested that the moisture absorption and trapping into the HTM is further aided by the presence of Li-TFSI making the HTM hydrophilic.

TABLE 4.1 $\tau_{1/2}$ as a Function of Relative Humidity for Different Test Structures Reported by Yang et al. [3] $\tau_{1/2}$ is the Time it Takes for the Absorbance of the Perovskite Film to Deteriorate Half of its Initial.

Structure	Relative Humidity (%) at 22.9 ± 0.5 °C	$\tau_{1/2}$ (hours)	
$CH_3NH_3PbI_3$	20	10,000 (extrapolated)	
$CH_3NH_3PbI_3$	50	1000	
$CH_3NH_3PbI_3$	80	34	
$CH_3NH_3PbI_3$	98	4	
$CH_3NH_3PbI_3$/ Spiro-OMeTAD	98	1 (with Li-TFSI)	4 (without Li-TFSI)
$CH_3NH_3PbI_3$/PTAA	98	7 (with Li-TFSI)	6 (without Li-TFSI)
$CH_3NH_3PbI_3$/P3HT	98	25 (with Li-TFSI)	6 (without Li-TFSI)

Yang et al. also suggested that $(CH_3NH_3)_4PbI_6 0.2H_2O$ or a related compound consists of isolated PbI_6^{4-} octahedra which has been predicted to have an XRD peak at q ≈ 8 nm^{-1} in a previous work[17] as an intermediate in the decomposition pathway of $CH_3NH_3PbI_3$ according to Equation 4.1. This suggestion was based on the new features (which cannot be assigned to either $CH_3NH_3PbI_3$, PbI_2, or CH_3NH_3I) found in the XRD pattern at 5 and 7 nm^{-1} after subjecting the perovskite film to 80% RH for 2.5 h. The authors discussed that this phase is partially reversible from the results of dehydrating a 24-hour-98%-RH-exposed perovskite-turned-yellow film with dry nitrogen, regenerating some of the dark brown color of $CH_3NH_3PbI_3$. Further decomposition of CH_3NH_3I to CH_3NH_2 and hydroiodic acid (HI) would ultimately leave PbI_2 as the only by-product of the reaction.

$$4CH_3NH_3Pb I_3 + 2H_2O \Leftrightarrow (CH_3NH_3)_4 Pb I_6 0.2H_2O + 3 Pb I_2 \qquad (4.1)$$

Christians et al. carried out similar absorbance measurements to ones like Yang et al. by monitoring.[4] It was observed that perovskite films stored in the dark underwent faster degradation. Christians et al. arrived at similar conclusions to those by Yang et al. with regards to the hydrated phase during $CH_3NH_3PbI_3$ decomposition from the new XRD peaks 7.93, 8.42, 10.46, and 16.01 that appear after exposing the $CH_3NH_3PbI_3$ film to 90% RH for 7 days in the dark. Partial recovery was also observed (from XRD and absorbance) when degraded perovskite films were stored in a vacuum or low RH environment.

While femtosecond transient absorption measurements were carried out on the degraded film, no significant changes in ultrafast carrier dynamics could be observed possibly due to defects being predominantly shallow trap states and the associated trap-mediated recombination that is largely absent in the nanosecond timescale. In terms of device degradation, the performance of FTO/c-TiO_2/mp-$TiO_2/CH_3NH_3PbI_3/spiro$-OMeTAD device dropped from the initial 12% to less than 1% efficiency in only 3 days of 90% RH exposure. Under 50% RH, the efficiency of the device drops by 10% in 13 days and to 40% of the initial value in 21 days. At 0% RH, the efficiency of the device drops by 5% in 21 days. Longer-term stability of this device was not reported.

TABLE 4.2 $\tau_{1/2}$ as a Function of Relative Humidity for Different Test Structures Reported by Christians et al.[4] $\tau_{1/2}$ is the Time it Takes for the Absorbance of the Perovskite Film to Deteriorate Half of its Initial.

Structure	Relative Humidity (%) at 22–24°C dark	$\tau_{1/2}$
$CH_3NH_3PbI_3/Al_2O_3$	0	76 days/1824 h
$CH_3NH_3PbI_3/Al_2O_3$	25	57 days/1368 h
$CH_3NH_3PbI_3/Al_2O_3$	50	26 days/624 h
$CH_3NH_3PbI_3/Al_2O_3$	90	26 days/111 h

In March 2015, Han et al. also studied the effect of higher temperature in the presence of moisture under one sun illumination on the encapsulated FTO/c-$TiO_2/CH_3NH_3PbI_3/spiro$-OMeTAD device's electrical performance.[18] Cross-sectional scanning electron microscopy (SEM) and powder XRD analyses were also carried out. Devices are encapsulated using UV-curable epoxy sandwiched between the device and glass cover (method A) or as an edge sealant (method B). In the latter, a desiccant is attached to the cover glass as a moisture absorber. Lifetimes of devices tested under different conditions are summarized in Table 4.3.

TABLE 4.3 Lifetimes of Encapsulated FTO/c-TiO$_2$/CH$_3$NH$_3$PbI$_3$/spiro-OMeTAD Devices Under One Sun Illumination at Controlled Temperatures and Relative Humidity.[18]

Encapsulation Method	Conditions: Environmental Temperature (Actual Cell Temperature), Relative Humidity	Lifetime (hours)	Performance Degradation (%)
A	-20°C (10°C), 0%	120	No longer-term data
A	55°C (85°C), 10%	50	10%
		80	50%
A	55°C (85°C), 80%	20	100%
B	55°C (85°C), 80%	20	70%
		40	80%
		60	90%

Test structures that are (i) perovskite-free, (ii) spiro-OMeTAD free, and (iii) Spiro-OMeTAD and silver-free are also tested under continuous illumination at 55°C (85°C), 50% RH for 500 h. The perovskite free structure underwent no degradation indicating that Spiro-OMeTAD was stable in contact with silver and TiO$_2$ under these conditions. In the spiro-OMeTAD-free structure, there are signs of Ag corrosion either through the chemical reaction in direct contact with the perovskite layer or gaseous by-products, most likely HI, from the perovskite decomposition. Ag corrosion by perovskite has been reported by Leijtens et al.[9] In the Spiro-OMeTAD and silver-free structure, an increase in the grain size of the perovskite was observed under SEM and an appearance of PbI$_2$ in the powder XRD pattern due to decomposition from the presence of moisture albeit a small amount. Therefore, some key conclusions were drawn in this work with regards to the failure modes of FTO/c-TiO$_2$/CH$_3$NH$_3$PbI$_3$/spiro-OMeTAD cells. Degradation of cells was accelerated by heat as the sealing becomes more permeable. The release of the HI and/or CH$_3$NH$_2$ vapors during decomposition will accelerate the decomposition further. Voids are observed in the Spiro-OMeTAD and perovskite layers as a result. Ag corrosion follows this, which consumes HI, further accelerating the perovskite decomposition.

4.3 UNSTABLE PEROVSKITE PHASE

In terms of thermal stability, $CsPbX_3$ (>300°C) is better than $FAPbX_3$ (>200°C), which in turn is better than $MAPbX_3$ (>100°C). However, in terms of device performance, Cs-based perovskite solar cells lag behind the organic counterparts part due to the higher (1.8–3.2 eV for $CsPbX_3$) than desired (1.1–1.4 eV) bandgap for single junction cell.[19]

$CsPbI_3$ and $FAPbI_3$ have the issue of transition to the undesirable non-perovskite (yellow) δ-phase with reduced photovoltaic attributes, see Figure 4.1. The desired perovskite α-phase (black in color) can be achieved via a high-temperature process and if left in the liquid, will transition to δ-phase upon cooling (e.g., to room temperature). Once in the solid form, the transition will be difficult especially for the black solid α-polymorph that has been previously transferred from yellow δ-phase via heating (above 120°C).[20] However, the black α-polymorph powder (previously yellow, before heated at 170°C) was observed to turn yellow again after 10 days of storage in air.[21]

Nevertheless, techniques to stabilize the α-phase and to eliminate the δ-phase have been demonstrated to be successful for formamidinium (FA) perovskite solar cells. They include (i) high-temperature annealing post-deposition;[5,22,23] (ii) sequential deposition of PbI_2 followed by FAI reducing the required temperature for post-annealing;[23,24–27] and (iii) intramolecular exchange[28] during sequential deposition, and (iv) mixing with other perovskite precursors such as MA- or/and Cs-halides.[21,29–32] The last strategy has been very effective in allowing stable perovskite (black) α or β-phase to form spontaneously. For $CsPbI_3$, a recent work by Luo et al.[33] has demonstrated that it may be possible to stabilize the black α-$CsPbI_3$ phase by sequential solvent engineering including the addition of HI and the subsequent isopropanol treatment although this first $CsPbI_3$ cell demonstrated experienced performance degradation during the course of light current-voltage measurement.

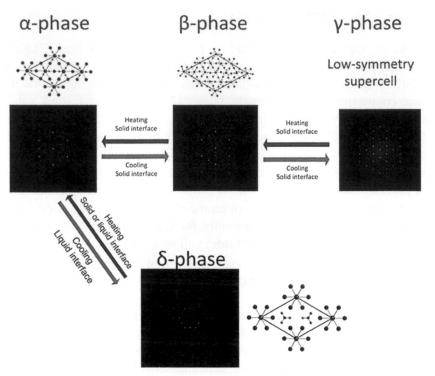

FIGURE 4.1   Phase transitions in the FA lead halide. Source: Reprinted/adapted with permission from ref. 20. © 2013 American Chemical Society.

4.4   OTHER FAILURE MODES AND SOLUTIONS PROPOSED

Apart from perovskite decomposition, different failure modes associated with the presence of or interaction with TiO_2;[9] Spiro-OMeTAD;[34] P3HT or PTAA, especially when additives are present;[3] PEDOT: PSS (acidity);[35,36] Ag,[18] Au[37], or Al[38] are also reported. The position of the interface layer (which determines the subsequent temperature treatment (s) it will receive) in the cell structure and the deposition method for the interface layer will determine its effect on the stability of the device. One example is ZnO, which has been reported to improve the stability of the "inverted" Glass/ITO/NiO/ $MAPbI_3$/ZnO/Al cell as ZnO acts as a barrier effect between the perovskite and Al.[38] However, when it is applied on the "standard polarity" cell structure on which the perovskite layer will be deposited, device degradation can accelerate.[34,39,40,41]

Other strategies proposed to improve device durability include:

- The use inorganic interface layers instead of organic ones although device performance remains to improve[38]
- The use of carbon to replace metal electrode[15,16]
- The use of barrier layers such as Cr_2O_3,[42] Cr,[37] Al_2O_3,[37,43,44] MoO_x[45] or PMMA with nano-carbon[46] between the carrier transport material and rear metal to stop metal migration through the transport layer (spiro-OMeTAD in most cases) causing device degradation
- Replacing mp-TiO_2 with mp-Al_2O_3;[9] or inserting Sb_2S_3[47] or $CsBr$[48] beneath the perovskite layer which appeared to improve the UV stability.

4.5 IMPROVED STABILITY WITH REDUCED DIMENSION

Most recently, it is shown that perovskites with reduced dimensions can have better moisture resistance and thus, stability.[1,49–52] The replacement of the small MA^+ or FA^+ cation with a much bulkier organic primary spacer cation confines the perovskite in two dimensions due to steric effects. It is believed that the better moisture resistance is due to the hydrophobicity of the long cation chain and the highly oriented and dense nature of the perovskite films prevent direct contact with water.[1,50–51] The introduction of multilayered perovskite structures may alter the mechanisms of electronic and ionic motion across the film, thus suppressing hysteresis.[52]

Thermal stability is still an issue of these devices and the best stability reported is two, 250 h under constant illumination for an encapsulated cell. Other techniques such as the addition of a phosphonic acid ammonium additive which is reported to cross-link the perovskite grains also shows better stability potential.[53] Recently, the addition of RbI to mixed-cation perovskite was found to increase the long-term stability and performance enhancement.[54,55]

Apart from environmental instabilities, reversible (ion migration) and irreversible (decomposition) behavior under operating conditions that include illumination, electrical bias and moisture have been reported. Their effects can be visualized using various characterization techniques such as photoluminescence microscopy and Kelvin probe force microscopy and are most visible at grain boundaries and interfaces.[12,56–61] Hysteresis also worsens over time as the cell degrades. Perovskite cells with stable electrical characteristics, free of hysteresis over time are yet to be developed, although increasing understanding will lead to effective strategies that either reduce

(i) the number of mobile ions, or (ii) the number of defect states, or (iii) the interation between the two if neither of them can be totally eliminated.

4.6 ENCAPSULATION SCHEMES

Encapsulating techniques from organic light emitter diode devices are commonly used for perovskite cell encapsulation.[62] For organic devices, an effective encapsulation scheme with a low water vapor transmission rate (WVTR), for example, 10^{-6}–10^{-5} g/m^2/day and a low oxygen transmission rate $\sim 10^{-3}$ cm^3/m^2/day will be required.[62] Strategies that may be compatible with perovskite solar cell may include multilayer stack using atomic layer deposited (ALD) Al$_2$O$_3$ with SiO$_2$ or SiN$_X$, which has been demonstrated to have such low transmission rates.[63,64] A water-resistant fluorous polymer CYTOP has been demonstrated to give similar results when compared to ALD aluminum oxide layer[65] on organic field-effect transistors. CYTOP has also been used on a CH$_3$NH$_3$PbI$_{3-X}$Cl$_X$ photodetector, extending its lifetime by 100 days in the air without additional glass protection.[65]

A more recent work by Shi et al.[66] demonstrated the effectiveness of using polyisobutylene (PIB) as an edge sealant and encapsulant for glass/FTO/c-TiO2/FAPbI3/PTAA/Au perovskite solar cells. PIB has been proven in the glazing industry. Its cost is comparable to ethylene vinyl acetate (EVA) commonly used for encapsulating silicon solar modules and is much lower than the UV-cured epoxy which is commonly used for packaging LEDs. Its WVTR is 2–3 orders lower than EVA (WVTR~28 g/m^2-day) and the UV-cured epoxy (WVTR~16 g/m^2-day)). In addition, PIB can be applied at a lower temperature compared to EVA and does not require UV curing, making it compatible with perovskite solar cells. In their work, Shi et al. also developed a feedthrough scheme that does not require metal feedthrough, which achieved a much tighter seal at the edge free of leakage pathways. By comparing three types of sealing and encapsulation schemes, Shi et al. showed that heat is the major factor for cell degradation after post-mortem studies of cells that went through International Electrochemical Commission (IEC) photovoltaic module thermal cycling (−45–85°C) and damp heat (85°C at 85% relative humidity) tests. In addition, the outgassing of degraded products accelerates further degradation. By using PIB as a "blanket" encapsulant, the outgassing is effectively stopped. Cells encapsulated by this method passed the IEC61215:2016 thermal cycling test (200 cycles) of experiencing no degradation in cell performance which was the best ever reported. In addition, these cells survived 500 h of damp heat tests as shown in Figure 4.2.

In the work by Bush et al.,[67] a glass/ITO/NiO/Cs$_{0.17}$FA$_{0.83}$Pb (Br$_{0.17}$I$_{0.83}$)$_3$/ LiF/PCBM/SnO$_2$/ZTO/ITO/Ag device was packaged between the top and bottom EVA encapsulants and two sheets of 3-mm-thick glass edge sealed by butyl rubber. The extra piece of glass (compared to Shi's work) and the presence of ITO providing the extra barrier, the use of the thermodynamically more favorably mixed CsFA perovskite, and the presence of the dense, atomic layer deposited SnO$_2$/ZTO stack are the reasons for improved stability allowing the packaged cell to pass the IEC61215:2016 damp heat test (<10% degradation after 1000 h of 85°C at 85% relative humidity).

With many researchers working in the field making progress, the durability will continue to be improved allowing cells to pass the critical IEC photovoltaic module environmental tests, including the humidity freeze and light soaking test making perovskite solar cells commercially viable.

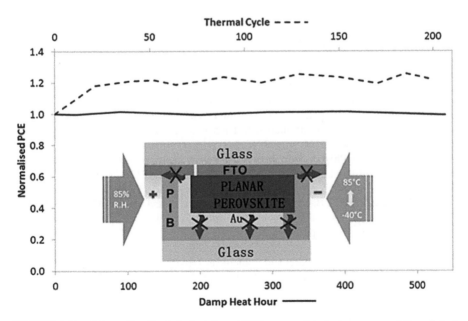

FIGURE 4.2 Schematic of polyisobutylene (PIB)-encapsulated planar perovskite cell that passes the IEC thermal stability test with results of the thermal cycling and damp heat tests. Reproduced with permission from ref 66. © 2017 American Chemical Society.

KEYWORDS

- perovskite cell stability
- perovskite cell encapsuation
- perovskite phase change
- perovskite cell durability
- water vapor transmission rate

REFERENCES

1. Tsai, H.; et al. High-Efficiency Two-Dimensional Ruddlesden–Popper Perovskite Solar Cells. Nature 2016, 536(7616), 312–316.
2. Noh, J. H.; et al. Chemical Management for Colorful, Efficient, and Stable Inorganic–Organic Hybrid Nanostructured Solar Cells. Nano Lett. 2013, 13(4), 1764–1769.
3. Yang, J.; et al. Investigation of $CH_3NH_3PbI_3$ Degradation Rates and Mechanisms in Controlled Humidity Environments Using in Situ Techniques. ACS Nano 2015, 9(2), 1955–1963.
4. Christians, J. A.; Miranda Herrera, P. A.; Kamat, P. V. Transformation of the Excited State and Photovoltaic Efficiency of $CH_3NH_3PbI_3$ Perovskite upon Controlled Exposure to Humidified Air. J. Am. Chem. Soc. 2015, 137(4), 1530–1538.
5. Eperon, G. E.; et al. Formamidinium Lead Trihalide: A Broadly Tunable Perovskite for Efficient Planar Heterojunction Solar Cells. Energy Environ Sci. 2014, 7(3), 982–988.
6. Leyden, M. R.; et al. Large Formamidinium Lead Trihalide Perovskite Solar Cells Using Chemical Vapor Deposition with High Reproducibility and Tunable Chlorine Concentrations. J. Mater. Chem. A 2015, 3(31), 16097–16103.
7. Wang, F.; et al. $HPbI_3$: A New Precursor Compound for Highly Efficient Solution-Processed Perovskite Solar Cells. Adv. Funct. Mater. 2015, 25(7), 1120–1126.
8. Aharon, S.; et al. Temperature Dependence of Hole Conductor Free Formamidinium Lead Iodide Perovskite Based Solar Cells. J. Mater. Chem. A 2015, 3(17), 9171–9178.
9. Leijtens, T.; et al. Overcoming Ultraviolet Light Instability of Sensitized TiO_2 with Meso-Superstructured Organometal Tri-Halide Perovskite Solar Cells. Nat. Commun. 2013, 4, 2885.
10. Misra, R. K. et al. Temperature- and Component-Dependent Degradation of Perovskite Photovoltaic Materials under Concentrated Sunlight. J. Phys. Chem. Lett. 2015, 6(3), 326–330.
11. Xiao, Z.; et al. Giant Switchable Photovoltaic Effect in Organometal Trihalide Perovskite Devices. Nat. Mater. 2015, 14(2), 193–198.
12. Leijtens, T.; et al. Mapping Electric Field-Induced Switchable Poling and Structural Degradation in Hybrid Lead Halide Perovskite Thin Films. Adv. Energy Mater. 2015, 5(20), 1500962-n/a.
13. Kim, H. S.; et al. Lead Iodide Perovskite Sensitized All-Solid-State Submicron Thin Film Mesoscopic Solar Cell with Efficiency Exceeding 9%. Sci. Rep. 2012, 2. DOI: 10.1038/srep00591.

14. Burschka, J.; et al. Sequential Deposition as a Route to High-Performance Perovskite-Sensitized Solar Cells. Nature 2013, 499(7458), 316–319.
15. Mei, A.; et al. A Hole-Conductor–Free, Fully Printable Mesoscopic Perovskite Solar Cell with High Stability. Science 2014, 345(6194), 295–298.
16. Zhou, H,; et al. Hole-Conductor-Free, Metal-Electrode-Free TiO$_2$/ CH$_3$NH$_3$PbI$_3$ Heterojunction Solar Cells Based on a Low-Temperature Carbon Electrode. J. Phys. Chem. Lett. 2014, 5(18), 3241–3246.
17. Vincent, B. R.; et al. Alkylammonium Lead Halides. Part 1. Isolated PbI$_6^{4+}$ Ions in (CH$_3$NH$_3$)$_4$PbI$_6$·2H$_2$O. Can. J. Chem. 1987, 65(5), 1042–1046.
18. Han, Y. et al. Degradation observations of encapsulated planar CH$_3$NH3PbI$_3$ Perovskite Solar Cells at High Temperatures and Humidity. J. Mater. Chem. A 2015, 3(15), 8139–8147.
19. Bremner, S. P.; et al. Optimum Bandgap Combinations to Make Best Use of New Photovoltaic Materials. Sol. Energy 2016, 135, 750–757.
20. Stoumpos, C. C.; Malliakas, C. D.; Kanatzidis, M. G. Semiconducting Tin and Lead Iodide Perovskites with Organic Cations: Phase Transitions, High Mobilities, and Near-Infrared Photoluminescent Properties. Inorg. Chem. 2013, 52(15), 9019–9038.
21. Jeon, N. J.; et al. Compositional Engineering of Perovskite Materials for High-Performance Solar Cells. Nature 2015, 517(7535), 476–480.
22. Lee, J. -W.; et al. High-Efficiency Perovskite Solar Cells Based on the Black Polymorph of HC(NH$_2$)$_2$PbI$_3$. Adv. Mater. 2014, 26(29), 4991–4998.
23. Lv, S.; et al. One-Step, Solution-Processed Formamidinium Lead Trihalide (FAPbI$_{(3-x)}$ Cl$_x$) for Mesoscopic Perovskite-Polymer Solar Cells. Phys. Chem. Chem. Phys. 2014, 16(36), 19206–19211.
24. Hanusch. F. C.; et al. Efficient Planar Heterojunction Perovskite Solar Cells Based on Formamidinium Lead Bromide. J. Phys. Chem. Lett. 2014, 5(16), 2791–2795.
25. Pellet, N. et al. Mixed-Organic-Cation Perovskite Photovoltaics for Enhanced Solar-Light Harvesting. Angew. Chem. 2014, 126(12), 3215–3221.
26. Binek, A.; et al. Stabilization of the Trigonal High-Temperature Phase of Formamidinium Lead Iodide. J. Phys. Chem. Lett. 2015, 6(7), 1249–1253.
27. Hu, M.; et al. Efficient Hole-Conductor-Free, Fully Printable Mesoscopic Perovskite Solar Cells with a Broad Light Harvester NH$_2$CH=NH$_2$PbI$_3$. J. Mater. Chem. A 2014, 2(40), 17115–17121.
28. Yang, W. S.; et al. High-Performance Photovoltaic Perovskite Layers Fabricated Through Intramolecular Exchange. Science 2015, 348(6240), 1234–1237.
29. Chiang, Y.-H.; et al. Mixed Cation Thiocyanate-Based Pseudohalide Perovskite Solar Cells with High Efficiency and Stability. ACS Appl. Mater. Interfaces 2017, 9(3), 2403–2409.
30. Saliba, M.; et al. Cesium-Containing Triple Cation Perovskite Solar Cells: Improved Stability, Reproducibility And High Efficiency. Energy Environ. Sci. 2016, 9(6), 1989–1997.
31. Rehman, W.; et al. Photovoltaic Mixed-Cation Lead Mixed-Halide Perovskites: Links Between Crystallinity, Photo-Stability and Electronic Properties. Energy Environ. Sci. 2017, 10(1), 361–369.
32. Kulbak, M.; et al. Cesium Enhances Long-Term Stability of Lead Bromide Perovskite-Based Solar Cells. J. Phys. Chem. Lett. 2016, 7(1), 167–172.

33. Luo, P.; et al. Solvent Engineering for Ambient Air-Processesd, Phase Stable CsPbI$_3$ Perovskite Solar Cells. J. Phys. Chem. Lett. 2016, 7(18), 3603–3608.
34. Dkhissi, Y.; et al. Stability Comparison of Perovskite Solar Cells Based on Zinc Oxide and Titania on Polymer Substrates. ChemSusChem. 2016, 9(7), 687–695.
35. Lee, D. Y.; et al. Graphene oxide/PEDOT:PSS Composite Hole Transport Layer for Efficient and Stable Planar Heterojunction Perovskite Solar Cells. Nanoscale 2016, 8, 1513–1522.
36. Igbari, F.; et al. A Room-Temperature CuAlO$_2$ Hole Interface Layer for Effieincy and Stable Planar Perovskite Soalr Cells. J. Mater. Chem. A 2016, 4, 1326–1335.
37. Domansk,i K.; et al. Not all that Glitters is Gold: Metal-Migration-Induced Degradation in Perovskite Solar Cells. ACS Nano 2016, 10(6), 6306–6314.
38. You, J.; et al. Improved Air Stability of Perovskite Solar Cells Via Solution-Processed Metal Oxide Transport Layers. Nat. Nanotechnol. 2016, 11(1), 75–81.
39. Yang, J.; et al. Origin of the Thermal Instability in CH$_3$NH$_3$PbI$_3$ Thin Films Deposited on ZnO. Chem. Mater. 2015, 27(12), 4229–4236.
40. Cheng, Y.; et al. Decomposition of Organometal Halide Perovskite Films on Zinc Oxide Nanoparticles. ACS Appl. Mater. Interfaces 2015, 7(36), 19986–19993.
41. Kumar, S.; et al. Acclerated Thermal-Aging-Induced Degradation of Organometal Triiode Perovskite on ZnO Nanostructures and its Effect on Hybrid Photovoltaic Devices. ACS Appl. Mater. Interfaces 2016, 8(28), 18309–18320.
42. Kaltenbrunner, M.; et al. Flexible High Power-Per-Weight Perovskite Solar Cells with Chromium Oxide-Metal Contacts for Improved Stability in Air. Nat. Mater. 2015, 14(10), 1032–1039.
43. Guarnera, S.; et al. Improving the Long-Term Stability of Perovskite Solar Cells with a Porous Al$_2$O$_3$ Buffer Layer. J. Physical. Chem. Lett. 2015, 6(3), 432–437.
44. Dong, X.; et al. Improvement of the Humidity Stability of Organic-Inorganic Perovskite Solar Cells Using Ultrathin Al$_2$O$_3$ Layers Prepared by Atomic Layer Deposition. J. Mater. Chem. A 2015, 3(10), 5360–5367.
45. Sanehira, E. M.; et al. Influence of Electrode Interfaces on the Stability of Perovskite Solar Cells: Reduced Degradation Using MoO$_x$/Al for Hole Collection. ACS Energy Lett. 2016, 1(1), 38–45.
46. Habisreutinger, S. N.; et al. Carbon Nanotube/Polymer Composites as a Highly Stable Hole Collection Layer in Perovskite Solar Cells. Nano Lett. 2014, 14(10), 5561–5568.
47. Ito, S.; et al. Effects of Surface Blocking Layer of Sb$_2$S$_3$ on Nanocrystalline TiO$_2$ for CH$_3$NH$_3$PbI$_3$ Perovskite Solar Cells. J. Phys. Chem. C 2014, 118(30), 16995–17000.
48. Li, W.; et al. Enhanced UV-Light Stability of Planar Heterojunction Perovskite Solar Cells with Caesium Bromide Interface Modification. Energy Environ. Sci. 2016, 9(2), 490–498.
49. Smith, I. C.; et al. A Layered Hybrid Perovskite Solar-Cell Absorber with Enhanced Moisture Stability. Angew. Chem. Int. Ed. 2014, 53(42), 11232–11235.
50. Cao, D. H.; et al. 2D Homologous Perovskites as Light-Absorbing Materials for Solar Cell Applications. J. Am. Chem. Soc. 2015, 137(24), 7843–7850.
51. Yao, K.; et al. Multilayered Perovskite Materials Based on Polymeric-Ammonium Cations for Stable Large-Area Solar Cell. Chem. Mater. 2016, 28(9), 3131–3138.
52. Quan, L. N.; et al. Ligand-Stabilized Reduced-Dimensionality Perovskites. J. Am. Chem. Soc. 2016, 138, 2649–2655.

53. Li, X.; et al. Improved Performance and Stability of Perovskite Solar Cells by Crystal Crosslinking with Alkylphosphonic Acid ω-Ammonium Chlorides. Nat. Chem. 2015, 7(9), 703–711.
54. Duong, T.; et al. Structural Engineering Using Rubidium Iodide as a Dopant Under Excess Lead Iodide Conditions for High Efficiency and Stable Perovskites. Nano. Energy 2016, 30, 330–340.
55. Zhang, M.; et al. High-Efficiency Rubidium-Incorporated Perovskite Solar Cells by Gas Quenching. ACS Energy Lett. 2017, 2(2), 438–444.
56. Deng, X.; Wen, X.; Lau, C. F. J.; Young, T.; Yun, J.; Green, M. A.; Huang, S.; Ho-Baillie, A. W. Y. Electric Field Induced Reversible and Irreversible Photolumines-cence Responses in Methylammonium Lead Iodide Perovskite. J. Mater. Chem. C 2016, 4, 9060–9068.
57. Soufiani, A. M.; Tayebjee, M. J. Y.; Meyer, S.; Ho-Baillie, A.; Yun, J. S.; McQueen, R.; Spiccia, L.; Green, M. A.; Hameiri, Z. Lessons Learnt from Spatially Resolved Electro- and Photoluminescence Imaging: Interfacial Delamination in $CH_3NH_3PbI_3$ Planar Perovskite Solar Cells Upon Illumination. Adv. Energy Mater. 2016, DOI: 10.1002/aenm.201602111.
58. Yun, J. S.; Ho-Baillie, A.; Huang, S.; Woo, S.; Heo, Y.; Seidel, J.; Huang, F.; Cheng, Y. –B.; Green, M. Benefit of Grain Boundaries in Organic–Inorganic Halide Planar Perovskite Solar Cells. J. Phys. Chem. Lett. 2015, 6, 875–880.
59. Kim, Y. C.; Jeon, N. J.; Noh, J. H.; Yang, W. S.; Seo, J.; Yun, J. S.; Ho-Baillie, A.; Huang, S.; Green, M. A.; Seidel, J.; Ahn, T. K.; Seok, S. I. Beneficial Effects of PbI_2 Incorporated in Organo-Lead Halide Perovskite Solar Cells. Adv. Energy Mater. 2015, 1502104. DOI: 10.1002/aenm.201502104.
60. Yun, J. S.; et al. Critical Role of Grain Boundaries for Ion Migration in Formamidinium and Methylammonium Lead Halide Perovskite Solar Cells. Adv. Energy Mater. 2016, 6(13), p 1600330-n/a.
61. Yuan, Y.; et al. Photovoltaic Switching Mechanism in Lateral Structure Hybrid Perovskite Solar Cells. Adv. Energy Mater. 2015, 5(15), 1500615-n/a.
62. Yu, D.; et al. Recent Progress on Thin-Film Encapsulation Technologies for Organic Electronic Devices. Opt. Commun. 2016, 362, 43–49.
63. Koushik, D.; et al. High-Efficiency Humidity-Stable Planar Perovskite Solar Cells Based on Atomic Layer Architecture. Energy Environ. Sci. 2017, 10(1), 91–100.
64. Zardetto, V.; et al. Atomic Layer Deposition for Perovskite Solar Cells: Research Status, Opportunities and Challenges. Sustainable Energy Fuels 2017, 1(1), 30–55.
65. Lu, H.; et al. A Self-Powered and Stable All-Perovskite Photodetector–Solar Cell Nano-system. Adv. Funct. Mater. 2016, 26(8), 1296–1302.
66. Shi, L.; Young, T. L.; Kim, J.; Sheng, Y.; Wang, L.; Chen, Y.; Feng, Z.; Keevers, M. J.; Hao, X.; Verlinden, P. J.; Green, M. A.; Ho-Baillie, A. W. Y. Accelerated Lifetime Testing of Organic–Inorganic Perovskite Solar Cells Encapsulated by Polyisobutylene. ACS Appl. Mater. Interfaces 2017, 9(30), 25073–25081.
67. Bush, K. A.; Palmstrom, A. F.; Yu, Z. J.; Boccard, M.; Cheacharoen, R.; Mailoa, J. P.; McMeekin, D. P.; Hoye, R. L. Z.; Bailie, C. D.; Leijtens, T.; Peters, I. M.; Minichetti, M. C.; Rolston, N.; Prasanna, R.; Sofia, S.; Harwood, D.; Ma, W.; Moghadam, F.; Snaith, H. J.; Buonassisi, T.; Holman, Z. C.; Bent, S. F.; McGehee, M. D. 23.6% Effi-cient Monolithic Perovskite/Silicon Tandem Solar Cells with Improved Stability. Nat. Energy. 2017, 2, 17009.

PART II

ELECTRON-TRANSPORTING LAYERS IN PEROVSKITE SOLAR CELLS

PHAM THI THU TRANG and HEMANT KUMAR MULMUDI

CHAPTER 5

METAL OXIDES AS ELECTRON SELECTIVE CONTACTS

5.1 INTRODUCTION

Mesoporous metal oxide scaffold plays an important role in achieving highly efficient perovskite devices. Various metal oxides such as TiO_2, SnO_2, ZnO, Zn_2SnO_4, $SrTiO_3$, and so forth[1-4] have been used in perovskite solar cells which cater as electron transporting materials.[5-24] Various nanostructures such as nanorods (NRs), nanofibers (NFs), nanoplatelets, and so forth of these metal oxides have been demonstrated in working perovskite solar cells. This section will briefly review the various metal oxides used in perovskite solar cells.

5.2 TITANIUM OXIDE (TIO$_2$)

Kim et al. reported highly efficient perovskite solar cells based on TiO_2 rutile NRs grown through hydrothermal means.[25] The X-ray diffraction (XRD) pattern of the vertically grown rutile NRs is shown in Figure 5.1a, which correspond to the reference data as shown in Figure 5.1b. The (101) diffraction peak is the most prominent peak compared to the (110) peak. The highly intense (101) peak and the enhanced (002) peak suggest that the rutile NRs are aligned in this direction. The height of the NRs was varied by changing the time duration of growth of NRs from 2 to 3.15 h. The length could be controlled from 0.56 to 1.58 um. Effect of the length of NRs on the device performance is shown in Figure 5.1c. The PCE of 0.56-um NRs device is 9.4% with J_{sc} of 15.6 mA/cm², V_{oc} of 0.95 V and FF of 0.63. Incident photon-to-electron conversion efficiency (IPCE) of more than 50% is observed over the entire wavelength range with a maximum of 71% between 420 and 500 nm (Figure 5.1d). As the length of NRs increases, the PV performance came down which was attributed to the lowering of J_{sc} and V_{oc}.

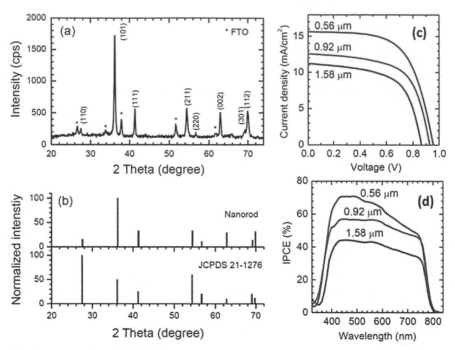

FIGURE 5.1 (a) X-ray diffraction (XRD) pattern of TiO_2 nanorods (NRs) grown on fluorine-doped tin oxide (FTO), (b) Comparison of diffraction pattern between rutile TiO_2 NRs and reference, Effect of TiO_2 nanorod length on (c) photovoltaic J–V curves of perovskite solar cells and (d) incident photon-to-electron conversion efficiency (IPCE) of the devices. Source: Adapted with permission from ref 25. © 2013 American Chemical Society.

Figure 5.2 shows the top view and cross section of the TiO_2 rutile NRs used in this study. The NRs have rectangular cross section and the length can be varied from 0.5 to 1.6 um. As the duration of the reaction increases, the dimension of the NRs increase from 80 to 150 nm. Vertically aligned and homogeneous NRs were obtained for shorter reaction time while the aligned with reduced degree was observed for longer durations.

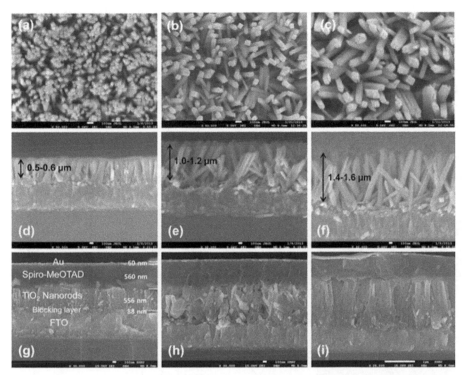

FIGURE 5.2 a–c) Surface and (d–f) cross sectional FESEM images of rutile TiO$_2$ NRs grown on FTO substrate. (g–i) Cross sectional scanning electron microscope (SEM) images of solid-state DSSCs based on perovskite CH3NH3PbI3-sensitized rutile TiO$_2$ nanorod photoanode, the spiro-MeOTAD hole transporting layer, and the Au cathode. Scale bars in panels (a–h) are 100 nm and that in panel (i) is 1 μm.

Source: Adapted with permission from ref 25. © 2013 American Chemical Society.

Shao et al. studied the effect of pore size and porosity of mesoporous TiO$_2$ films (TFs) on the device performance.[26] TiO$_2$ paste using block copolymer P123 and 2-butoxyethyl acetate as solvent was used to modulate the pore size by adjusting the ratios in which P123 and solvent were added. Figures 5.3a–c show the top-view scanning electron microscope (SEM) images of TF-16, TF-25, and TF-34; where -16,-25, and -34 denote the actual pore size of the TFs. TF-16 is film in which block copolymer P123 was not added and P-25 and P-34 represent films in which the weight ratio of P123: TiO$_2$ is 1:3 and 2:1, respectively. As seen from the SEM images, as the P123 content increases in the paste, larger voids are left in the films leading to larger pore size and porosity. Figure 5.3d shows the cross section of the TF-34 sample after sintering with voids which will aid in pore-filling of perovskite. Figures 5.3e–f show nitrogen adsorption–desorption isotherms. The pore size distribution curve in Figure 5.3e shows that the pore size increases with the P123 content. The wider pore size distribution and sharp increase in the relative pressure (Figure 5.3f) indicate more pores exist in those films. This is definite evidence that adding P123 in the films increases the pore size. The largest average pore size of the TFs obtained by this method was 34.2 nm and the corresponding porosity was found to be 73.5%, of which the porosity is the highest parameter that has ever been achieved for thick TFs. The thickness of TF with porosity of 73.5% obtained was up to 10 μm without cracking by screen-printing process.

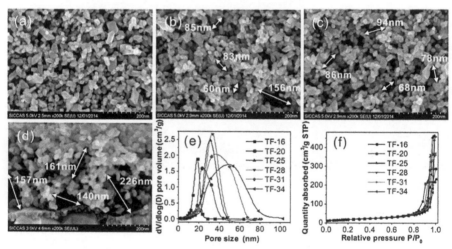

FIGURE 5.3 Top-view SEM images of porous TiO_2 films (TF)-16 (a), TF-25 (b) and TF-34 (c) and the corresponding cross section SEM image of TF-34 (d). Pore size distribution curves (e) and nitrogen adsorption–desorption isotherms (f) for sintered TFs.

Source: Reprinted with permission from Ref. 26. © 2016 American Chemical Society.

The current density–voltage characteristics for perovskite solar cells were tested based on TF-16, TF-25, and TF-34, consisting of a device structure of FTO/bl-TiO_2/mp-TiO_2–$CH_3NH_3PbI_3$/$CH_3NH_3PbI_3$/spiro-OMeTAD/Ag, where spiro-OMeTAD is a hole transport material of (2,2′,7,7′-tetrakis-N, N-di-4-methoxyphenylamino)-9,9′-spirobifluorene, and Ag is the silver contact. The effect of varying pore size of TFs on the hysteresis extents of J–V curves under reverse scan (from open circuit to short circuit, RS) and the opposite way forward scan (from short circuit to open circuit, FS) was studied. The hysteresis effect was quantified by measuring the hysteresis index (HI) given by the following formula:

$$HI = \frac{J_{RS}(0.6V_{OC}) - J_{FS}(0.6V_{OC})}{J_{RS}(0.6V_{OC})}$$

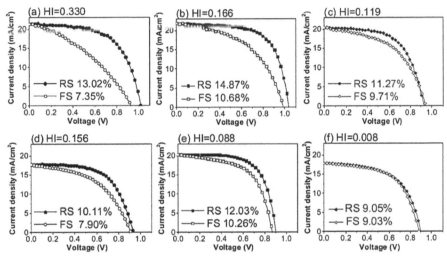

FIGURE 5.4 Scan direction dependent J–V curves for the PSCs based on 1L 250-nm thick TF-16 (a, d), 2L 450-nm-thick TF-25 (b, e) and 2L 450-nm thick TF-34 (c, f) through one-step solution method (a, b, and c) and two-step VASP method (d, e, and f).

Source: Reprinted with permission from Ref. 26. © 2016 American Chemical Society.

Figure 5.4 clearly shows the effect of pore size on the HI of the perovskite devices. It is also seen that the type of deposition for perovskites has an effect on hysteresis. With increasing pore size and thickness of the TFs, it is observed that the HI vanishes. This effect is attributed to the fast charge extraction by improved infiltration of perovskite into mp-TiO2 that suppresses charge accumulation on the interlayer of $TiO_2/CH_3NH_3PbI_3$. Perovskite solar cells fabricated by the VASP method show much lower HI due to the fact that this method would facilitate the pore-filling of perovskite into mp-TiO$_2$. This is because the CH_3NH_3I vapor can easily penetrate into the bottom of the film. Employing one-step solution procedure retards fast-evaporation of DMF, and crystallization occurs during diethyl ether dripping and subsequent annealing on hot plate which is in the time scale of minutes. Immediate formation of perovskite due to strong ionic interactions between the metal cations and halogen anions hinders excellent infiltration of perovskite into the bottom of the TiO$_2$ scaffold. By contrast, the VASP procedure avoids the extremely high reaction rate of perovskite often observed in the solution-processed method. Ultimately, a hysteresis-free PSC is achieved by depositing perovskite on 450-nm thick TF-34 through the two-step VASP method with power conversion efficiency (PCE) of 15.47%. This is an easy methodology for the preparation of highly porous TFs which provides a new way to fabricate hysteresis-free PSCs.

Another approach to facilitate pore-filling of perovskite is to use one-dimensional structure such as NFs. Sabba et al. employed electrospinning as technique to synthesise TiO_2 NFs. Though the NFs exhibit several advantages over nanoparticles in terms of charge transport and reduced charge recombination, they suffered from being limited by the film thickness when spun directly on conducting substrates. Thicker films (>2 μm) undergo shrinkage leading to delamination from the substrate. Additionally, the formation of cracks is predominant when the composite film is sintered to remove the organic precursors and to attain the crystalline semiconductor. For this purpose, several researchers employ additional steps to make pastes from the electrospun NFs and then obtain thick photoanodes by either screen printing or by spin-coating. The second approach employs sensitizers with high light absorption coefficients. In this sense, $CH_3NH_3PbI_3$ (an organic–inorganic halide perovskite) is a suitable sensitizer whose absorption coefficient is 1.5×10^4/cm at 550 nm. Thus, coupling the high absorption coefficient of the perovskite material and the good charge transport properties and porous structure of the NFs is a route toward high photovoltaic performance through a non-expensive process.

FIGURE 5.5   FESEM images showing: (a) small diameter and discontinuous nanofibers (NFs), (b) optimized diameter and continuous NFs and (c) NFs with large diameter but with closed pores, for similar electrospinning times.

Source: Reprinted with permission from Ref. 27. © 2014 Royal Society of Chemistry.

The synthesis of NF was performed using a solgel solution comprising of 0.8-g PVP (Mw = 1,300,000), 4-g titanium (IV) butoxide (97%), 1.18-g acetylacetone (≥99%) in 10-mL methanol. The optimized electrospinning conditions were: 25 kV with a feed-rate of 0.3 mL/h. Various morphologies of NFs can be attained by varying extrinsic parameters such as applied voltage, solution feed-rate, distance travelled by the solution to reach the substrate and solution viscosity. When the solution is diluted and is electrospun at 25 kV, with a feed rate of 0.1 mL/h, the electrospun NFs are thin but are discontinued as shown in Figure 5.5a. Upon increasing the solution viscosity and with voltage of 25 kV and feed the rate of 0.3 mL/h, NFs are less broken and the pores between them are more pronounced (Figure 5.5b). If the viscosity of the solution is further enhanced, with higher feed rate of 0.5 mL/h and voltage of 18 kV, the diameter of the NFs increases which leads to reduction in pores in-between the NFs, as seen in Figure 5.5c. When the solution viscosity changes, it is essential to change other parameters like feed rate and voltage to acquire the required shape and size of the NFs.

In addition to the porosity of the films, the total thickness of the final TF also has a critical effect on the device performance. In this respect, the photovoltaic characteristics for different nanofiber film thickness are illustrated in Figure 5.6a. The optimum thickness of nanofiber films to achieve the best cell performance was found to be 413 nm. With thicker films of 844 nm and 1215 nm, V_{oc} dropped to 0.78 and 0.74 V, respectively. This could be attributed to a higher recombination rate in the thicker films, due to the increase in surface area and the consequent increase of trap-assisted recombination in TFs. This higher recombination is also expressed as an early onset of the dark current for thicker TFs as seen in Figure 5.6a.

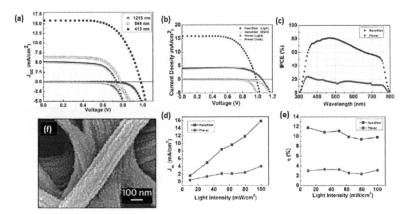

FIGURE 5.6 (a) Effect of different nanofiber film thicknesses on the photovoltaic performance, (b) Current density versus voltage plots, (c) IPCE action spectra for nanofiber (represented by squares) and planar devices (represented by circles). Effect of different light intensities on (d) J_{sc} and (e) efficiency, for planar and nanofiber cells, (f) high magnification SEM image of TiO_2 NFs, which were employed in the perovskite solar cells. *Source*: Reprinted with permission from Ref. 27. © 2015 American Chemical Society.

Given the limitation of the electrospinning technique and the diameter of the optimum fibers, homogenous films thinner than 413 nm could not be obtained. Therefore, planar devices were fabricated to get a reference device. The photovoltaic characteristics (J–V plots) of the devices fabricated on nanofiber and planar substrates are presented in Figure 5.6b. The nanofiber-based perovskite cell has exceedingly high efficiency (9.82%) compared to the planar (3.11%) device. The 413-nm thick mesoporous nanofiber film enables more loading of the perovskite material as compared to the compact structure of the planar device. This effect is reflected in terms of the higher J_{sc}: 15.88 mA/cm for the nanofiber device while the planar device exhibits only 4.02 mA/cm. The V_{oc} of the nanofiber devices dropped in comparison to the planar devices. This is expected since the trap-assisted recombination will be lower for the latter, following the same trend observed in the nanofiber devices with different film thickness. The high J_{sc} obtained for the nanofiber devices is also validated by the IPCE spectra shown in Figure 5.6c. The J_{sc} calculated from the IPCE data concurs well with the J_{sc} calculated from the J–V plots, and also the shape of the IPCE spectrum for nanofiber devices is in good agreement with the shapes reported for nanoparticle devices. The photovoltaic parameters (J_{sc}, η) of the nanofiber and planar devices as a function of light intensity are shown in Figures 5.6d and e. Short circuit current density showed linear dependence on light flux. The linear behavior of short circuit current density with light intensity demonstrates that charge collection efficiency is independent of light intensity. The nanofiber devices systematically show higher short circuit current densities and higher power conversion efficiencies compared to planar devices at all light intensities. Remarkably, the nanofiber cell yielded 11.79% PCE at 0.1 sun while the planar device yielded 3.03%. It is found that the efficiencies are mainly determined by the open porosity of the electrospun nanofiber network which varies with TiO_2 nanofiber photoanode thicknesses and fiber diameters. High magnification scanning electron image of the sample is shown in Figure 5.6(f) which shows voids and porosity in the nanofiber network aiding in charge transport.

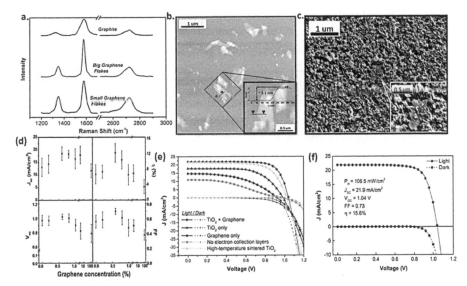

FIGURE 5.7 (a) Raman spectra of starting graphite flakes and its exfoliated graphene deposited on the SiO₂ substrate with increasing sonication time. (b) AFM image of exfoliated graphene scanned with 5 × 5 μm size and enlargement of the graphene flakes and its height profile of the lines shown in the inset figure. (c) SEM images of graphene-TiO₂ nanocomposites with enlargement micrographs in the inset, (d) Graphene-concentration dependence in the nanocomposites of the parameters extracted from a single batch J–V measurements of solar cells, (e) Current–voltage characteristics of different electron collection layers under simulated AM 1.5, 100 mW/cm² solar irradiation (solid line), and in the dark (dotted line). (f) Best performing (η = 15.6%) solar cell based on graphene-TiO₂ nanocomposites under simulated AM 1.5, 106.5 mW/cm² solar irradiation (solid line), and in the dark (dotted line), which processed at temperatures not exceeding 150°C. Solar cell performance parameters are given in the inset.

Source: Reprinted with permission from Ref. 28. © 2014 American Chemical Society.

Wang et al. used nanocomposites of graphene and TiO₂ as electron transporter.[28] The graphene nanoflakes proved to be excellent electron transporter and enabled solution processing at temperatures less than 150°C. The group observed reduced series resistance and reduced recombination losses with the low-temperature composite. Detailed material characterization was performed to unravel the reasons behind such astonishing results for the nanocomposites. Graphene was produced from graphite flakes using liquid-phase exfoliation using isopropanol as the solvent and sonicating for 70 h.

Raman spectroscopy and atomic force microscopy (AFM) was used to probe the quality of the graphene film and to know the number of graphene layers in each flake. Graphene was drop cast on silicon substrates and heated at 100°C for this study. Figure 5.7a shows the Raman spectrum for samples with decreasing flake size, which is done by increasing sonication time. The two most prominent peaks are the G band ($\sim$ 1580/cm) and the two-dimensional (2D) band ($\sim$ 2700/cm) as expected. We observe a difference in the contribution to the 2D band's shape from multiple peaks between exfoliated graphene flakes and starting graphite, suggesting that the prepared samples are few-layer graphene. The AFM image of exfoliated graphene flakes is shown in Figure 5.7b. It gives information on the graphene morphology and its number of layers. It is seen that the graphene flakes are randomly distributed over the substrate. The lateral dimension of the flakes varies between 0.1 and 1 μm. Cross sectional analysis of the graphene flakes revealed a height of $\sim$ 3.1 nm (equivalent to 5 layers of graphene) as shown in the inset to Figure 5.7b. To prepare the TiO_2 photoanode, anatase TiO_2 particles of 25 nm were blended with the graphene dispersion. The SEM image in Figure 5.7c, we show the exposed surface of the nanocomposite highlighting TiO_2 nanoparticles anchored upon a graphene nanoflake. Graphene's remarkable conductivity can serve as a highway for efficient electron transportation. A series of solar cells with varying amounts of graphene flakes were fabricated to assess its impact on the series resistance and overall device performance. Figure 5.7d shows a clear trend in the enhancement of PV parameters with the introduction of graphene in the nanocomposites. For the pristine TiO_2 nanoparticles, the average short-circuit current density (J_{sc}) and fill factor (FF) were around 13 mA/cm^2 and 0.56, respectively. With the increase in graphene concentration, both the J_{sc} and FF increased to average values of around 18.5 mA/cm^2 and 0.7, respectively, peaking at 0.6 wt.%, followed by a decrease with further increases in graphene content. However, in contrast to the J_{sc} and FF, the average open circuit voltage (V_{oc}) only increased slightly with graphene's addition, with an average V_{oc} of around 1 V for 0.8 wt.%. V_{oc} begins to decrease as the graphene concentration increases over 0.8 wt.%, probably due to direct contact of graphene with perovskite, leading to recombination directly between the electrons in the graphene and holes in the perovskite. Figure 5.7e, shows the J–V curves of the best-performing cells of each series, including cells with pure TiO_2, pure graphene, high-temperature sintered TiO_2, and the substrate with pure fluorine-doped tin oxide (FTO). As can be seen from the J–V curves, devices on FTO (without any electron selective contact) have inferior performance. One of the main losses for the perovskite device coated directly upon the FTO conducting glass is FF, this could be due to poor electronic contact at the FTO/perovskite interface. The electron collection layers with either pure graphene or low-temperature processed TiO_2 nanoparticles show improvements in measured J_{sc} and FF compared to devices fabricated directly on FTO. Further improvements are observed when graphene and TiO_2 nanoparticles are employed as nanocomposites. The most efficient low-temperature processed perovskite solar cells (Figure 5.7f) based on a graphene–TiO_2 nanocomposite electron collection layer had the following photovoltaic parameters: $J_{sc} = 21.9$ mA/cm², $V_{oc} = 1.05$ V, and FF=0.73 yielding an efficiency of 15.6%.

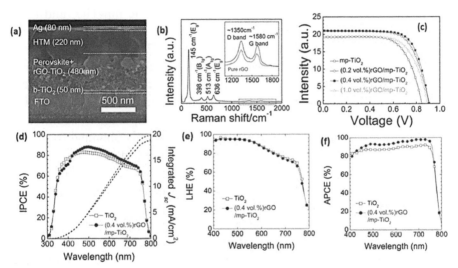

FIGURE 5.8 (a) Cross section SEM image of 0.4 vol.% rGO/mp-TiO_2 nanocomposite based perovskite solar cell, (b) Raman characterization of rGO/mp-TiO_2 nanocomposite films after 450°C annealing, (c) J–V curve of rGO/mp-TiO_2 nanocomposite-based perovskite solar cell with varying rGO contents, (d) IPCE, (e) light-harvesting efficiency (LHE), and (f) absorbed photon-to-current conversion efficiency (APCE) of the devices fabricated in this study. *Source:* Reprinted with permission from Ref. 29. © 2015 American Chemical Society.

Han et al. employed nanocomposite of reduced graphene oxide (rGO) and mesoporous (mp) TiO_2 in perovskite solar cells which had improved electrical properties.[29] Figure 5.8a shows the cross section of a typical device fabricated in this study. Raman spectroscopy was performed to determine the amount of rGO in 0.4% sample. Figure 5.8b shows the Raman peaks located at 143 (E_g), 199 (E_g), 396 (B_{1g}), 514 (A_{1g}), and 636/ cm (E_g) are attributed to the anatase TiO_2 phase. The Raman peaks of rGO located at 1350/cm (D band) and 1580/cm (G band) confirm the existence of rGO in the rGO–TiO$_2$ nanocomposites. The G band is related to the optical E$_{2g}$ phonons of sp^2 carbon atoms, and the D band corresponds to the breathing mode of the sp^2 atoms in rings. The Raman intensity ratio (I$_D$/I$_G$) slightly increases from 1.06 (pure rGO) to 1.16 after the annealing with TiO$_2$, indicating the rGO/mp-TiO$_2$ nanocomposite has a slightly higher amount of defects due to the conversion of sp^2 to sp^3. Figure 5.8d shows the IPCEs of the mp-TiO$_2$ and rGO/mp-TiO$_2$ nanocomposite perovskite solar cells. The IPCE of rGO/mp-TiO$_2$ nanocomposite is higher than the mp-TiO2 NPs in the wavelength region 400–700 nm. The integrated J$_{sc}$ obtained from the IPCE spectra for the mp-TiO$_2$ and (0.4 vol.%) rGO/mp-TiO$_2$ nanocomposite-based perovskite solar cells were 18.8 and 19.9 mA/cm^2 with maximum IPCEs of 82.99% (at 480 nm) and 88.02% (at 500 nm), respectively (Figure 5.8c). In order to understand these differences in photovoltaic properties, light harvesting efficiency (LHE) and absorbed photon to current conversion efficiency (APCE) were compared. Figure 5.8e shows the LHE of the mp-TiO$_2$ and rGO (0.4 vol.%)/mp-TiO$_2$ nanocomposites perovskite solar cells obtained from the relation LHE = (1—R)(1 —10^{-A}), where R is the reflectance and A is the absorbance. Though the rGO/mp-TiO$_2$ nanocomposite based perovskite solar cell shows slightly lower LHE than the mp-TiO$_2$ based perovskite solar cell at the longer wavelength, the integrated light absorption values (∼ 84%) in both the mp-TiO$_2$ and (0.4 vol.%) rGO/mp-TiO$_2$ perovskite solar cells are comparable. APCE of the nanocomposite based perovskite solar cells were ∼ 7% higher than that of the mp-TiO$_2$-based perovskite solar cells throughout the entire action spectra (Figure 5.8f). This is an indication of improved electron collection efficiency for the rGO/mp-TiO$_2$ nanocomposite. Introduction of rGO into the TiO$_2$-based electron transport layer (ETL) was found to be one of the simplest ways to mitigate large recombination at the TiO$_2$ grain boundaries.

Huang et al. used solgel technique to synthesize TiO$_2$ sub-micron spheres and used it as electron transporting scaffold.[30] The TiO$_2$ sub-micron spheres were synthesized by a two-step process, where titanium isopropoxide was added into ethanol solution containing DI water and KI solution, followed by the step of autoclave heating procedure.[30] The as obtained TiO$_2$ spheres was made into paste using ethylcellulose and α-terpineol. This paste was further diluted with ethanol (1:5 weight ratio) to spin coat on substrates. Figure 5.9a shows the top-view SEM image of the film spin-coated on FTO substrate. As seen clearly, the film is quite different from the conventional

20-nm film obtained by spin coating. There are areas of FTO which are still open when using sub-micron spheres as scaffold. The average diameter of the particles was found to be 250 nm. Figure 5.9b shows the diffuse reflectance of the TiO_2 sub-microsphere films and the conventional nanoparticle film measured using an integrated sphere. The light scattering effect of the sub-microsphere film is quite evident from the data. This will enhance the path length of the light and hence might be beneficial for light harvesting in perovskite solar cells. Apart from the optical advantage, the scaffold layer of sub-microsphere film should also have an effect on the surface morphology of the perovskite film, which in turn drastically change the absorption properties of the films. Figure 5.9c shows the optical absorption of TiO_2/PbI_2 films, where the optical absorption at 500 nm is slightly higher for the sub-microsphere film. After the perovskite film formation, the absorption of sub-microsphere film is increased in the longer wavelength region (550–800 nm). LHE was also calculated using the following formula:

$$LHE(\lambda) = (1-R) \times \left(1 - 10^{-A(\lambda)}\right)$$

where R is the total reflectance and A (λ) is the absorbance. The LHE is similar for both the films in the shorter wavelength region (400–500 nm), but in the longer wavelengths (550–800 nm), the sub-microsphere films have a higher LHE compared to the conventional nanoparticle film, Figure 5.9d. The current–voltage characteristics of the perovskite solar cells fabricated using the two types of films is shown in Figure 5.9e. The J_{sc}, V_{oc}, and FF are 18.01 mA/cm^2, 1.05 V, and 70.70% for the solar cells based on the nanoparticle film, while 19.41 mA/cm^2, 1.05 V, and 73.65% are the PV parameters for the device based on the sub-microsphere scaffold layer. The open circuit voltage of both the devices remain same, while the J_{sc} and FF show remarkable enhancement for the film based on sub-microsphere film resulting in the overall efficiency increasing from 13.37 to 15.01%. The IPCE action spectra in Figure5.9f show enhanced conversion efficiency in the longer wavelength region (500–800 nm). It was also found out that the strong light scattering and superior charge transfer contribute to the improved IPCE. Hence, designing novel scaffold layer microstructure is another way of boosting device performance of perovskite based solar cells.

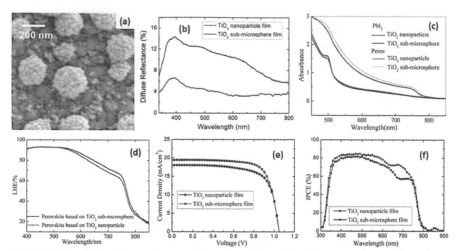

FIGURE 5.9 (a) SEM top view of TiO_2 microsphere film, (b) comparison of the diffuse reflectance of the nanoparticle and sub-microsphere films, (c) Absorbance of both PbI_2 and perovskite on nanoparticle and sub-microsphere films, (d) Light harvesting efficiency of the films, (e) J–V curves of the perovskite-based solar cells fabricated on both the films, (f) IPCE action spectra of the devices fabricated in the study.

Source: Reprinted with permission from Ref. 30. © 2016 American Chemical Society.

Peng et al. employed a simple and single-step solution processed method to fabricate high-quality indium doped titanium oxide as ETL.[31] The conductivity of compact In-TiO_x thin films was found to increase due to the indium dopant, thus dramatically improving the FF of perovskite solar cells compared to cells with pure TiO_2 ETLs. Furthermore, indium-doping allows to tune the work function (WF) of the ETL to improve the band alignment at the ETL/perovskite interface. In-TiO_x thin films for XPS/UPS characterization were prepared for the best performing indium-doping condition (3%-In-TiO_x). XPS measurements were conducted to elucidate the chemical compositions of TiO_2 and In-TiO_x deposited on FTO substrates. Figures 5.10a and b compares the XPS spectra of the Ti 2p and In 3d peaks, respectively, for the two films. The Ti 2p1 and Ti 2p3 peaks of the TF at binding energies of ≈ 464.3 and ≈ 458.5 eV. For the In-TiO_x films, the Ti 2p3 peak is shifted slightly higher to 458.8 eV. This shift can be explained by the Pauling electronegativity theory; the electronegativity value of Ti is 1.5 and In is 1.7, which indicates negative charge transfer toward indium in the Ti–O–In bond, thereby increasing the Ti 2p core level binding energy. Figure 5.10b shows the XPS spectrum of In 3d5 for the In-TiO_x film, with a single peak centered at ≈ 444.3 eV that can be ascribed to indium oxide. As expected, the TiO_2 does not have a corresponding peak. The surface stoichiometry of the samples was calculated by comparing the elements' relative peak areas and their corresponding relative sensitivity factors.

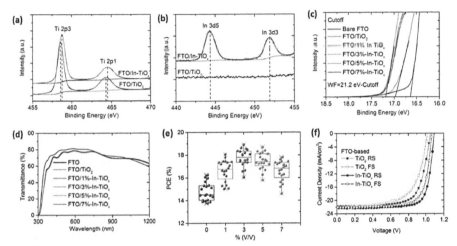

FIGURE 5.10 (a) XPS spectra of Ti 2p peaks, (b) XPS spectra of In 3d peaks, (c) UPS spectra of bare FTO, FTO/TiO$_2$, and FTO/In-TiO$_x$, (d) Transmittance spectra of bare FTO, FTO/TiO$_2$, and FTO/In-TiO$_x$, (e) Statistical distribution of the photovoltaic conversion efficiencies for cells with different indium doping concentrations, with a structure of FTO/In-TiO$_2$/meso-TiO$_2$/MAPbI$_3$/Spiro-OMeTAD/Au, (f) Current density–voltage curves of the ITO/In-TiO$_x$ (or TiO$_2$)/meso-TiO$_2$/MAPbI$_3$/Spiro-OMeTAD/Au champion device.
Source: Reprinted with permission from Ref. 31. © 2016 John Wiley & Son.

UPS measurements were also performed to estimate the WF of FTO/TiO$_2$ and FTO/In-TiO$_x$ samples. Figure 5.10c shows that the photoemission cut-off of bare FTO is at 16.65 eV corresponding to a WF of ≈4.55 eV, while the WF of FTO/TiO$_2$ is ≈4.22 eV. To further investigate the WF of the FTO/In-TiO$_x$, films were prepared with four different indium doping concentrations varying from 1% to 7% (v/v), labeled as 1%-In-TiO$_x$ to 7%-In-TiO$_x$. The WF initially decreases with increasing indium concentration from ≈4.06 eV for 1%-In-TiO$_x$ to ≈4.00 eV for 3%-In-TiO$_x$. With additional indium content, the WF increases again, to ≈4.02 eV for 5%-In-TiO$_x$ and ≈4.06 eV for 7%-In-TiO$_x$, which is the same as for 1%-In-TiO$_x$. For all doping levels, the WF of In-TiO$_x$ is closer to the conduction band of MAPbI$_3$ perovskite (≈3.9 eV). This translates into higher V$_{oc}$ values for MAPbI$_3$-based solar cells with In-TiO$_n$ ETLs compared to those with TiO$_2$ ETLs. Figure 5.10d demonstrates that the transmittance of FTO/In-TiO$_x$ samples is almost identical to that of FTO/TiO$_2$. All of the coated samples have lower transmittance than the bare FTO due to increased reflectance resulting from the high refractive index TiO$_2$/In-TiO$_x$ films. Therefore, it can be concluded that the indium doping has a negligible effect on the optical properties of the TiO_2/In-TiO_x films. To optimize the performance of In-TiO_x-based cells, perovskite solar cells on FTO substrates using indium doping concentrations varying from 0% to 7% (v/v), labeled as TiO_2, 1%-In-TiO_x, 3%-In-TiO_x, 5%-In-TiO_x, and 7%-In-TiO_x were fabricated. Figure 5.10e shows the PCE of 100 different cells as a function of indium concentration. It was found that even 1% indium doping can dramatically improve the V_{oc} from ≈ 1.04 V (the median V_{oc} for TiO_2) to ≈ 1.07 V (median V_{oc} for 1%-In-TiO_x). The V_{oc} reached a maximum median value of 1.09 V for the 3%-In-TiO_x. A slightly decreased V_{oc} (≈ 1.085 V) was observed with the 5% indium doping concentration before the V_{oc} drops back to ≈ 1.07 V for the 7%-In-TiO_x. This trend is consistent with the UPS measurements of the WF of In-TiO_x and TiO_2 as discussed before. A median FF of 0.665 was obtained from TiO_2-based control cells, whereas the 3%-In-TiO_x-based cells exhibited a median FF of ≈ 0.750, corresponding to a $\approx 13\%$ improvement. Comparing to the current density versus voltage (J–V) curve of In-TiO_x-based and TiO_2-based cells shown in Figure 5.10f, it appears that the improved FF is largely due to a reduced series resistance in the In-TiO_x-based cells. Therefore, Hall Effect measurements were employed to further study the electronic properties of the ETL layer. The results (data not provided here) show that the conductivity of In-TiO_x is higher than that of TiO_2, which is possibly caused by the increased carrier density and hall mobility, thus reducing the series resistance in In-TiO_x-based cells. The median J_{sc} for In-TiO_x-based cells showed only a weak variation with indium content, where the median J_{sc} of 22.09 mA/cm^2 for 3%-In-TiO_x-based cells is slightly ($\approx 4\%$) higher than that of the TiO_2-based control devices ($J_{sc} \sim 21.3$ mA/cm^2). The combination of higher V_{oc} and FF, and J_{sc} results in an absolute efficiency increase of more than $\approx 3.3\%$ for 3%-In-TiO_x-based cells ($PCE_{median} \approx 17.8\%$) as compared to the TiO_2-based control cells ($PCE_{median} \approx 14.5\%$), as seen in Figure 5.10f.

Liu et al. used a low-temperature approach to fabricate a uniform and pinhole-free compact TiO_2 layer for enhancing the photovoltaic performance of perovskite solar cells.[32] $TiCl_4$ treatment was employed to modify TiO_2 for efficient charge generation and significantly reduced recombination loss. The TiO_2 nanoparticles were prepared by sol-gel method. The diameter of these nanoparticles was found to be 5 nm. They were used to prepare the low-temperature TiO_2 compact layer (ltc-TiO_2). These nanoparticles were deposited on ITO to form the blocking layer. This film was treated with 200-mM $TiCl_4$ solution at 70°C for different time durations. The surface morphologies of the films obtained by AFM after different time durations of $TiCl_4$ treatment is shown in Figure 5.11. The roughness of ltc-TiO_2 reduced from 7.8 to 6.3 nm after 20 min of $TiCl_4$ treatment. It was found that the untreated TFs had pinholes with size varying from 5 to 10 nm, but after the $TiCl_4$ treatment, the pinholes were found to vanish making the films smooth. This could be attributed to the lower roughness values obtained for the films which were treated with $TiCl_4$. After 30 min of TiCl4 treatment, cracks were also found on the films which would be unfavorable for device performance.

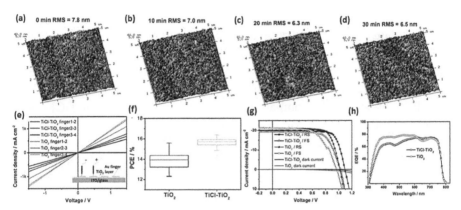

FIGURE 5.11 AFM image of TF with $TiCl_4$ treatments for (a) 0, (b) 10, (c) 20, and (d) 30 min, respectively, (e) J–V curve of devices with ITO/TiO_2/Au configurations based on TiCl–TiO_2 layer (solid line) and TF (dashed line). Inset: the schematic diagram of ITO/TiO_2/Au device, (f) Statistical distribution of photovoltaic PCE under simulated AM 1.5 (100 mW/cm^2) illumination, (g) Forward scan (FS, from −0.2 to 1.2 V) and reverse scan (RS, from 1.2 V to −0.2 V) of current–voltage (J–V) curves for TiCl–TiO_2-based devices and untreated ltc-TiO_2-based devices under simulated AM 1.5 (100 mW/cm^2) illumination with a step size of 20 mV and a delay time of 1 ms. Dark current measurements were performed with a scan range from −1.5 to 1.5 V and a scan rate of 0.05 V per step, (h) External quantum efficiency (EQE) curve of the device based on $TiCl_4$-treated TF (TiCl–TiO_2) and untreated TF.
Source: Reprinted with permission from Ref. 32. © 2016 American Chemical Society.

To study the effect of $TiCl_4$ treatment on the uniformity of films, devices were fabricated with the structure of ITO/ltc-TiO_2/Au. AS cracks appeared in the 30-min $TiCl_4$ treatment sample, 20-min $TiCl_4$ treatment sample was used for the study. The J–V curves are shown in Figure 5.11e. All devices show resistive behavior. The TiCl–TiO_2-based devices exhibit lower current density, that is, larger resistance, than that of the TiO_2 devices (2.25 (TiCl-TiO2) versus 1.18 (TiO2) ohm cm^2). In the case of high-quality films without any pinholes, the resistance derived from the J–V characteristics should depend only on the conductivity of the TFs. However, the dramatic difference in resistance observed could be attributed to the pinhole effects, which result in direct contact between ITO and Au. For the TiCl–TiO$_2$ layer, nanoscale pinholes were partially filled after TiCl$_4$ treatments, which can effectively prevent the direct contact of ITO and Au. Fewer pinholes indicate that the TiCl–TiO$_2$ should act as an effective blocking layer that physically separates the photovoltaic active layer and metallic electrode. According to this observation, 20-min TiCl$_4$-treated ltc-TF was chosen as optimal condition to fabricate photovoltaic cells. The ltc–TiO$_2$ layer in a planar device configuration with the following stack: ITO/TiO$_2$/Perovskite/Spiro-OMeTAD/Au were fabricated. Device performance was characterized by measuring J–V curves under simulated AM 1.5G (100 mW/cm^2) solar illumination with statistical distribution as shown in Figure 5.11f. As seen in Figure 5.11g, the devices based on TiO$_2$ without TiCl$_4$ treatment shows a short circuit current density (J$_{SC}$) of 21.0 mA/cm^2, an open circuit voltage (V$_{OC}$) of 1.02 V, a FF of 72.9% and a PCE of 15.6%. In comparison, the best device based on TiCl–TiO$_2$ shows a slightly lower J$_{SC}$ of 19.7 mA/cm^2, a V$_{OC}$ of 1.09 V, a higher FF of 75.9%, and a higher PCE of 16.4%. It is well-known that an anomalous hysteresis is frequently observed in perovskite solar cells, which has been attributed to the charge selective layers, ionic movement or ferroelectric effects. The hysteresis effects for both the ltc-TiO$_2$ and TiCl–TiO$_2$ devices are shown in Figure 5.11g. The J$_{SC}$ was not significantly affected; however, the FF and V$_{OC}$ were significantly changed with respect to the scan direction. It was found that both devices show hysteresis which could originate from the effects discussed before. Figure 5.11h shows the external quantum efficiency (EQE) of both the devices. The EQE of TiCl–TiO$_2$-based device shows obvious decrease in the wavelength region of 300–600 nm, which could originate from the absorption of TiCl–TiO$_2$. Hence, a simple TiCl$_4$ treatment can be used to obtain uniform and pinhole-free TiO$_2$ compact layer via low-temperature processing for planar heterojunction perovskite solar cells.

5.3   ZINC OXIDE (ZNO)

Son et al. prepared perovskite solar cells on zinc oxide NRs grown on ZnO seed layer grown from solution as shown in the schematic illustration (Figure 5.12).[33] The seed layer was prepared from an ethanolic solution of zinc acetate dehydrate (5 mM). This solution was spun coated on FTO substrates and then annealed at 350°C for 15 min. The solution for growing ZnO NRs was prepared by dissolving equimolar zinc nitrate hexahydrate, and hexamethylenetetramine in deionized (DI) water. The solution concentration was varied from 20 to 35 mM for controlling the diameter of the ZnO NRs.

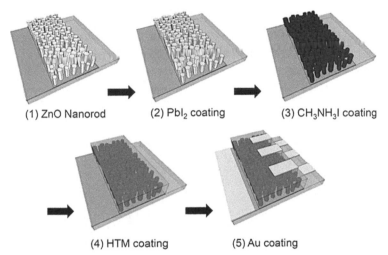

(1) ZnO Nanorod   (2) PbI₂ coating   (3) CH₃NH₃I coating

(4) HTM coating   (5) Au coating

FIGURE 5.12 Fabrication procedure of perovskite solar cell based on the ZnO nanorod electrode.

Source: Reprinted with permission from Ref. 33. © 2014 American Chemical Society.

FIGURE 5.13 Top-view SEM images of hexagonal ZnO NRs grown at (a) 20, (b) 25, (c) 30, and (d) 35 mM of the precursor solution containing equimolar zinc nitrate hexahydrate and hexamethylenetetramine. The ZnO seed layer deposited FTO substrates were immersed in the precursor solution at 90°C for 180 min. Insets represent the distribution of diameters of ZnO NRs.

Source: Reprinted with permission from Ref. 33. © 2014 American Chemical Society.

Figure 5.13 shows that average diameter (d) increases from 54 to 61, 73, and 82 nm as the precursor concentration increases from 20 to 25, 30, and 35 mM, respectively, when the immersion time is fixed at 180 min. The length of the grown ZnO NRs remains to 1 µm irrespective of the precursor concentration, which indicates that change in the precursor concentration at the fixed immersion time affects only the diameter of the ZnO NRs. The rate of increase in diameter of the ZnO NRs is estimated to be ≈2.4 nm/mM. Length of ZnO NRs can also be varied when the immersion time changes at the fixed concentration. It was found that the average length of the ZnO NRs increases from 440 to 620 nm, 820 nm, and 1 µm as the immersion time increases from 90–120, 150, and 180 min, respectively, when the precursor concentration was kept constant.

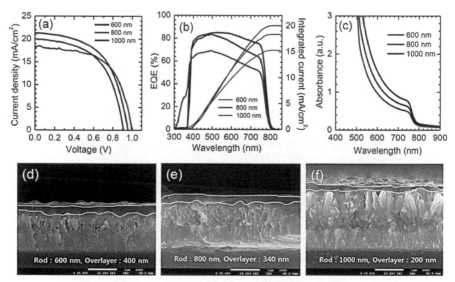

FIGURE 5.14   (a) Current density–voltage curves and (b) EQE spectra together with EQE data-based integrated current density for the perovskite solar cells based on ZnO NRs with different lengths. (c) Absorption spectra of the perovskite-coated ZnO NRs with different lengths. Cross sectional SEM images of perovskite-coated (d) 600, (e) 800, and (f) 1000-nm long ZnO NRs. Different ZnO nanorod lengths were grown for 120 min (600 nm), 150 min (800 nm), and 180 min (1000 nm) in the 35-mM precursor solution.
Souce: Reprinted with permission from Ref. 33. © 2014 American Chemical Society.

The PV performance and IPCE of the devices fabricated with different NRs lengths are shown in Figures 5.14a and b. The ZnO nanorod-based perovskite solar cell demonstrates short circuit current density J_{sc} of

20.08 mA/cm^2, open circuit voltage V$_{oc}$ of 991 mV, FF of 0.56 and PCE of 11.13% at AM 1.5G 1 sun (100 mW/cm^2) illumination. The highest photovoltaic performance was observed for the NRs of 1000 nm. It was also found that increase in NRs length improves the absorbance of perovskite as seen in Figure 5.14c, which could be attributed to increase in the overall thickness of the perovskite layer. The cross sectional SEM images of devices fabricated with different growth times shown in Figures 5.14d–f.

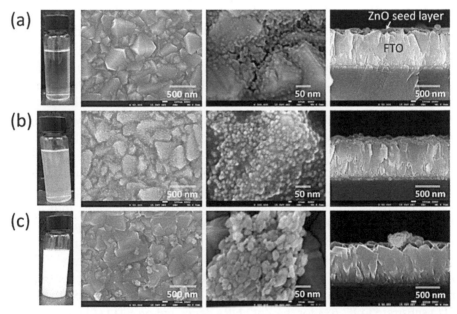

FIGURE 5.15 Top-view and cross sectional SEM images of the seed layers for growth of ZnO NRs formed by (a) an ethanolic solution of zinc acetate (solution), (b) a colloidal solution of zinc acetate in the presence of KOH (colloidal), and (c) a ZnO nanopowders dispersed in ethanol (powder).
Source: Reprinted with permission from Ref. 34. © 2015 American Chemical Society.

Son et al. studied the effect of the seed layer on the growth of ZnO NRs and its effect on perovskite photovoltaic performance.[34] Three different precursor solutions were used in this work to make the seed layer. The three solutions used were made of zinc acetate solution (called as solution), powdered zinc oxide dispersed in solvent (called as powder) and a colloidal zinc oxide solution (called as colloidal). In Figure 5.15, each precursor solution in the glass vial is shown on the left. The "solution" sample is colorless because zinc acetate completely dissolves in ethanol, while the "colloidal" sample is bluish due to the presence of ZnO colloids. ZnO colloids are formed by the addition of alkaline KOH solution in acidic zinc acetate solution. The "powder" sample shows the white opaque solution, although 20 nm sized ZnO powders are dispersed in ethanol, which is probably due to aggregation of the primary nanoparticles. Top view SEM images show that the coverage of the deposited seed layer is better using the colloidal coating solution than the solution or powder coating solutions. It is also clear from the cross sectional SEM images that FTO surface is almost fully covered with ZnO seed layer when the colloidal coating solution is used, whereas FTO is partly and locally covered by the solution and the powder coating solution, respectively. The deposited seed layers are seen to be composed of ZnO nanoparticles regardless of coating solutions. It was also seen from the cross section SEM image that the colloidal-based seed layer resulted in uniform and vertically aligned NRs compared to the powder based seed solution. The diameter of the NRs grown using ZnO nanopowder-based seed solution was also broader than the colloidal and solution-based seed solutions.

The optical properties of the seed solution-based ZnO layers and the NRs grown was studied using transmittance spectra as shown in Figures 5.16a and b. The transmittance of all the seed layers remain the same, indicating that there is not much loss or change of the incident light. On the other hand, ZnO NRs exhibit strong band edge absorption near 370 nm, which is indicative of the optical band gap of ~3.3 eV. The ZnO nanorod grown on the colloidal-based seed layer shows enhanced transmittance, where the maximum transmittance at 600 nm is enhanced from 82.9% for the bare FTO substrate to 87.6% for the ZnO NRs deposited on FTO (Figure 5.16b). The transmittance of 83.9% at 600 nm for the ZnO nanorod grown on the solution-based seed layer is almost the same as that of the bare FTO, but transmittance of the ZnO nanorod on the powder-based seed layer is decreased to 74.3%. The enhanced transmittance of the ZnO nanorod grown on the colloidal-based seed layer is related to vertical alignment, while the reduced transmittance and the pronounced reduction in transmittance at the blue light for the ZnO nanorod grown on the powder-based one are related to the Rayleigh scattering of light because of the titled structure. The Rayleigh scattering is wavelength dependent, hence, the much lowered transmittance at shorter wavelength than that at longer wavelength shown in the ZnO nanorod on the powder-based seed layer results from Rayleigh scattering. The J–V curves and the IPCE for the perovskite ZnO NRs solar cells based on the three seed solutions are shown in Figures 5.16c and d. The open circuit voltage remarkably changes as the seed layer changes. V_{oc} is highest for the colloidal-based seed layer ($V_{oc} = 0.956$ V), while the solution-based one ($V_{oc} = 0.808$ V) exhibits intermediate value and the powder-based one shows lowest V_{oc} of 0.526 V. The J_{sc} for all the three devices ranges between 20.4 and 20.9 mA/cm². PCE of 11.68% is obtained for the colloidal-based seed layer, whereas lower PCEs of 8.03 and 4.66% are observed for the solution- and powder-based ones, respectively. Not much difference is observed in the IPCE over the entire wavelength, except for the wavelength below 370 nm, and the integrated IPCE concurs well with the J_{sc} measured using J–V curves. The LHE is calculated based on the reflectance and absorbance spectra from Figure 5.16e.

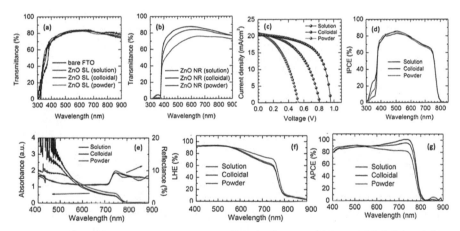

FIGURE 5.16 Transmittance spectra of (a) three different seed layers (SLs) formed from solution, colloidal, and powder precursor solutions and (b) ZnO NRs grown on the three different seed layers, (c) J–V curves and (d) IPCE spectra for the perovskite solar cells based on ZnO NRs grown on the three different seed layers. Active area was 0.12 cm², (e) Absorbance and reflectance, (f) LHE, and (g) APCE spectra for the MAPbI₃ film deposited on ZnO nanorod films formed on three different seed layers using solution, colloidal, and powder precursor solutions.
Source: Reprinted with permission from Ref. 34. © 2015 American Chemical Society.

The LHE in Figure 5.16f is higher for the powder-based seed layer in the longer wavelength compared to the other two seed solutions. For the powder case, this higher LHE at the long wavelength is due to the less dense ZnO nanorod layer, and the similar LHE at short wavelength is related to the strong blue light scattering effect by the tilted ZnO NRs. Despite higher LHE, lowest APCE (APCE = IPCE/LHE) is observed for the powder-based seed layer (Figure 5.16g), which indicates that photoexcited electrons by red light are not effectively collected. The APCE for the solution-based seed layer case is higher than the powder case but still lower than the colloidal case. In this study, it was found that the seed layers play a vital role in the alignment of the NRs and also the photovoltage of the devices.

Cheng et al. studied the chemistry and crystal growth of perovskite on ZnO nanoparticles.[35] In this study, it was found that having a buffer layer between ZnO nanoparticles and perovskite retards the formation of PbI_2 on annealing. ZnO nanoparticles of 5 nm in diameter were dissolved in chloroform and spun coat on ITO. Figure 5.17a show the photovoltaic devices fabricated in this study which have a general structure of ITO/ZnO (30 nm)/$CH_3NH_3PbI_3$ perovskite (280–300 nm)/spiro-OMeTAD (100 nm)/ Au. Figure 5.17b shows the current density–voltage (J–V) curves of the perovskite solar cell under AM 1.5G illustration at 1 sun, which has an open circuit voltage (V_{oc}) of 0.84 V, short circuit current density (Jsc) of 7.3 mA/ cm2, FF of 47.6%, and resulting PCE of 2.9%.

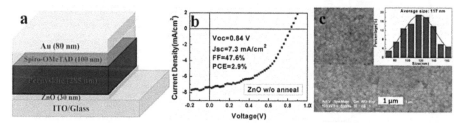

FIGURE 5.17 (a) Device structure of ZnO-based perovskite solar cell, (b) current density– voltage (J–V) curve of perovskite solar cell-based on ZnO without thermal annealing, and (c) SEM image and grain size distribution of perovskite film (without thermal annealing) deposited on ZnO (without thermal annealing).

Source: Reprinted with permission from Ref. 35. © 2015 American Chemical Society.

As shown in Figure 5.17c, the average grain size of the perovskite is only 117 nm. Such small grain size and large number of grain boundaries could have significant recombination of photocurrent and detrimental effect on the overall device performance. To study the effect of annealing on the grain size of the perovskite, samples were annealed at 100°C.

As shown in the inset in Figure 5.18a, the perovskite film on ZnO-NPs exhibits completely different behavior: it turns from dark brown to yellow even after being annealed at 100°C just for a few minutes, and the film becomes completely yellowish in 30 min. This suggests that the CH3NH3PbI3 perovskite on ZnO layer could easily decompose into PbI2 during the thermal annealing process. The decomposition of perovskite crystal was confirmed by the small-angle XRD, and the results are shown in Figure 5.18a. The diffraction peaks located at 14.1° (110), 28.4° (220), and 42.1° (330) correspond to the perovskite phase and diminish after annealing. Meanwhile, the intensity of the diffraction peak at 12.7° that corresponding to PbI_2 significantly increased after annealing. This indicates that the perovskite film on ZnO–NP layer completely decomposed into PbI_2 after annealing treatment at 100°C for 30 min. As seen in Figure 5.18b, the O 1 score level spectrum can be resolved into two main peaks located at 531.2 and 530.0 eV, which correspond to the O 1 s level in ZnO and other chemisorbed oxygen species such as hydroxide (OH–) on the surface of the ZnO-NPs, respectively. Perovskite films were also deposited on annealed ZnO to avoid the detrimental effects of hydroxide on the perovskite decomposition. As seen in Figure 5.18c, after annealing of the perovskite film at 100°C for 30 min, although there is notable increase of the PbI_2 diffraction peak at 12.7°, the perovskite diffraction peaks located at 14.1°, 28.4°, and 42.1° are also enhanced. Such reduced decomposition with annealed ZnO is also shown in the inset of Figure 5.18c; the majority of the perovskite film remains dark brown after annealing, compared to the completely yellowish PbI_2 film on the untreated ZnO.

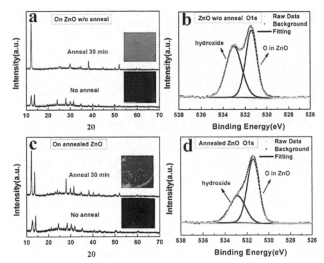

FIGURE 5.18 (a) XRD patterns and photographs of perovskite film on ZnO (without annealing) without thermal annealing and annealing at 100°C for 30 min; high-resolution of O 1 s XPS core level spectra of ZnO (b) without annealing and (d) with annealing at 200°C for 1 h in air. (c) XRD patterns and photographs of perovskite film on annealed ZnO without thermal annealing and annealing at 100°C for 30 min.

Source: Reprinted with permission from Ref. 35. © 2015 American Chemical Society.

From the above results, it is clear that the presence of hydroxide groups on the metal oxide is detrimental to the device performance, as decomposition of perovskite is promoted and leads to the formation of PbI_2. Hence, this study investigated the use of buffer layers between the electron transporting layer (ETL) and the perovskite layer. To investigate whether the $PC_{61}BM$ and PEI buffer layers can effectively prevent the perovskite from direct contact with the ZnO layer during the annealing process, perovskite films were deposited on the $PC_{61}BM$ (30 nm) and PEI (10 nm) coated ZnO and annealed at 100°C. As seen in Figure 5.19a, the XRD results and optical images of perovskite film with $PC_{61}BM$ as the buffer layer, it can be clearly seen that $PC_{61}BM$ can reduce the decomposition but cannot completely avoid it.

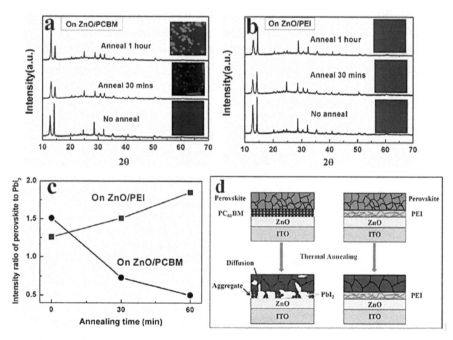

FIGURE 5.19 XRD patterns and photographs of perovskite film deposited on (a) $PC_{61}BM$-coated ZnO (b) and PEI-coated ZnO with different thermal annealing treatment, (c) intensity ratio of perovskite to PbI_2 on $PC_{61}BM$-coated ZnO and PEI-coated ZnO, and (d) schematics of the growth of perovskite deposited on $PC_{61}BM$-coated ZnO and PEI-coated ZnO during thermal annealing.
Source: Reprinted with permission from Ref. 35. © 2015 American Chemical Society.

The perovskite film in some areas started to turn yellow at 30 min and increase in area after 1 h. Furthermore, as seen in Figure 5.19c, the XRD diffraction peak intensity ratio of perovskite to PbI_2 decreased from 1.51 for no annealing to 0.73 and 0.5 for annealing 30 min and 1 h, respectively. On the other hand, as shown in Figure 5.19b, the PEI buffer layer showed superb performance to separate the ZnO-NPs and perovskite. The perovskite on PEI-coated ZnO remained dark brown even after annealing for 1 h. This is confirmed by the XRD results, as seen in Figure 5.19c: the intensity ratio of perovskite to PbI_2 kept increasing from 1.26 for no annealing to 1.51 and 1.85 for annealing 30 min and 1 h, respectively.

Figure 5.20 (a) shows the J–V curves of perovskite solar cells using $PC_{61}BM$ and PEI as the buffer layers under AM 1.5G illumination at 100 mW/cm². The perovskite solar cell with $PC_{61}BM$ as buffer layer had $V_{oc} = 0.88$ V, $J_{sc} = 16.0$ mA/cm², FF=46.0%, and PCE=6.4%. The device with PEI as the buffer layer had significant improvement in FF and resulting PCE, with $V_{oc} = 0.88$ V, Jsc=16.8 mA/cm², FF=69.0%, and PCE=10.2%. As seen in Figure 5.20b, the device with PEI as the buffer layer exhibited a broad EQE from 400 to 800 nm with an integrated J_{sc} of 15.6 mA/cm² at AM 1.5G illumination at 100 mW/cm², which concurs well with the measured J_{sc} from J-V curves. This study shows that using a buffer layer such as PEI could help in boosting the device performance due to decrease in the decomposition of perovskite into the PbI_2 phase.

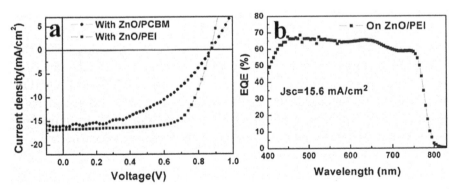

FIGURE 5.20 (a) J–V curves of perovskite solar cell based on PC61BM-coated and PEI-coated ZnO; (b) EQE spectra of perovskite solar cell based on PEI-coated ZnO. *Source:* Reprinted with permission from Ref. 35. © 2015 American Chemical Society.

5.4 OTHER METAL OXIDE

Qin et al. employed a low-temperature solution processed method to deposit the indium oxide electron selective contact in perovskite solar cells.[36] PCBM was also employed over the In_2O_3-electron selection layer to enhance the performance of perovskite solar cell. The In_2O_3 films were fabricated on FTO substrates by sol-gel method followed by an annealing process at 200°C. The thickness of the In_2O_3 films was varied by changing the concentration of the precursor solution. Different precursor concentrations (0.05, 0.10, 0.15, and 0.20 M) were obtained by adding different amount of In $(NO_3)_3 \cdot 4.5H_2O$ into a fixed volume of ethanol. Here, the resultant In_2O_3 films were named after $0.05\text{-}In_2O_3$, $0.10\text{-}In_2O_3$, $0.15\text{-}In_2O_3$, and $0.20\text{-}In_2O_3$. The J–V characteristics on different In_2O_3 ESLs are presented in Figure 5.21a. It is clear that the performance of the PSCs with In_2O_3 ESLs is much better than that of the PSC without an In_2O_3 ESL. The best performance is obtained from the PSC with the $0.10\text{-}In_2O_3$ thin film with thickness of 120 nm, achieving a PCE of 13.01%, with $V_{OC}=1.07$ V, $J_{SC}=$of 17.90 mA/cm², and FF$=0.679$. The solar cells fabricated without an In_2O_3 ESL, had a low PCE of 4.44% with the following PV parameters: $V_{OC}=0.99$ V, $J_{SC}=10.61$ mA/cm², and FF$=0.421$. For the $0.05\text{-}In_2O_3$ thin film, due to a non-continuous compact layer caused more serious recombination between electrons and holes. The In_2O_3 films obtained from a high concentration precursor solution, such as 0.15 and 0.20 M, are too thick to act as an efficient ESL. Those films cannot extract electrons quickly and suppress the charge recombination effectively, which are partially responsible for the lower J_{SC} and V_{OC}. Hence, 0.10 M was found to be the optimal molarity of precursor to produce high-efficiency PSCs. The IPCE was conducted to confirm the trend of J_{SC} in the J–V curves, which are consistent with the J_{SC} of the PSCs with different In_2O_3 ESLs (Figure 5.21b).

The transparency of the ETL is also an important factor which influences the light absorption in PSCs. The transmission spectra of different InO_3 films and bare FTO were measured and is shown in Figure 5.21c. It is seen that the $0.05\text{-}In_2O_3$ and $0.10\text{-}In_2O_3$ thin films coated on FTO substrates are antireflective in nature, and such optical properties can be beneficial to increase the light transmittance and facilitate the generation of electron–hole pairs. For the In_2O_3 films fabricated from high concentration precursor solution such as 0.15 and 0.20 M, the optical absorption edges shift to longer wavelengths, and the transmittance is much lower because the indium oxide films are too thick. The poor optical properties of both the $0.15\text{-}In_2O_3$ and $0.20\text{-}In_2O_3$ films are responsible for the mediocre performance of the In_2O_3-based PSCs. Figure 5.21d shows the dependence of $(\alpha h\upsilon)^2$ versus hυ for the different films fabricated in this study, and the optical band gap energy (E_g) was extracted from the intercept of the linear portion of the curve to the energy axis. The band gap was found to be 3.75 eV irrespective of the thickness of the film.

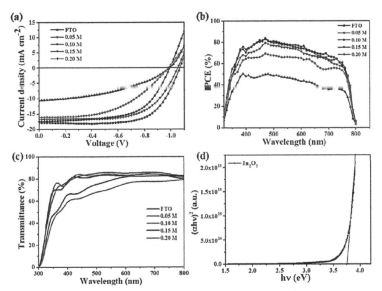

FIGURE 5.21 (a) J–V curves of the PSCs based on the In_2O_3 ESLs prepared by different precursor concentrations. The corresponding (b) IPCE spectra and (c) transmission spectra. (d) tauc plot for In_2O_3 film.
Source: Reprinted with permission from Ref. 36. © 2016 American Chemical Society.

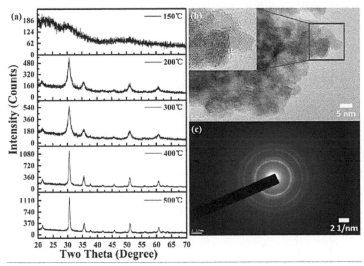

FIGURE 5.22 (a) XRD patterns of In_2O_3 films coated on glass substrates after annealing at 150°C, 200°C, 300°C, 400°C, and 500°C, respectively. (b) Transmission electron microscopy (TEM) and (c) selective area electron diffraction (SAED) images of an In_2O_3 nanocrystalline film.
Source: Reprinted with permission from Ref. 36. © 2016 American Chemical Society.

The XRD patterns of the In_2O_3 films as a function of annealing tempera-ture (150°C, 200°C, 300°C, 400°C, and 500°C) are presented in Figure 5.22a. When annealed at 150°C, In_2O_3 films are amorphous, and there are no distinct peaks in the XRD pattern. However, the characteristic peaks of the In_2O_3 film appear when the annealing temperature is increased to 200°C, which indicates the formation of In_2O_3 crystals. The XRD patterns of the In_2O_3 films, annealed at 200°C or higher, all match with pure In_2O_3 (PDF card number $1-071-2194$) and reveal strong orientation along (222) direction at $\sim 30.65°$. It is seen that the crystallization of the In_2O_3 film is enhanced with the increment of annealing temperature. The In_2O_3 film annealed at 200°C has the highest FWHM value; therefore, the crystallite size of In_2O_3 is the smallest. The crystallite size of In_2O_3 annealed at 200°C was found to be about 10 nm from the images of the transmission electron micros-copy (TEM) (Figure 5.22b), which is close to the value of 11.74 nm calcu-lated from XRD measurements. Furthermore, the In_2O_3 films, formed by the same process, are demonstrated to be nanocrystalline according to the selec-tive area electron diffraction (SAED) images shown in Figure 5.22c. The small In_2O_3 particles are expected to form a compact and uniform blocking layer. When the annealing temperature is increased, the value of FWHM will decrease, indicating that the crystallite size of In_2O_3 will become larger. Large In_2O_3 crystalline grains formed in this study may increase the rough-ness of In_2O_3 film, which could be detrimental to the device fabrication.

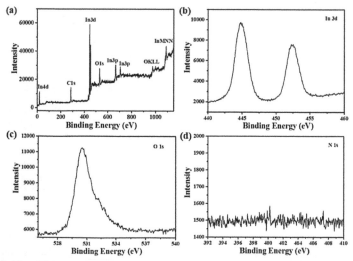

FIGURE 5.23 XPS spectra of (a) survey, (b) In 3d, (c) O 1 s, and (d) N 1 s peaks for an In2O3 film coated on a glass substrate.

Source: Reprinted with permission from Ref. 36. © 2016 American Chemical Society.

The composition of the In_2O_3 film annealed at 200°C was found using XPS, as shown in Figure 5.23. By resolving the XPS spectra, the composition of the films prepared by In $(NO_3)_3 \cdot 4.5H_2O$ precursor at a low temperature was found to be pure In_2O_3. Figure 5.23a presents the full XPS spectrum survey, which reveals the presence of In and O. Figure 5.23b shows the binding energies of 445.1 and 452.5 eV which are attributed to the In 3d5/2 and In 3d3/2 peaks, respectively. The main binding energy of 530.1 eV corresponds to the O 1 s peak, which is the O^{2-} state in In_2O_3 (Figure 5.23c). There was no residual N found in the In_2O_3 films as seen in Figure 5.23d. Therefore, it can be concluded that the In $(NO_3)_3 \cdot 4.5H_2O$ is converted to In_2O_3 completely after annealing at 200°C for 40 min. Some unexpected pinholes and cracks were found along In_2O_3 grain boundaries, and it could possibly arise due to the separation during the annealing process. This imperfection on the surface could lead to direct contact of the perovskite layer and FTO, which could lead to severe recombination. To avoid such recombination, the surface of the In_2O_3 film was modified with the introduction of a PCBM layer between the In_2O_3 ESL and the perovskite layer. To gain insight into the effect of introducing a PCBM layer in the In_2O_3-based PSCs, optimized the thickness of PCBM layer was first optimized. It was found that the highest PCE was achieved when using the PCBM precursor that 20 mg PCBM dissolved in 1 mL of chlorobenzene. Data pertaining to the optimization of this layer is not presented here. For further details please refer to ref. 36.

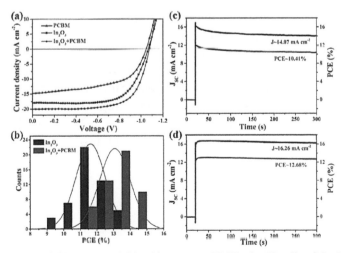

FIGURE 5.24 (a) J–V curves and (b) a histogram of PCEs for 50 cells of the In_2O_3-based PSCs with and without a PCBM layer. Steady-state efficiency of the In_2O_3-based PSC (c) without and (d) with a PCBM layer at a constant bias voltage of 0.74 and 0.78 V, respectively.
Source: Reprinted with permission from Ref. 36. © 2016 American Chemical Society.

The performance of In_2O_3-based PSCs with and without a PCBM layer, and the corresponding J–V curves are shown in Figure 5.24a. The perovskite solar cell with a PCBM layer has a higher PCE of 14.83% with the following PV parameters: $J_{SC} = 20.06$ mA/cm^2, $V_{OC} = 1.08$ V, and FF $= 0.685$, demonstrating that PCBM has a positive effect on In_2O_3-based PSCs. For comparison, fabrication of PSCs with only PCBM layer was carried out, which had a low PCE of 8.74% with a poor FF, demonstrating the importance of In_2O_3 layer in electron selection. The reproducibility of the devices was checked by fabricating 50 cells with and without a PCBM layer, and the results are presented in the histogram shown in Figure 5.24b. It is seen clearly that the overall performance of the In_2O_3-based PSCs with a PCBM layer is better than that without a PCBM layer. The steady-state efficiencies of the In_2O_3-based PSCs without and with a PCBM layer are also shown in Figures 5.24c and d, respectively. For the In_2O_3-based PSC without a PCBM layer, the J_{SC} first decreased quickly and then stabilized when measured at a constant bias voltage of 0.74 V. In contrast, the J_{SC} of the PSC containing a PCBM layer decreased very little for 300 s. A steady-state current density of 16.26 mA/cm^2 and a steady-state efficiency of 12.68% were achieved under a constant bias voltage of 0.78 V. The constant bias voltages of 0.74 and 0.78 V were found to be consistent with the voltage at the maximum power points of J–V curves of the cells.

Ren et al. investigated solution processed Nb-doped tin oxide as electron transporter in planar perovskite solar cells.[37] To prepare the doped oxide films, $SnCl_2 \cdot 2H_2O$ in ethanol solution with different Nb contents was spin-coated onto the glass/FTO substrates, followed by thermal annealing at 190°C for 60 min. Electrical conductivity is a critical figure of merit for the ETL. Figure 5.25a shows the electrical conductivity of ETLs deposited on FTO substrates. It is apparent that the electrical conductivity of the SnO_2 film is significantly increased by the Nb-doping regardless of the concentration of Nb in the film. Figure 5.25b shows the optical transmission spectra of the SnO_2 and Nb: SnO_2, both having excellent transmittance in the wavelength range of 400–800 nm.

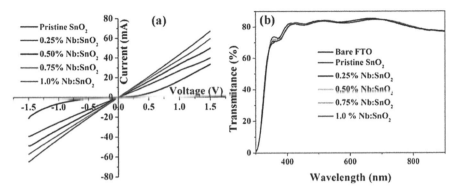

FIGURE 5.25 (a) I–V characteristics of SnO_2 and Nb: SnO_2 films. (b) Transmission spectra of bare FTO, pristine SnO_2 and Nb: SnO_2 with different Nb contents.
Source: Reprinted with permission from Ref. 37. © 2017 American Chemical Society.

The electron mobility of the films was also found to double by Nb doping. The low mobility of the undoped ETL indicates more traps existed leading to charge accumulation at the interface and inferior charge transport. The electrical measurements show that the SnO_2 properties are significantly improved by the Nb-doping including enhanced electron mobility, smoother surface, and larger electrical conductivity. To test their PV performance, fabrication of PSCs using the Nb: SnO_2 ETL. The device structure consisted of FTO as anode, SnO_2 or Nb: SnO_2 film as the ETL, the $(FAPbI_3)_{0.85}(MAPbBr_3)_{0.15}$ as the absorber layer, the Spiro-OMeTAD as the HTL, and the gold layer as the cathode. It was also found that the PSC performance was affected by the SnO_2 ETL thickness. The device with a thick ETL led to a high series resistance (R_s) and a small shunt resistance (R_{sh}), reducing the J_{sc} and FF. However, when the ETL is thinner, pin-holes were seen in the film, leading to direct contact between the perovskite and the FTO electrode and serious carrier recombination. In this study, the SnO_2 film thickness was adjusted by controlling $SnCl_2 \cdot 2H_2O$–ethanol solution concentration. The optimized of SnO_2 was found to be 60 nm formed by controlling the precursor concentration at 0.07 mol/mL.

The PCE increased from 15.13 to 17.57% when the Nb content changed from 0–0.50%. When the Nb content was further increased to 2.00%, the PCE drops to 15.11%. Figure 5.26a shows the J–V curves of the champion devices based on both SnO_2 and Nb: SnO_2 ETLs measured under reverse and forward scan directions. The device based on the pristine SnO_2 ETL under reverse scan direction gives a PCE of 15.13% with $J_{sc} = 21.65$ mA/cm^2, $V_{oc} = 1.06$ V and FF = 0.659. While with Nb doping, the PCE increased to 17.57%. Comparing to the control device, all key J–V parameters are significantly improved. The higher J_{sc} and FF may be ascribed to the high electron mobility and electrical conductivity of the Nb: SnO_2 ETL. The high V_{oc} is likely due to the reduced charge recombination and improved electron extraction. Figure 5.26b shows the EQE and integrated current based on various ETLs. The EQE integrated current density for the pristine SnO_2-based cell is 21.11 mA/cm^2 and it is increased to 21.79 mA/cm^2 for the Nb: SnO_2-based device, which is in good agreement with the J–V curves. Such doping strategies in metal oxides can be used to build devices with higher efficiency.

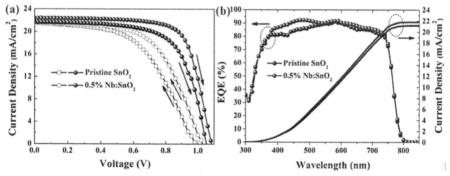

FIGURE 5.26   (a) The J–V curves of PSC devices based on pristine SnO_2 and Nb: SnO_2 under reverse and forward scanning directions. (b) EQE of the champion devices based on SnO_2 and Nb: SnO_2 ETLs.
Source: Reprinted with permission from Ref. 37. © 2017 American Chemical Society.

Zhao et al. demonstrated highly efficient planar perovskite solar cells based on aluminum-doped zinc oxide (AZO) films, where the AZO not only serves as the ETL but also the transparent conduction substrate, thus simplifying the cell structure.[2] The cell exhibited outstanding photovoltaic performance with a PCE of 12.6% and also extraordinarily good thermal stability compared with those of devices based on intrinsic ZnO. Such simple architecture and good performance bear great potential in flexible plastic substrates and thus in the future commercialization.

Ke et al. worked on developing low-temperature solution-processed tin oxide (SnO_2) as an alternative ETL for perovskite solar cells.[38] The SnO_2 ETLs were synthesized by spin-coating process using $SnCl_2 \cdot 2H_2O$ as precursor prepared at room temperature and followed by thermal annealing in air at 180°C for 1 h. The ETLs were treated by ultraviolet ozone for

15 min before perovskite deposition. The device structure was as follow FTO/planar-SnO$_2$/perovskite/spiro-OMeTAD/Au.

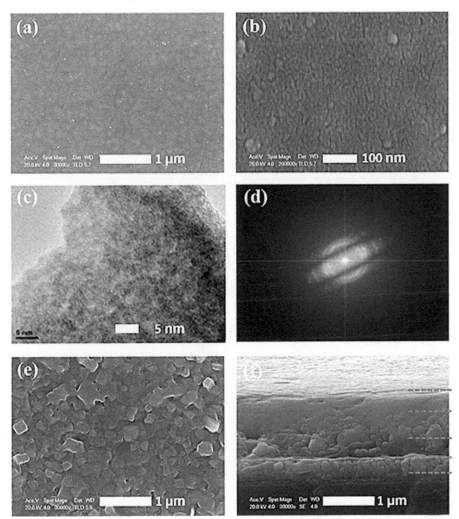

FIGURE 5.27 (a) Top-view SEM images of a SnO$_2$ nanocrystalline film coated on FTO at (a) low and (b) high magnifications. (c) TEM and (d) SAED images of a SnO$_2$ nanocrystalline film. (e) Top-view SEM image of a perovskite CH$_3$NH$_3$PbI$_3$ film coated on the SnO$_2$ ETL. (f) Cross sectional SEM image of the device.
Source: Reprinted with permission from Ref. 38. © 2015 American Chemical Society.

The low temperature processed SnO$_2$ films were found to be nanocrystalline in nature. This enabled smooth and conformal coating of the SnO$_2$ thin layers on the FTO substrates. Figures 5.27a and b shows SEM images of an FTO substrate fully covered with a SnO_2 nanocrystalline film at low and high magnifications, respectively. Figure 5.27a shows only the grains of the FTO substrate; however, at high magnification, the SnO_2 nanocrystallites are seen. TEM and SAED images confirm that the low temperature processed SnO_2 films are nanocrystalline in nature, Figures 5.27c and d. The low-temperature processed SnO_2 ETLs are fully compatible with the growth of $CH_3NH_3PbI_3$ perovskite absorbers. As shown in Figure 5.27e, the $CH_3NH_3PbI_3$ perovskite films coated on the SnO_2 ETLs have smooth surfaces and good coverage. The cross section of the fully fabricated device is presented in Figure 5.27f showing good coverage and no voids in the films.

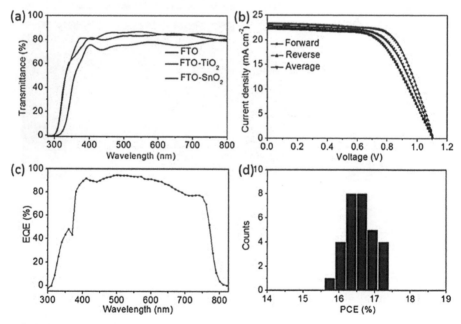

FIGURE 5.28 (a) Transmission spectra of FTO substrates without and with a compact 60 nm TiO_2 film or a 60 nm SnO_2 nanocrystalline film. (b) J–V curves of the best-performing perovskite $CH_3NH_3PbI_3$-based solar cell using a 60 nm SnO_2 ETL measured under reverse and forward voltage scans. (c) EQE spectrum of the best-performing cell using a SnO_2 ETL. (d) Histograms of PCEs measured for 30 cells using the SnO_2 ETLs. *Source:* Reprinted with permission from Ref. 38. © 2015 American Chemical Society.

The thickness of SnO_2 was found to influence the cell performance. In this study, SnO_2 films with varying thicknesses were prepared by using solutions with different $SnCl_2 \cdot 2H_2O$ concentrations but a fixed spin rate and annealing temperature. Optimum thickness for the SnO_2 ETL was about 60 nm, which was achieved by using a 0.1 M $SnCl_2 \cdot 2H_2O$ solution. The transmittance spectra of the SnO_2 film and TiO_2 film for comparison have been shown in Figure 5.28a. The SnO_2 nanocrystalline film shows much higher transmission and a wider band gap than the TiO_2 compact film, suggesting an important way to achieve improved J_{sc} and, therefore, higher PCE for lead halide perovskite solar cells. Figure 5.28b shows the J–V curves of the best performing perovskite cell using a 60 nm SnO_2 ETL measured both in reverse and forward voltage scan directions with a scan rate of 0.1 V/s. The cell achieved a PCE of 17.21%, with the following PV parameters: V_{oc}=1.11 V, J_{sc}=23.27 mA/cm², and FF=0.67 when measured under a reverse voltage scan. The same cell achieved a PCE of 14.82%, V_{oc}=1.11 V, a J_{sc}=22.39 mA/cm², and FF=0.60 when measured under a forward voltage scan. The PCE, V_{oc}, J_{sc}, and FF values averaged from the J–V curves measured under different scan directions are 16.02%, 1.11 V, 22.83 mA/cm², and 0.64, respectively. It is seen that the FF reduces drastically when scanned in the forward direction. The reduction of J_{sc} was marginal, and there was almost no reduction in V_{oc}. Therefore, the perovskite solar cells using the SnO_2 ETLs developed in this study have a low hysteresis. Perovskite solar cells using the SnO_2 ETLs have a fast electron injection process and, therefore, a low carrier recombination. The results suggest that good charge transport at the ETL/perovskite interface could be partially responsible for the low hysteresis observed in perovskite solar cells with SnO_2 as ETL. Figure 5.28c represents the EQE spectrum of the best-performing perovskite solar cell using a 60 nm SnO_2 ETL. The EQE spectrum shows a broad peak value of above 80% in the range from 400–760 nm. The current matching from these results in within 10% of the current density obtained from the J–V curves. To check the reproducibility of the performance of the planar perovskite solar cells using the SnO_2 ETLs, 30 devices were fabricated. The statistics of the PCEs measured under reverse voltage scan and a scan rate of 0.1 V/s are shown in Figure 5.28d. The average efficiency for the 30 cells using low-temperature solution-processed SnO_2 ETLs is 16.44%, with the following PV parameters V_{oc}=1.09 V, a J_{sc}=23.10 mA/cm², and FF=0.65. The outstanding performance of the perovskite solar cells using the SnO_2 ETL originates from the unique properties of nanocrystalline SnO_2 films such as good antireflection, high electron mobility, and wide band gap. This study opens new directions to push the performance of organic–inorganic lead halide perovskite solar cells.

Yin et al. used compact films of ternary oxides in the TiO_2–ZnO system by spray pyrolysis and applied them as ETLs for perovskite solar cells.[39] ZnO ETL was deposited on FTO substrate by spraying 0.2 mol/L Zn (OAc)$_2$ at 350°C. In the case of ternary oxides, Ti (i-OPr)$_2$(acac)$_2$ was added into the Zn^{2+} isopropyl alcohol precursor solution with different molar ratios of Ti^{4+} to Zn^{2+}. The thickness of the compact oxide layers was controlled to be ~ 60 nm by varying the spraying cycles. Three different ternary oxide layers were investigated in this study: $TiZn_2O_4$, $TiZnO_3$, and $Ti_3Zn_2O_8$ which were named $TiZnO_{12}$, $TiZnO_{11}$, and $TiZnO_{32}$, respectively. Figure 5.29a shows the planar device structure used in this study and energy alignment diagram is shown in Figure 5.29b. As seen from the energy level diagram (measured from UPS and optical bandgap data), TiZnO11 and TiZnO32 are not suitable to be applied as ETLs for perovskite solar cells in terms of their conduction band position, whereas TiZnO12 seems to be a good choice due to its favorable conduction band position.

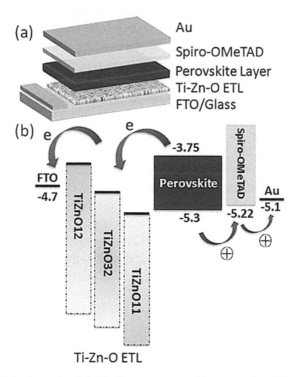

FIGURE 5.29   (a) schematic of the device structure and (b) energy level diagram of device in the study.
Source: Reprinted with permission from Ref. 39. © 2016 American Chemical Society.

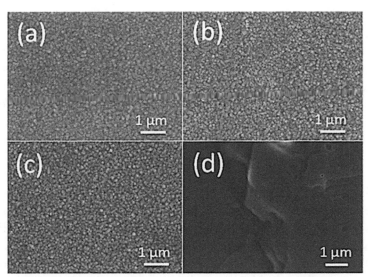

FIGURE 5.30   SEM images of ternary oxide ETLs: (a) $TiZnO_{11}$, (b) $TiZnO_{32}$, (c) $TiZnO_{12}$, and (d) perovskite film on $TiZnO_{12}$ layer.
Source: Reprinted with permission from Ref. 39. © 2016 American Chemical Society.

The SEM morphology of the ternary oxide layers are shown in Figure 5.30. All three kinds of ternary oxide layers were found to be smooth and homogenous. These ETLs seem to have formed by the stacking of nanoparticles. There was no obvious difference in morphology observed among these three kinds of ternary oxide layers. SEM image of the perovskite formed on the top of ternary oxide layers is shown in Figure 5.30d. It was also found that the perovskite layer formed on the oxide films was smooth and compact, which is favorable for the superior performance of solar cells.

To investigate the rectifying behavior of the contact between FTO and ternary oxide films, simple devices with a structure of "FTO/ternary oxides/hole-transport layer/Au electrode" were fabricated. The typical J–V curves measured under the dark for all the three ternary oxide layers are shown in Figure 5.31a. In the case of a blank FTO (absence of an oxide layer), a linear J–V relationship was observed, indicating an ohmic contact. When the compact layer of TiZnO12 is deposited on the FTO substrate, the strong rectifying characteristic is observed due to the Schottky contact between the metallic FTO and semiconducting TiZnO12. Rectifying behaviors were also observed in the cases of TiZnO11 and TiZnO32, although the rectification ratios are much worse than that observed for TiZnO12. Hence, in terms of the rectifying characteristic, the compact layer of TiZnO12 is the most efficient among the three ternary oxides in blocking the electrons from recombination. Figure 5.31b shows the representative J–V curves of devices using ternary oxides as ETLs. The device using TiZnO12 as the ETL exhibited the best photovoltaic performance with the following PV parameters V_{oc}=0.969 V, J_{sc}=21.13 mA/cm^2, and FF=0.742, resulting in a PCE of 15.10%. A poor photovoltaic performance was observed for the device based on TiZnO32 ETL with V_{oc}=0.903 V, J_{sc}=18.29 mA/cm^2, and FF=0.713, leading to a PCE of 11.78%. The TiZnO11-based device had the lowest PCE of 5.05%, resulting from a J_{sc}=14.45 mA/cm^2, V_{oc}=0.760 V, and FF=0.460. The IPCE curves of these three devices are presented in Figure 5.31c. The J_{sc} values obtained by integrating IPCE curves are in good agreement with those determined from the J–V measurements. Histograms of PCEs for each group of 25 devices using TiZnO12 and TiZnO32 ETLs, respectively, are shown in Figure 5.31d. Results obtained in this study indicate that ternary materials can be quite different in terms of their optoelectronic properties from the binary members, and these properties can be tuned over a wide range by simply altering the composition.

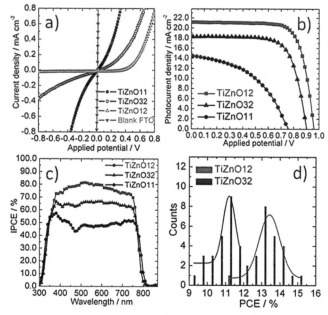

FIGURE 5.31   (a) Current–voltage curves for devices with three types of ternary oxide ETLs. (b) Photocurrent–voltage characteristic curves and (c) IPCE spectra of perovskite solar cells with three types of ternary oxide ETLs. (d) Histogram of power conversion efficiencies for each group of 25 devices using TiZnO$_{12}$ and TiZnO$_{32}$ ETLs, respectively. *Source:* Reprinted with permission from Ref. 39. © 2016 American Chemical Society.

Shin et al. reported Zn_2SnO_4 (nanoparticles and quantum dots) as ETL which was synthesized at low-temperatures for fabricating flexible perovskite solar cells.[40] The group also demonstrated performance improvement for flexible PSCs by employing an energy level-graded oxide ETL using oxide quantum dots. The layer was formed by sequential deposition of Zn_2SnO_4 QDs and NPs on a flexible substrate at low temperature below 100°C and exhibited effective electron collection. 16.5% PCE was reported which is highest for a flexible perovskite solar cell.

KEYWORD

- **nanorods**
- **nanofibers**
- **X-ray diffraction**
- **incident photon-to-electron conversion efficiency**
- **TiO$_2$ films**

REFERENCES

1. Wang, H.-H.; et al. Improving the TiO2 Electron Transport Layer in Perovskite Solar Cells Using Acetylacetonate-Based Additives. *J. Mater. Chem. A* **2015**, *3*(17), 9108–9115.
2. Zhao, X.; et al. Aluminum-Doped Zinc Oxide as Highly Stable Electron Collection Layer for Perovskite Solar Cells. *ACS Appl. Mater. Interfaces* **2016**, *8*(12), 7826–7833.
3. Bera, A.; et al. Perovskite Oxide SrTio$_3$ as An Efficient Electron Transporter For Hybrid Perovskite Solar Cells. *J. Phys. Chem. C* **2014**, *118*(49), 28494–28501.
4. Anaraki, E. H.; et al. Highly Efficient and Stable Planar Perovskite Solar Cells by Solution-Processed Tin Oxide. *Energy Environ. Sci.* **2016**, *9*(10), 3128–3134.
5. AitDads, H.; et al. Structural, Optical and Electrical Properties of Planar Mixed Perovskite Halides/Al-doped Zinc Oxide Solar Cells. *Sol. Energy Mater. Sol. Cells* **2016**, *148*, 30–33.
6. Bai, Y.; et al. Low Temperature Solution-Processed Sb:SnO$_2$ Nanocrystals for Efficient Planar Perovskite Solar Cells. *ChemSusChem.* **2016**, *9*(18), 2686–2691.
7. Dong, Q.; et al. In$_2$O$_3$ Based Perovskite Solar Cells. *Proceedings of SPIE – The International Society for Optical Engineering.* **2016**, 97491S-1 to 97491S-2.
8. Islavath, N.; et al. Seed Layer-Assisted Low Temperature Solution Growth of 3D ZnO Nanowall Architecture for Hybrid Solar Cells. *Mater. Des.* **2017**, *116*, 219–226.
9. Kim, D. H.; et al. Niobium Doping Effects on TiO2 Mesoscopic Electron Transport Layer-Based Perovskite Solar Cells. *ChemSusChem.* **2015**, *8*(14), 2392–2398.

10. Lee, K.; et al. Morphology-Controlled Mesoporous SiO_2 Nanorods for Efficient Scaffolds in Organo-Metal Halide Perovskite Solar Cells. *Chem. Commun.* **2016,** *52*(22), 4231–4234.

11. Li, Y.; et al. Mesoporous SnO_2 Nanoparticle Films as Electron-Transporting Material in Perovskite Solar Cells. *RSC Adv.* **2015,** *5*(36), 28424–28429.

12. Liu, X.; et al. A Low-Temperature, Solution Processable Tin Oxide Electron-Transporting Layer Prepared by the Dual-Fuel Combustion Method for Efficient Perovskite Solar Cells. *Adv. Mater. Interfaces* **2016,** *3*(13), 1600122.

13. Mali, S. S.; Shim, C. S.; Hong, C. K. Highly porous Zinc Stannate (Zn_2SnO_4) Nanofibers Scaffold Photoelectrodes for Efficient Methyl Ammonium Halide Perovskite Solar Cells. *Sci. Rep.* **2015,** *5*.

14. Mali, S. S.; et al. Reduced Graphene Oxide (rGO) Grafted Zinc Stannate (Zn_2SnO_4) Nanofiber Scaffolds for Highly Efficient Mixed-Halide Perovskite Solar Cells. *J. Mater. Chem. A* **2016,** *4*(31), 12158–12169.

15. Numata, Y.; Sanehira, Y.; Miyasaka, T. Impacts of Heterogeneous TiO_2 and Al_2O_3 Composite Mesoporous Scaffold on Formamidinium Lead Trihalide Perovskite Solar Cells. *ACS App. Mater. Interfaces* **2016,** *8*(7), 4608–4615.

16. Park, M.; et al. Low-Temperature Solution-Processed Li-Doped SnO_2 as an Effective Electron Transporting Layer for High-Performance Flexible and Wearable Perovskite Solar Cells. *Nano Energy* **2016,** *26,* 208–215.

17. Rao, H. S.; et al. Improving the Extraction of Photogenerated Electrons with SnO_2 Nanocolloids for Efficient Planar Perovskite Solar Cells. *Adv. Funct.* **2015,** *25*(46), 7200–7207.

18. Singh, T.; Singh, J.; Miyasaka, T. Role of Metal Oxide Electron-Transport Layer Modification on the Stability of High Performing Perovskite Solar Cells. *ChemSusChem.* **2016,** *9*(18), 2559–2566.

19. Wang, C.; et al. Low-Temperature Plasma-Enhanced Atomic Layer Deposition of Tin Oxide Electron Selective Layers for Highly Efficient Planar Perovskite Solar Cells. *J. Mater. Chem. A* **2016,** *4*(31), 12080–12087.

20. Yang, J.; Fransishyn, K. M.; Kelly, T. L. Comparing the Effect of Mesoporous and Planar Metal Oxides on the Stability of Methylammonium Lead Iodide Thin Films. *Chem. Mater.* **2016,** *28*(20), 7344–7352.

21. Yue, Y.; et al. Selective Deposition of Insulating Metal Oxide in Perovskite Solar Cells with Enhanced Device Performance. *ChemSusChem.* **2015,** *8*(16), 2625–2629.

22. Zhang, C.; et al. Influence of Different TiO2 Blocking Films on the Photovoltaic Performance of Perovskite Solar Cells. **2016,** *Appl. Surf. Sci 388,* 82–88

23. Zhou, L.; et al. Influence of Insulating Oxide Coatings on the Performance of Perovskite Solar Cells and the Interface Charge Recombination Dynamics. *Wuli Huaxue Xuebao/Acta Physico-Chimica Sinica* **2016,** *32*(5), 1207–1213.

24. Zhou, P.; et al. Tin Oxide Nanosheets as Efficient Electron Transporting Materials for Perovskite Solar Cells. *Sol. Energy* **2016,** *137,* 579–584.

25. Kim, H.-S.; et al. High Efficiency Solid-State Sensitized Solar Cell-Based on Submicrometer Rutile TiO_2 Nanorod and $CH_3NH_3PbI_3$ Perovskite Sensitizer. *Nano Lett.* **2013,** *13*(6), 2412–2417.

26. Shao, J.; et al. Pore Size Dependent Hysteresis Elimination in Perovskite Solar Cells Based on Highly Porous TiO_2 Films with Widely Tunable Pores of 15–34 nm. *Chem. Mater.* **2016,** *28*(19), 7134–7144.

27. Dharani, S.; et al. High Efficiency Electrospun TiO_2 Nanofiber Based Hybrid Organic-Inorganic Perovskite Solar Cell. *Nanoscale* **2014,** *6*(3), 1675–1679.

28. Wang, J. T.-W.; et al. Low-Temperature Processed Electron Collection Layers of Graphene/TiO_2 Nanocomposites in Thin Film Perovskite Solar Cells. *Nano Lett.* **2014,** *14*(2), 724–730.

29. Han, G. S.; et al. Reduced Graphene Oxide/Mesoporous TiO_2 Nanocomposite Based Perovskite Solar Cells. *ACS Appl. Mater. Interfaces* **2015,** *7*(42), 23521–23526.

30. Huang, Y.; et al. TiO_2 Sub-microsphere Film as Scaffold Layer for Efficient Perovskite Solar Cells. *ACS Appl. Mater. Interfaces* **2016,** *8*(12), 8162–8167.

31. Peng, J.; et al. Efficient Indium-Doped TiO_x Electron Transport Layers for High-Performance Perovskite Solar Cells and Perovskite-Silicon Tandems. *Adv. Energy Mater.* **2017,** *7,* 1601768.

32. Liu, Z.; et al. Low-Temperature TiO_x Compact Layer for Planar Heterojunction Perovskite Solar Cells. *ACS Appl. Mater. Interfaces* **2016,** *8*(17), 11076–11083.

33. Son, D.-Y.; et al. 11% Efficient Perovskite Solar Cell Based on ZnO Nanorods: An Effective Charge Collection System. *J. Phys. Chem. C* **2014,** *118*(30), 16567–16573.

34. Son, D.-Y.; et al. Effects of Seed Layer on Growth of ZnO Nanorod and Performance of Perovskite Solar Cell. *J. Phys. Chem. C* **2015,** *119*(19), 10321–10328.

35. Cheng, Y.; et al. Decomposition of Organometal Halide Perovskite Films on Zinc Oxide Nanoparticles. *ACS Appl. Mater. Interfaces* **2015,** *7*(36), 19986–19993.

36. Qin, M.; et al. Perovskite Solar Cells Based on Low-Temperature Processed Indium Oxide Electron Selective Layers. *ACS Appl. Mater. Interfaces* **2016,** *8*(13), 8460–8466.

37. Ren, X.; et al. Solution-Processed Nb:SnO_2 Electron Transport Layer for Efficient Planar Perovskite Solar Cells. *ACS Appl. Mater. Interfaces* **2017,** *9*(3), 2421–2429.

38. Ke, W.; et al. Low-Temperature Solution-Processed Tin Oxide as an Alternative Electron Transporting Layer for Efficient Perovskite Solar Cells. *J. Am. Chem. Soc.* **2015,** *137*(21), 6730–6733.

39. Yin, X.; et al. Ternary Oxides in the TiO_2–ZnO System as Efficient Electron-Transport Layers for Perovskite Solar Cells with Efficiency over 15. *ACS Appl. Mater. Interfaces* **2016,** *8*(43), 29580–29587.

40. Shin, S. S.; et al. Tailoring of Electron-Collecting Oxide Nanoparticulate Layer for Flexible Perovskite Solar Cells. *J. Phys. Chem. Lett.* **2016,** *7*(10), 1845–1851.

CHAPTER 6

ORGANIC N-TYPE MATERIALS

6.1 INTRODUCTION

The first few perovskite solar cells (PSCs) were developed from the architecture of solid-state dye-sensitized solar cell with TiO_2 as the electron transporting material (ETM) and spiro-OMeTAD as hole-transporting material. Researches to replace these materials to enhance device performance or to employ in various architectures, or to facilitate low-temperature fabrication have been widely studied. Although started later, n-type organic ETM have seen a rapid growth in the past few years compared to their p-type counterparts. In this chapter, different n-type organic ETMs will be discussed and categorized into groups based on the nature of their molecular formulae. Chemical and physical properties, synthesis, and deposition method and their employment as electron transporting layer (ETL) in PSCs will also be discussed.

To efficiently accept electron from perovskite and block holes, desired organic materials for ETL should have suitable energy levels and wide band gap. The lowest unoccupied molecular orbital (LUMO) level should align with the conduction band of perovskite while the highest occupied molecular orbital (HOMO) level should be lower than the perovskite valence band. Full coverage of the ETL on top of perovskite in inverted p-i-n configuration is essential and high electron mobility ETM can allow a thicker film without increasing too much series resistance. A thicker film with a good hole blocking property can reduce shunting and current leakage as well as charge recombination at the interfaces. Wide optical band gap ETM can also ensure maximize photon absorption by the perovskite layer.[1,2,3]

6.2 FULLERENE AND ITS DERIVATIVES

Unlike n-type inorganic semiconductor employed in PSCs being developed from dye-sensitized solar cell background, the usage of n-type organic semiconductor was based on the working principle of the organic solar cell. Therefore, fullerene (C_{60}) and its derivatives became the easiest choice. Fullerene is a spherical bulky ball with 60 carbon atoms that has widely been used as electron acceptor in bulk heterojunction organic photovoltaics.[4] Owing to its poor solubility in many common solvents, some of its derivatives are often employed. These derivatives include $PC_{61}BM$ (phenyl-C_{61}-butyric acid methyl ester) and ICBA (indene-C_{60} bisadduct) have high solubility in organic solvent and are suitable to be deposited by spin-coating or printing methods. These derivatives also have distinctive energy levels and electron mobility compared to fullerene[1,5]. In a preliminary study, photoluminescence (PL) measurement of $CH_3NH_3PbI_3$ showed efficient quenching to $PC_{61}BM$ to prove charge transfer from perovskite to phenyl-c61-butyric acid methyl ester (PCBM)[5]. Since the choice of fullerene comes from the fact that it was employed as electron acceptor in the organic solar cell, it is first understood that the function of fullerene is to extract and transport charge for the charge-carriers formed in the perovskite layer. On the other hand, other debates suggest that perovskite has excitonic behavior and fullerene's main role is to dissociate photogenerated excitons formed in the active layer of perovskite.[1]

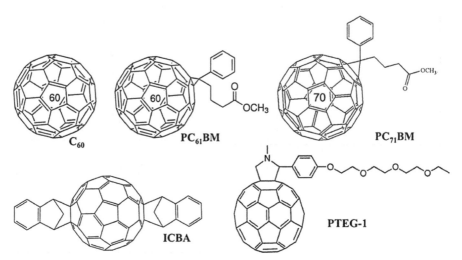

FIGURE 6.1   Chemical structures of fullerene C_{60} and some of its derivatives.
Source: Adapted from ref. 1.

The first paper reported using C_{60} and its derivatives $PC_{61}BM$ and ICBA demonstrated a rather low power conversion efficiency (PCE) of 3.0, 3.9, and 3.4%, respectively[6] with device configuration with ITO/PEDOT: PSS/

(fullerene or derivative) bathocuproine (BCP)/Ag (PEDOT: PSS is the hole transporting layer (HTL) poly (3,4-ethyl-enedioxythiophene) poly (styrene-sulfonate) and BCP is hole blocking layer BCP) (Figure 6.1). Nonetheless, at the early stage of developing perovskite-based solar cell, this was considered a breakthrough not only in terms of device architecture of planar configuration but also of the choice of materials: both the HTL and ETL in these devices (energy level alignment, Figure 6.2) were organic and promising for low-temperature deposition.

Liang et al. prepared similar comparison among ICBA, fullerene C_{60} and $PC_{61}BM$[7] in a later study with a slightly modified device configuration with ITO/PEDOT: PSS/(fullerene or derivative)/BisC$_{60}$/Ag and better control of perovskite deposition. The devices exhibited much higher PCE of 8.06, 13.37, and 15.44% for ICBA, $PC_{61}BM$, and C_{60}, respectively. It is worth noting that all three of these materials were spincoated on perovskite film from a solution of 15 mg/ml in dichlorobenzene. The PSC with ICBA often has higher V_{oc} (0.95 V) due to its higher LUMO level (−3.6 eV) compared to $PC_{61}BM$ and C_{60}. However, for a difference in the energy level of 0.2–0.3 eV (LUMO of $PC_{61}BM$ and C_{60} are −3.8 and -3.9 eV), the increment of ICBA's V_{oc} is not sufficiently high. This is attributed to ICBA's lower electron mobility that leads to increasing recombination (6.9×10^{-3} cm^2 V/s for IC$_{60}$BA, 6.1×10^{-2} cm^2 V/s for $PC_{61}BM$, and 1.6 cm^{2}V/s for C_{60}). Furthermore, the higher electron mobility of $PC_{61}BM$ and C_{60} facilitates charge dissociation and charge transport at the perovskite/fullerene interface and in the bulk of fullerene, resulting in significantly higher J_{sc} of $PC_{61}BM$, and C_{60} devices. Besides, PL quenching efficiency of perovskite to fullerene follows the trend of electron mobility $C_{60} > PC_{61}BM \geq ICBA$ proving the poorer performance of ICBA-based PSC.

The control of perovskite surface roughness is critical as the ETL has to fully cover the perovskite but should not be too thick due to increasing resistance, especially in the case of ICBA. Therefore, a strategy was applied to cover the pores or even out the roughness of the perovskite layer with PCBM or ICBA and then deposited a C60 thin film on top for better coverage. Beside the above-mentioned study, Wang et al. also successfully improved the ICBA-based device by incorporating it with C_{60} in the configuration of ITO/PEDOT: PSS/CH$_3$NH$_3$PbI$_3$/ICBA/C$_{60}$/BCP/Al.[8] The best device exhibited 15.7% efficiency.

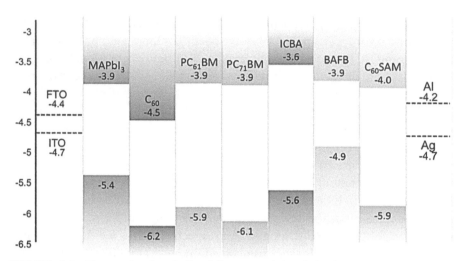

FIGURE 6.2 Energy levels of C_{60} fullerene and its derivatives as respect to the transparent conducting electrode (FTO or ITO), metallic electrode (Al, Ag) and perovskite methylammonium lead iodide $MAPbI_3$.

An early report employing fullerene's derivative $PC_{61}BM$ as the hole-blocking layer was performed by Malinkiewicz et al.[9] in an inverted planar configuration of ITO/PEDOT: PSS/PolyTPD/$CH_3NH_3PbI_3$/$PC_{61}BM$/Au. $PC_{61}BM$ was chosen because its LUMO level matches that of perovskite. The authors achieved a remarkable efficiency of 12.04%. In another report at the same time, Decampo et al. measured Steady-state PL spectra of $CH_3NH_3PbI_3$-$_xCl_x$ coated with $PC_{61}BM$ as quencher and the calculated quenching efficiency was 91% indicating that there was certain charge transfer from perovskite to the fullerene derivative.[5] The same group employed $PC_{61}BM$ in the inverted planar device and achieved 9.8% efficiency, but more notably, they also fabricated flexible device on polyethylene substrate with 6.4%. With this result, the possibility of manufacturing PSC in low-temperature with the printable method is highly desirable.

In order to compare fullerene C_{60} with its derivative $PC_{61}BM$ by Jeng et al. utilized them in PSC and also demonstrated the optimization of device performance by modulating the film thickness of the perovskite $CH_3NH_3PbI_3$ layer through spincoating.[10] Both C_{60} and $PC_{61}BM$ were deposited by thermal evaporation on the spincoated perovskite layer, followed by a thin film of BCP and Al as the n contact. A comparison between C_{60} and $PC_{61}BM$ using the optimized condition of perovskite $CH_3NH_3PbI_3$ and HTL NiO_x showing a much higher V_{oc} (0.74 V for C_{60} and 0.92 V for $PC_{61}BM$). The improvement of V_{oc} is attributed to the higher LUMO energy level of $PC_{61}BM$ (−3.9 eV) compared to that of C_{60} (−4.5 eV) that minimizes the energy loss at the $CH_3NH_3PbI_3$/ETL interface.

A major concern when depositing a thin layer of organic ETM to construct a device is the incapability to cover the rough surface of perovskite that will lead to shunting and poor performance of the device. By solvent engineering the deposition of perovskite layer, Seo et al. could create highly uniform active layer with the roughness of 10 nm. They then demonstrated the variation of $PC_{61}BM$ film thickness deposited by spincoating from 55 nm to 140 nm. The device performance showed a gradual decrease of all parameters V_{oc}, J_{sc}, fill factor (FF), and PCE when the $PC_{61}BM$ thickness decreased from 55 to 140 nm due to increasing resistance of the ETL layer. This also depicted that 55 nm was sufficient to cover the perovskite layer if the roughness can be well controlled.

In a more recent work, also inspired by the heterojunction structure in the organic solar cell, Chiang and Wu introduced a new method to incorporate PCBM in perovskite with an upgraded architecture and deposition process. The device architecture is ITO/PEDOT: PSS/($CH_3NH_3PbI_3$: PCBM) blend/ Ca/Al.[11] The fabrication process resembles the two-step sequential deposition method in which a PbI_2 film was first spincoated, after treatment, it was then dipped in CH_3NH_3I solution to be converted into perovskite.[12] However, in this work, the PbI_2 solution was mixed with PCBM before spincoating. Since the PbI_2 film is not completely compact but has some physical vacancies or holes, they can be filled with the PCBM molecules. After perovskite conversion, the interface between $CH_3NH_3PbI_3$ and PCBM is greatly enhanced and the active layer was composed of densely packed large grains with high durability upon impact. Observation of the cross section under high-resolution SEM revealed a film so dense that no defect could be seen at ×100,000 magnification. Electrons and holes can diffuse to PCBM and PEDOT:PSS, respectively, was within one single grain neither passing through any grain boundaries nor losing energy due to interparticle resistance, leading to a high fill factor of 0.82 and a PCE of 16%. This same research group has also reported substituting $PC_{61}BM$ with $PC_{71}BM$ because of its similar energy level but higher electron mobility and better optical absorption in the ultraviolet (UV)-visible region. The interface of perovskite/$PC_{71}BM$ and crystallinity of $PC_{71}BM$ was enhanced by solvent annealing of the $PC_{71}BM$ layer for 24 h. Compared to $PC_{61}BM$, the PSC was higher in both V_{oc}, J_{sc}, and FF with an efficiency of 16.31%.[13]

Interlayer is often integrated between the ETL and the top electrode and serves as several roles: (1) as hole-blocking layer, (2) transport electrons, (3) protect the ETL, and (4) improve the optical field. Several organic n-type interlayers are often used in p-i-n planar devices including fullerene C_{60} itself and BCP, bathophenanthroline (BPhen). These materials have often been used previously as interlayer in all-organic solar cells. BPhen has higher electron mobility than BCP and although it serves as similar function as BCP, it can transport electrons more effectively across the interlayer to the metal electrode. C_{60} has also been employed as interlayer when it is deposited on top of its derivative to form a double fullerene layer to reduce current leakage. Double fullerene layer C_{60}/PCBM was also found to passivate the defects at the perovskite surface and grain boundaries. Liang et al. added solution-processed bis-C60 surfactant layer in between the PCBM and Ag electrode to facilitate the band alignment at the ELT/electrode interface.[14] Zhang et al. also applied two polyelectrolytes PEIE and P3TMAHT poly [3-(6-trimethylammoniumhexyl) thiophene] to raise the WF of Ag from -4.7 to -3.96 eV and -4.13 eV to reduce the barrier between the ETL and the electrode. In a device configuration of ITO/PEDOT:PSS/$CH_3NH_3PbI_{3-x}Cl_x$/$PC_{61}BM$/interlayer/Ag, the efficiency without interlayer was 8.5% and improved to 12.0% with PEIE and 11.3% with P3TMAHT.[15]

Another example recently being introduced is benzoic acid fullerene bis-adducts (BAFB) which are expected to passivate the defects on the surface and grain boundaries of perovskite with the dicarboxylic group on the fullerene.[16] Although the preliminary report by Erten-Ela et al. still showed a slightly lower performance of BAFB compared to $PC_{61}BM$ control sample (9.21 vs. 10.13%, respectively), this is an interesting material to be further pursued.

Fullerene and its derivatives are also employed to assist the inorganic ETL such as TiO_2.[17] An example is self-assembled monolayer of fullerene (C_{60}-SAM) on metal oxide can passivate charge traps and enhance electron extraction[18] in normal n-i-p PSC (Figure 6.3). Furthermore, by coupling C_{60}-SAM with TiO_2, the fullerene unit close to the perovskite layer can reduce non-radiative recombination channels at the interface and lower hysteresis, increase device performance and stability (Figure 6.3).[19] Another example is Li et al.'s work on PCBB-2CN-2C8 with multiple functionalities assisting TiO_2 ETL resulting in a significant increase of V_{oc} from 0.98 (TiO_2 only) to 1.06 (TiO_2 with PCBB-2CN-2C8) and FF from 0.72 to 0.79, leading to a high efficiency of 17.35% (Figure 6.4).[20]

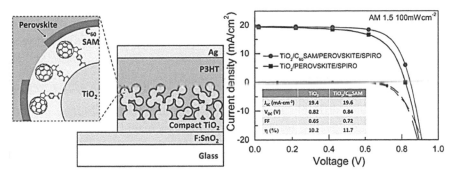

FIGURE 6.3   Coupling C_{60}-SAM fullerene with TiO_2 can passivate defects on the surface of perovskite and enhance device performance.
Source: Reprinted with permission from ref 19. © 2013 American Chemical Society.

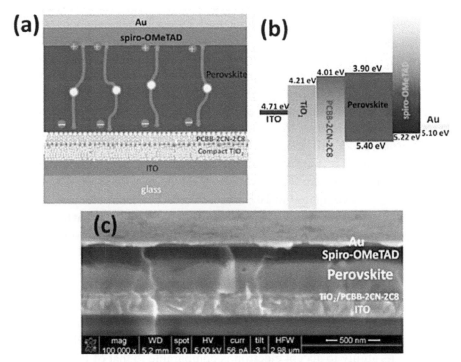

FIGURE 6.4 TiO_2/PCBB-2CN-2C8/perovskite/spiro/Au, (b) Energy level diagram, (c) Cross section scanning electron microscope (SEM) image of the device.
Source: Reprinted with permission from Ref. 20. © 2015 American Chemical Society.

Yoon et al.[21] recently fabricated PSC with $CH_3NH_3PbI_3$ and thermally evaporated C60 as ETL on ITO with a surprisingly high efficiency of 19.1% on rigid and 16.0% on flexible substrate. The device architecture is as follows $ITO/C_{60}/CH_3NH_3PbI_3/spiro\text{-}MeOTAD/Au$. With a C_{60} layer of 35 nm, the FF and V_{oc} of PSC are enhanced due to reduced current leakage and better electron extraction of C_{60}.

Fullerenes with different dielectric constant and their influence on the device performance of PSC upon light-soaking was investigated by Loi et al. The high dielectric constant of PTEG-1 (dielectric constant 5.9) can help reduce electron trapping due to recombination in the extraction layer and passivate defects in perovskite better than $PC_{61}BM$ (dielectric constant 3.9). A PCE of 15.7% was obtained with PTEG-1.[22]

6.3 NON-FULLERENE SMALL MOLECULES AND OLIGOMERS

Several other alternatives to fullerene and its derivatives such as small molecules or oligomers have also been investigated as ETL or HTL in PSCs. 3TPYMB (tris (2,4,6-trimethyl-3-pyridin-3-yl) phenyl) borane is an example of organoborane compounds introduced by Malinkiewicz et al. in PHJ PSCs.[23] 3TPYMB has a higher LUMO level compared to $PC_{61}BM$ and hence causes a mismatch between its LUMO level and the conduction band of perovskite, leading to poorer electron transfer and inferior J_{sc} and V_{oc}. Therefore, 3TPYMB is more suitable to function as a hole blocking layer rather than ETL. Furthermore, the organoborane layer might not be sufficiently compact to provide coverage of the perovskite layer. The PCE obtained was 5.5% compared to 10% of its $PC_{61}BM$ counterpart.

Di-perylene diimide (diPDI) is a non-fullerene molecule which has been studied as electron acceptor by Kim et al.[24] diPDI was doped with DMBI (1,3-dimethyl-2-phenyl-2,3-dihydro-1H- benzoimidazole) which has successfully been introduced as n-type dopant for PCBM to be used as ETL in inverted perovskite (Figure 6.5).[25] diPDI was shown to be effectively doped with DMBI by increasing absorption when measured UV-Vis absorption spectra of different doping level and up-shifting of Fermi level by measuring UV photoelectron spectroscopy. When examining its PL and it is shown that its ability to accept electron is comparable to that of PCBM. The device fabricated from diPDI with 1% doping of DMBI exhibited the highest efficiency of 10.0% with an impressive J_{sc} of 21.6 mA/cm^2. The up-shifted Fermi level after doping (from −4.95 eV to −4.71 eV) has resulted in higher V_{oc} but it is still lower compared to PCBM.

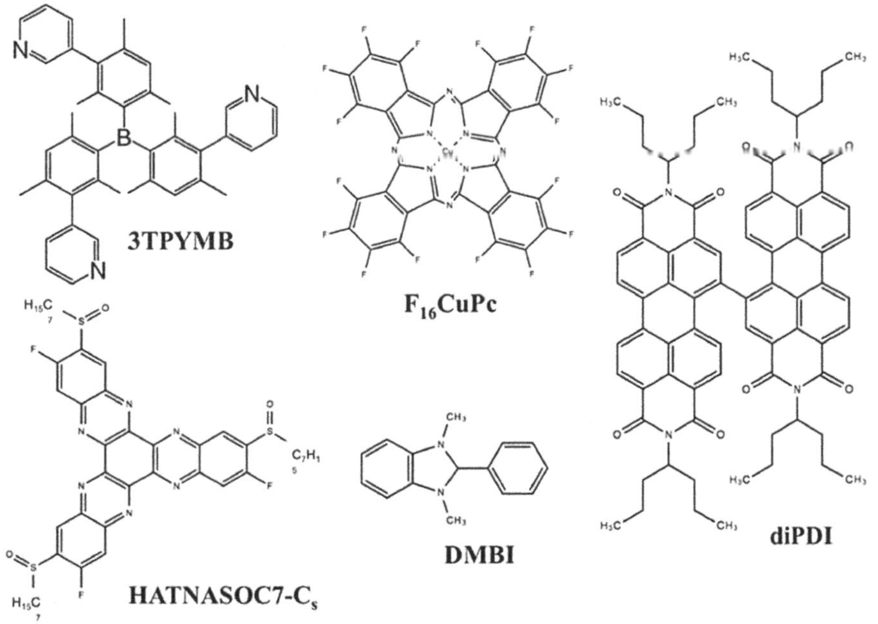

FIGURE 6.5 Chemical structures of several non-fullerene molecules used as the electron transporting materials in perovskite solar cells.
Source: Adapted from ref. 23-27.

A copper complex hexadecafluorophthalocyaninatocopper (F_{16}CuPc) has previously been employed in n-channel organic thin film transistors was also investigated as electron acceptor in perovskite solar cell[26] with configuration ITO/PEDOT: PSS/$CH_3NH_3PbI_3$/F_{16}CuPc/BPhen/Al. F_{16}CuPc has low absorption in the visible and UV region and shows effective quenching of PL indicating that it can effectively accept electron from perovskite. By varying film thickness of the F_{16}CuPc, the highest efficiency of 12.6% was achieved at 40-nm thick. As the interface of $CH_3NH_3PbI_3$/F_{16}CuPc is rough, thick F_{16}CuPc film is necessary to prevent shunting due to direct contact of perovskite and Al.

Although hexaazatrinaphthylene (HATNA) derivatives have been known for their low cost, high charge mobility and large band gap for electronic devices, they have high LUMO levels compared to $CH_3NH_3PbI_3$ making it difficult to be applied in PSC directly. Jen et al. were the first to introduce HATNA-based ETM in PSC by replacing the three F groups in HATNA-F_6 core with three alkylsulfanyl chains.[27] The modification improved material solubility and passivate the surface traps of perovskite significantly. However, as alkylsulfanyl chains raise the LUMO level, the authors introduce different oxidation states for the sulfur atoms to sure the LUMO aligns properly with the conduction band of perovskite. Among the five isomers studied, sulf-oxide HATNASOC7-C$_s$ provided the best efficiency of 17.6% with remarkable V$_{oc}$ of 1.08V and J$_{sc}$ 20.73%. HATNASOC7-C$_s$ not only has a matching LUMO with perovskite (-4.16 eV) but it also has high electron mobility (5.13×10^{-3} cm^2/V.s). The larger band gap compared to PCBM means that it is a better hole blocking layer and recombination can be reduced. Moreover, this material has hydrophobic nature that can prevent moisture approaching the perovskite layer; hence it also acts as a protection layer and improves the device stability significantly.

6.4   POLYMERS

Conducting polymers have been widely used in organic solar cell and organic thin film transistors, are recently attracting attention to be employed as ETM in perovskite solar cell. One of the most explored classes of polymers used as n-type semiconductor is naphthalene diimide whose charge transport and electronic properties have been proved.

Three naphthalene diimide-based polymers have been investigated by Wang et al. for inverted planar PSC ITO/PEDOT: PSS/CH$_3$NH$_3$PbI$_3$/Polymer/ZnO/Al. They are N2200—poly {[N, N′-bis (2-octyldodecyl)-1,4,5,8-naphthalene diimide-2,6-diyl]-alt-5,5′-(2,2′-bithiophene)}, PNVT-8—poly ({N, N′-bis (alkyl)-1,4,5,8-naphthalene diimide-2,6-diyl-alt-5,5′-di (thiophen-2-yl)-2,2′-(E)-2-[2-(thiophen-2-yl) vinyl] thiophene}) and PNDI2OD-TT—poly {[N, N′-bis (2-octyldodecyl)-1,4,5,8-naphthalene diimide-2,6-diyl]-alt-5,5′-[thieno (2,3-b) thiophene]} [28] (Figure 6.6). The LUMO level of N2200, PNVT-8 and PNDI20OD-TT are −3.93 eV, −3.91 eV, and −3.87 eV, respectively, well matching of perovskite to be used as electron acceptors. The optimized devices obtained efficiency of 8.5, 7.13, and 6.11% compared to 8.51% of their PCBM counterpart. Electrochemical impedance spectroscopy revealed a faster recombination rate of N2200/perovskite interface compared to PCBM, this is the main reason for lower FF.

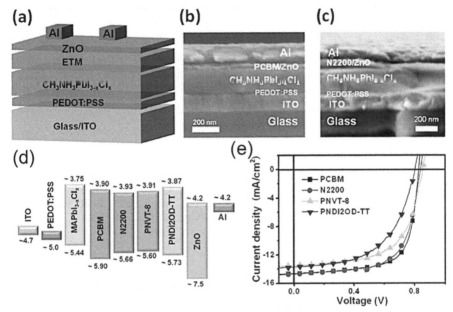

FIGURE 6.6 Three naphthalene diimide-based polymers used as the electron transporting layer in PSCs in comparison to $PC_{61}BM$: (a) device architecture, (b) and (c) cross section SEM images, (d) energy band levels and (e) photovoltaic performance.
Source: Reprinted with permission from Ref. 28. © 2015 American Chemical Society.

The naphthalene diimide family continues to attract attention from researchers as Sun et al. reported a new design of n-type conjugated amino-functionalized polymer with excellent charge transport and interfacial property. The backbone of these polymers consisted of fluorine and naphthalene diimide unites conjugated by two thiophene spacers (PF-2TNDI).[29] The fluorine incorporation lowers the HOMO level of the polymer to ensure good hole blocking property and the thiophene spacers improve polymer packing and charge transport property. To further enhance the interfacial property, amine functional groups were introduced to the side chains of the fluorine unit to form new polymer PFN-2TNDI. The modification was expected to form interfacial dipoles with metal electrode leading to a reduction of the metal work function and increase charge collection at the cathode. The resulted device with PFN-2TNDI showed a significant improvement of 16.7% compared to PCBM-based PSC of 12.9% PCE while the polymer without amine group PF-2TNDI showed almost no photovoltaic effect. Several techniques were used to explain the effect of interfacial property including Kelvin probe, PL and time-resolved PL decay, transient photovoltage, and impedance spectroscopy. The authors found that PFN-2TNDI not only modified the cathode WF but also passivated the surface traps and grain boundaries on the perovskite, leading to suppressed surface recombination and reduced interfacial resistance. Furthermore, due to good electron transport property of PFN-2TNDI, a wide range of ETL thickness could be used without disturbing the device performance. Table 6.1 provides a detailed processing conditions and materials used in different configurations.

TABLE 6.1 Device Performance of Various Organic n-Type Materials Employed as Electron Transporting Layer in Perovskite Solar Cell

n-type material	Device architecture	Deposition method	Voc (V)	Jsc (mA/cm²)	FF	PCE (%)	References
C60	ITO/PEDOT:PSS/ CH3NH3PbI3/ C60/BCP/Al	Thermal evaporation	0.55	10.32	0.61	3.0	6
C60	ITO/NiOx/ CH3NH3PbI3/ C60/BCP/Al	Thermal evaporation	0.74	12.95	0.6	5.7	10
C60	ITO/PEDOT:PSS/ C60/BisC60/Ag	Spincoat	0.92	21.07	0.8	15.44	7
PC61BM	ITO/PEDOT:PSS/ PC61BM/BisC60/ Ag	Spincoat	0.89	18.85	0.8	13.37	7
PC61BM	ITO/PEDOT:PSS/ PolyTPD/ CH3NH3PbI3/ PC61BM/Au	Spincoating	1.05	16.12	0.67	12.04	9
PC61BM	FTO/ PEDOT:PSS/ CH3NH3P-bI3xClx/ PC61BM/TiOx/Al	Spincoating	0.94	15.8	0.66	15.8	5
PC61BM	ITO/NiOx/ CH3NH3PbI3/ PC61BM/ BCP/Al	Thermal evaporation	0.92	12.43	0.68	7.8	10
PC61BM	ITO/PEDOT:PSS/ CH3NH3PbI3/ PC61BM/LiF/Al	Spincoating	0.866	20.7	0.78	14.1	30

TABLE 6.1   *(Continued)*

n-type material	Device architecture	Deposition method	Voc (V)	Jsc (mA/cm^2)	FF	PCE (%)	References
PC61BM	ITO/PEDOT·PSS/ (CH3NH3PbI3: PC61BM) blend/ Ca/Al	Spincoat blend of PbI2: PCBM, then dipcoated in CH3NH3I for conversion to perovskite.	0.96	20.2	0.82	16.0	11
PC71BM	ITO/PEDOT:PSS/ (CH3NH3PbI3: PC71BM) blend/ Ca/Al	Spincoat and solvent annealing	1.05	19.98	0.78	16.31	13
ICBA	ITO/PEDOT:PSS/ CH3NH3PbI3/ ICBA/BCP/Ag	Thermal evaporation	0.58	10.03	0.58	3.4	6
ICBA	ITO/PEDOT:PSS/ CH3NH3PbI3/ ICBA/Bis-C60/ Ag	Spincoating	0.95	11.27	0.75	8.06	7
BAFB	ITO/PEDOT:PSS/ CH3NH3PbI3/ BAFB/ZnO/Ag	Spincoating	0.822	16.2	0.69	9.21	16
3TPYMB	ITO/PEDOT:PSS/ PolyTPD/ CH3NH3PbI3/ 3TPMYB/Au	Spincoating	0.7	14.2	0.56	5.5	23
C60	ITO/C60/ CH3NH3PbI3/ spiro-MeOTAD/ Au	Thermal evaporation	1.09	23.67	0.74	19.1	21
PTEG-1	ITO/PEDOT:PSS/ CH3NH3I3-xClx/ PTEG-1/Al	Spincoating	0.94	20.63	0.81	15.71	22
DMBI-doped diPDI	ITO/PEDOT:PSS/ CH3NH3PbI3/ DMBI-doped diPDI/Al	Spincoating	0.86	21.6	0.54	10.0	24

TABLE 6.1   *(Continued)*

n-type material	Device architecture	Deposition method	Voc (V)	Jsc (mA/cm^2)	FF	PCE (%)	References
F16CuPc	ITO/PEDOT:PSS/ CH3NH3PbI3/ F16CuPc/BPhen/ Al.	Spincoating	0.93	19.97	0.68	12.62	26
HATNASOC7-Cs	ITO/NiOx/ CH3NH3PbI3/ HATNASOC7-Cs/ Ag	Spincoating	1.08	20.73	0.78	17.62	27
N2200	ITO/PEDOT:PSS/ CH3NH3PbI3/ N2200/ZnO/Al	Spincoating	0.84	14.70	0.66	8.15	28
PNVT-8	ITO/PEDOT:PSS/ CH3NH3PbI3/ PNVT-8/ZnO/Al	Spincoating	0.85	13.53	0.62	7.13	28
PNDI2OD-TT	ITO/PEDOT:PSS/ CH3NH3PbI3/ PNDI2OD-TT/ ZnO/Al	Spincoating	0.81	13.71	0.55	6.11	28
PFN-2TNDI	ITO/PEDOT:PSS/ CH3NH3PbI3/ PFN-2TNDI/Ag	Spincoating	0.98	21.9	0.78	16.7	29
DMAPA-C60	ITO/PEDOT:PSS/ CH3NH3PbI3/ PC61BM/ DMAPA-C60/Al	Spincoating	0.97	17.9	0.77	13.4	31

KEYWORDS

- electron transporting layer
- lowest unoccupied molecular orbital
- highest occupied molecular orbital
- photoluminescence
- bathocuproine

REFERENCES

1. Salim, T.; Sun, S.; Abe, Y.; Krishna, A.; Grimsdale, A. C.; Lam, Y. M. Perovskite-Based Solar Cells: Impact of Morphology and Device Architecture on Device Performance. *J. Mater. Chem. A Mater. Energy Sustain.* **2015**, *3*, 8943–8969.
2. Völker, S. F.; Collavini, S.; Delgado, J. L. Organic Charge Carriers for Perovskite Solar Cells. *ChemSusChem.* **2015**, *8*(18), 3012–3028.
3. Ullah, F.; Chen, H.-Z.; Li, C.-Z. Organic Functional Materials Based Buffer Layers for Efficient Perovskite Solar Cells. *Chin. Chem. Lett.* **2017**, *28*(3), 503–511.
4. Chueh, C.; Li, C.; Jen, A. K. Recent Progress and Perspective in Solution-Processed Interfacial Materials for Efficient and Stable Polymer and Organometal Perovskite Solar Cells. *Energy Environ. Sci.*, **2015**, *8*, 1–30.
5. Docampo, P.; Ball, J. M.; Darwich, M.; Eperon. G. E.; Snaith, H. J. Efficient Organometal Trihalide Perovskite Planar-Heterojunction Solar Cells on Flexible Polymer Substrates. *Nat. Commun.* **2013**, *4*, 2761.
6. Jeng, J.; Chiang, Y.-F.; Lee, M.; Peng, S.-R.; Guo, T.-F.; Chen, P.; Wen, T. $CH_3NH_3PBI_3$ Perovskite/Fullerene Planar-Heterojunction Hybrid Solar Cells. *Adv. Mater.* **2013**, *25*(27), 3727–3732.
7. Liang, P.-W.; Chueh, C.-C.; Williams, S. T.; Jen, A. K.-Y. Roles of Fullerene-Based Interlayers in Enhancing the Performance of Organometal Perovskite Thin-Film Solar Cells. *Adv. Energy Mater.* **2015**, *5*(10), 1–7.
8. Wang, Q.; Shao, Y.; Dong, Q.; Xiao, Z.; Yuan, Y.; Huang, J. Large Fill-Factor Bilayer Iodine Perovskite Solar Cells Fabricated by a Low-Temperature Solution-Process. *Energy Environ. Sci.* **2014**, *7*(7), 2359–2365.
9. Malinkiewicz, O.; Yella, A.; Lee, Y. H.; Espallargas, G. M.; Graetzel, M.; Nazeeruddin, M. K.; Bolink, H. J. Perovskite Solar Cells Employing Organic Charge-Transport Layers. *Nat. Photonics* **2014**, *8*, 128–132.
10. Jeng, J.-Y.; Chen, K.-C.; Chiang, T.-Y.; Lin, P.-Y.; Tsai, T.-D.; Chang, Y.-C.; Guo, T.-F.; Chen, P.; Wen, T. C.; Hsu, Y.-J. Nickel Oxide Electrode Interlayer in $CH_3NH_3PBI_3$ Perovskite/PCBM Planar-Heterojunction Hybrid Solar Cells. *Adv. Mater.* **2014**, *26*(24), 4107–4113.
11. Chiang, C.-H.; Wu, C.-G. Bulk Heterojunction Perovskite–PCBM Solar Cells with High Fill Factor. *Nat. Photonics* **2016**, *10*(3), 196–200.
12. Burschka, J.; Pellet, N.; Moon, S. J.; Humphry-Baker, R.; Gao, P.; Nazeeruddin, M. K.; Grätzel, M. Sequential Deposition as a Route to High-Performance Perovskite-Sensitized Solar Cells. **2013**, *Nature* 3–7.
13. Chiang, C.-H.; Tseng, Z.-L.; Wu, C.-G. Planar Heterojunction Perovskite/$PC_{71}BM$ Solar Cells with Enhanced Open-Circuit Voltage via a (2/1)-step Spin-Coating Process. *J. Mater. Chem. A* **2014**, *2*(38), 15897–15903.
14. Liang, P.-W.; Liao, C.-Y.; Chueh, C.-C.; Zuo, F.; Williams, S. T.; Xin, X,-K.; Lin, J.; Jen A. K.-Y. Additive Enhanced Crystallization of Solution-Processed Perovskite for Highly Efficient Planar-Heterojunction Solar Cells. *Adv. Mater.* **2014**, *26*(22), 3748–3754.
15. Zhang, H.; Azimi, H.; Hou, Y.; Ameri, T.; Przybilla, T.; Spiecker, F.; Kraft, M.; Scherf, U.; Brabec, C. J. Improved High-Efficiency Perovskite Planar Heterojunction Solar Cells Via Incorporation of a Polyelectrolyte Interlayer. *Chem. Mater.* **2014**, *26*(18), 5190–5193.

16. Erten-Ela, S.; Chen, H.; Kratzer, A; Hirsch, A.; Brabec, C. J. Perovskite Solar Cells Fabricated Using Dicarboxylic Fullerene Derivatives. *New J. Chem.* **2016**, *40*(3), 2829–2834.

17. Li CZ; Huang, J.; Ju, H.; Zang, Y.; Zhang, J.; Zhu, J.; Chen, H.; Jen, A. K. Y. Modulate Organic-Metal Oxide Heterojunction via [1,6] Azafulleroid for Highly Efficient Organic Solar Cells. *Adv. Mater.* **2016**, *28*(33), 7269–7275.

18. Wojciechowski, K.; Stranks, S. D.; Abate, A.; Sadoughi, G.; Sadhanala, A.; Kopidakis, N.; Rumbles, G.; Li, C.-Z.; Friend, R. H.; Jen, A. K.-Y.; Snaith, H. J. Heterojunction Modification for Highly Efficient Organic - Inorganic Perovskite Solar Cells. *ACS Nano* **2014**, *8*(12), 12701–12709.

19. Abrusci, A.; Stranks, S. D.; Docampo, P.; Yip, H.; Jen, A. K.-Y.; Snaith, H. J. High Performance Perovskite-Polymer Hybrid Solar Cells via Electronic Coupling with Fullerene Monolayers. *Nano Lett.* **2013**, *13*(7), 3124–3128.

20. Li, Y.; Zhao, Y.; Chen, Q.; Yang, Y.; Liu, Y.; Hong, Z.; Liu, Z.; Hsieh, Y.-T.; Meng, L.; Li, Y.; Yang, Y. Multifunctional Fullerene Derivative for Interface Engineering in Perovskite Solar Cells. *J. Am. Chem. Soc.* **2015**, *137*(49), 15540–15547.

21. Yoon, H.; Seong, A.; Kang, M.; Lee, J.-K.; Choi, M. Hysteresis-Free Low-Temperature-Processed Planar Perovskite Solar Cells with 19.1% Efficiency. *Energy Environ. Sci.* **2016**, *9,* 2262–2266.

22. Shao, S.; Abdu-Aguye, M.; Qiu, L.; Lai, L.-H.; Liu, J.; Adjokatse, S.; Jahani, F.; Kamminga, M. E.; ten Brink, G. H.; Palstra, T. T. M.; Kooi, B. J.; Hummelen, J. C.; Antonietta Loi, M. Elimination of the Light Soaking Effect and Performance Enhancement in Perovskite Solar Cells Using a Fullerene Derivative. *Energy Environ. Sci.* **2016**, *9*(7), 2444–2452.

23. Malinkiewicz, O.; Roldán-Carmona, C.; Soriano, A.; Bandiello, E.; Camacho, L.; Nazeeruddin, M. K.; Bolink, H. J. Metal-Oxide-Free Methylammonium Lead Iodide Perovskite-Based Solar Cells: The Influence of Organic Charge Transport Layers. *Adv. Energy Mater.* **2014**, *4*(15), 1–9.

24. Kim, S. S.; Bae, S.; Jo, W. H. A perylene diimide-Based Non-Fullerene Acceptor as an Electron Transporting Material for Inverted Perovskite Solar Cells. *RSC Adv.* **2016**, *6*(24), 19923–19927.

25. Kim, S. S.; Bae, S.; Jo, W. H. Performance Enhancement of Planar Heterojunction Perovskite Solar Cells by N-Doping of the Electron Transporting Layer. *Chem. Commun.* **2015**, *51,* 17413–17416.

26. Jin, F.; Liu, C.; Hou, F.; Song, Q.; Su, Z.; Chu, B.; Cheng, P.; Zhao, H.; Li, W. Hexadeca-fluorophthalocyaninatocopper as an Electron Conductor for High-Efficiency Fullerene-Free Planar Perovskite Solar Cells. *Sol. Energy Mater. Sol. Cells* **2016**, *157,* 510–516.

27. Zhao, D.; Zhu, Z.; Kuo, M.-Y.; Chueh, C.-C.; Jen, A. K.-Y. Hexaazatrinaphthylene Derivatives: Efficient Electron-Transporting Materials with Tunable Energy Levels for Inverted Perovskite Solar Cells. *Angew. Chem. Int. Ed.* **2016**, *55*(31), 8999–9003.

28. Wang, W.; Yuan, J.; Shi, G.; Zhu, X.; Shi S; Liu, Z.; Han, L.; Wang, H.-Q.; Ma, W. Inverted Planar Heterojunction Perovskite Solar Cells Employing Polymer as the Electron Conductor. *ACS Appl. Mater. Interfaces* **2015**, *7*(7), 3994–3999.

29. Sun, C.; Wu, Z.; Yip, H.-L.; Zhang, H.; Jiang, X.-F.; Xue, Q.; Hu, Z.; Hu, Z.; Shen, Y.; Wang, M.; Huang, F.; Cao, Y. Amino-Functionalized Conjugated Polymer as an Efficient Electron Transport Layer for High-Performance Planar-Heterojunction Perovskite Solar Cells. *Adv. Energy Mater.* **2016**, *6*(5), 1–10.

30. Seo, J.; Park, S.; Kim, Y. C.; Jeon, N. J.; Noh, J. H.; Yoon, S. C.; Il Seok, S. Benefits of Very Thin PCBM and LiF Layer for Solution-Processed P-I-N Perovskite Solar Cells. *Energy Environ. Sci.* **2014,** *7,* **2642–2646.**

31. Azimi, H.; Ameri, T.; Zhang, H.; Hou, Y.; Quiroz, C. O. R.; Min, J.; Hu, M.; Zhang, Z.-G.; Przybilla, T.; Matt, G. J.; Spiecker, E.; Li, Y.; Brabec, C. J. A Universal Interface Layer Based on an Amine-Functionalized Fullerene Derivative with Dual Functionality for Efficient Solution Processed Organic and Perovskite Solar Cells. *Adv. Energy Mater.* **2015,** *5*(8), 6–11.

PART III

HOLE-TRANSPORTING LAYERS IN PEROVSKITE SOLAR CELLS

KUNWU FU and PHAM THI THU TRANG

The architecture of mesoscopic perovskite-based solar cells initially developed was similar to the solid-state dye-sensitized solar cells where perovskite material was considered as the light-absorbing material only and injects electrons to n-type electron transporting layer and holes to p-type hole-transporting layer (HTL) after photoexcitation.[1] Later as perovskite was discovered to have intrinsic capability of conducting charge carriers, the typical architecture of the perovskite solar cells (PSCs) has been in p-i-n structure.[2–5] In such cases, one layer of light-absorbing perovskite material is sandwiched between electron and HTLs. After photoexcitation, the electron–hole pair moves to perovskite HTL interface to separate. The role of the hole-conducting layer is to extract the positive hole carriers from light-absorbing perovskite layer and conduct them away.

Proper energy level alignment with perovskite materials, high charge carrier conductivity, good solution processability, and resistance towards moisture degradation of the perovskite material are the desired properties for a candidate to be a promising hole-transporting material (HTM). Typically, an organic small molecule material was employed for this purpose due to the initial structural similarity of PSCs to the solid-state dye-sensitized solar cells. Among various types of small molecule organic materials, Spiro-OMeTAD (2,2',7,7'-Tetrakis[N,N-di(4-methoxyphenyl)amino]-9,9'-spirobifluorene) has been commonly used as the HTM. Polymer hole conductors were also introduced later due to their typically higher charge carrier mobility. Meanwhile, over the years of continuous development and increasing understanding, a wide variety of HTMs, both organic and inorganic, has also been developed and employed in perovskite-based solar cells with the aim to enhance device performance, increase device stability, reduce fabrication cost, and so on. Hence, this section focuses on the introduction and discussion of the representative organic and inorganic HTMs that have been demonstrated to achieve high efficiency in PSCs.

CHAPTER 7

ORGANIC HOLE-TRANSPORTING MATERIALS

As mentioned previously in earlier chapters, mesoscopic perovskite solar cell (PSC) was initially considered as a sensitized solar cell. The hole-transporting materials (HTMs) that had been commonly used in solid-state dye-sensitized solar cells, such as spiro-OMeTAD and P3HT, were first used in the PSCs.[1,2,6,7] As more research revealed later that perovskite material has also good charge transport properties, thin film perovskite planar devices were realized and further optimized to have device performance beyond 15%.[8–10] As an example, in Figure 7.1a, the perovskite planar thin film developed through vapor-assisted process by Chen et al., has very smooth thin formation and large grains (Figure 7.1b), both of which are beneficial for the device performance. With the high quality, planar perovskite film fabricated, Chen et al. obtained the high power-conversion efficiency (PCE) of 12.1%, with the J-V curve shown in Figure 7.1c, the highest efficiency for planar architecture PSCs at that time.

In this planar architecture configuration, to achieve high performance, selective contacts need to be in good electronic contact with the perovskite layer to conduct away the positive and negative charge carriers after the charge carrier separation. Thus, a lot of research efforts have been focused on exploring various types of HTMs, both organic and inorganic, which form p-type selective contact with perovskite layer and transport the hole carriers. The main criteria that HTM candidates for PSCs should meet consist: (1) compatible energy level alignment relative to perovskite material for efficient charge injection, (2) good hole mobility and conductivity, (3) good solubility and processability, (4) good resistance to thermal and hydroscopic degradation, and (5) low-cost synthesis process. These current HTMs can be briefly categorized into three groups: (a) small-molecule HTMs, (d) conducting polymers, and (c) inorganic HTMs. This chapter will provide an overview of the research in small molecule and conducting polymer hole transport materials.

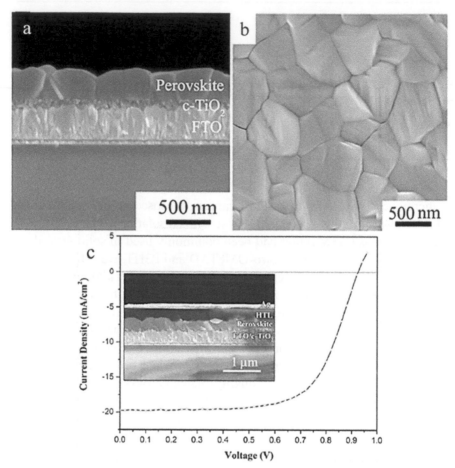

FIGURE 7.1   (a) Cross-sectional scanning electron microscope (SEM) image of perovskite film on the FTO/c-TiO$_2$ substrate obtained by reacting PbI$_2$ film and CH$_3$NH$_3$I vapor. (b) Top-view SEM images of the perovskite film by annealing ~ 200 nm PbI$_2$ film in the presence of CH$_3$NH$_3$I at 150°C in N$_2$ atmosphere after post stage at 4 h. (c) Current density–voltage (*J–V*) characteristics of the solar cell based on the as-prepared perovskite films under AM 1.5G illumination, and cross-sectional SEM images of the device (inset).
Source: Reprinted with permission from ref 9. © 2014 American Chemical Society..

Up to date, the most common p-type selective contact material that is adopted in PSCs and also gives the highest cell efficiency is spiro-OMeTAD.[11,12] Initially, it was incorporated to infiltrate into the mesoscopic structure of MAPbI$_3$/TiO$_2$ nanocomposite films and to have good interfacial contact with the perovskite material to transport away the hole charge carriers after hole injection.[1] However, as the device structure continues to develop and evolve, with the realization of thin-film perovskite layer and mesoscopic TiO_2 underlayer structure, the contact between spiro-OMeTAD has changed to more towards planar interfacial contact. Due to the continuous advancement in the perovskite thin film deposition techniques, recently the perovskite device demonstrated a maximum 20.8% PCE, with $MAPbI_3$- and $FAPbBr_3$-mixed perovskite film fabricated through single-step solvent engineering technique from a mixed solution of FAI, PbI_2, MABr, and $PbBr_2$ (where FA stands for formamidinium cations and MA stands for methylammonium cations), with mesoporous TiO_2 as the electron-transporting layer and spiro-OMeTAD as the optimized HTM layer.[11] More recently, the adoption of vacuum flash-assisted method enabled researchers to produce more uniform, smoother, and more crystalline thin film of $MAPbI_3$- and $FAPbBr_3$-mixed perovskite with high electronic quality, thus with spiro-OMeTAD as the HTM in the device, high efficiency of 20.5% (19.6% certified) over large area of more than 1 cm^2 was obtained.[12] In both cases, the high-efficiency devices were fabricated with spiro-OMeTAD as the hole-transporting layer (HTL).

Though high efficiency values have been achieved with spiro-OMeTAD, the charge mobility of this material has been lower as compared to many other p-type HTMs, especially polymer-type materials. Recent studies by Shi et al. have demonstrated that the single crystal form of spiro-OMeTAD has 3 orders of magnitude higher charge mobility as compared to its amorphous thin film form, which could show more potential of this material in the HTM application.[13] However, the synthesis process for this material is still rather more complex, and its cost is relatively higher and conductivity is generally lower. Thus, it becomes important to design and develop alternative HTMs with low-cost synthesis process and high carrier mobility to replace it. Various existing types of low-cost polymer materials and new types of small-molecule organic materials, with compatible highest occupied molecular orbital (HOMO)/lowest unoccupied molecular orbital (LUMO) energy levels relative to perovskite materials and high hole carrier mobility, have been selected to replace spiro-OMeTAD in solar cell devices.[14–18]

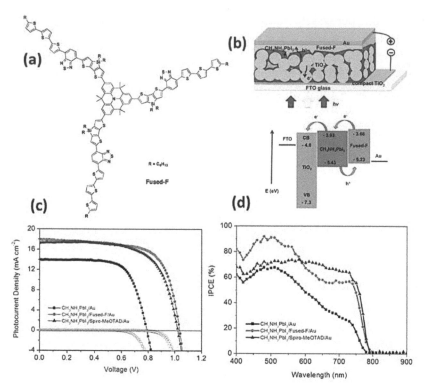

FIGURE 7.2 (a) Molecular structure of the HTM-Fused-F. (b) Device architecture of CH$_3$NH$_3$I$_3$ perovskite device fabricated with Fused-F as the HTM and the corresponding energy level diagram. (c)and(d) Current density–voltage characteristics and incident photon to converted electron (IPCE) spectra of the solar cells with Fused-F (red), spiro-OMeTAD (blue) as the HTM and without HTM (black) measured in the dark and under ~ 100 mW / cm photon flux (AM 1.5G).
Source: Reprinted with permission from ref 17. © 2014 American Chemical Society.

Qin et al. developed a new type of fused quinolizino acridine core-based small molecule with low bandgap values, effective hole mobility in the same range of the commonly used spiro-OMeTAD and an oxidation potential of 5.23 eV versus the vacuum level. The molecular structure of this material is shown in Figure 7.2a. With the appropriate oxidation potential level, the alignment between the HOMO level of Fused-F and the conduction band of CH$_3$NH$_3$I$_3$ perovskite are close range in Figure 7.2b, which will contribute to high output potential from the device. The PSC based on this new type of material as the HTM showed high PCE of 12.8% as compared to the spiro-OMeTAD-based device fabricated under the same conditions, illustrated by the J–V curves in Figure 7.2c.[17] Additionally, it is worth to note that the device with this HTM displays much higher incident photon to converted electron (IPCE) values in the blue region than that of spiro-OMeTAD, which indicates that it not only works as an HTM but also contributes to the overall photon harvesting; thus, the IPCE spectrum of the device with Fused-F as HTM has higher values in 400–570 nm wavelength range.

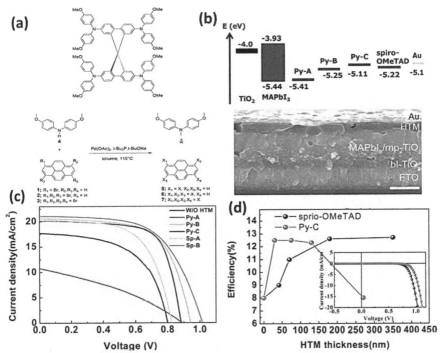

FIGURE 7.3 (a) and (b) Structures of spiro-OMeTAD and synthetic routes for pyrene arylamine derivatives. (b) Energy level diagram of the corresponding materials used in our devices and cross-sectional structure of the representative device. Scale bar=500 nm. (c). Current density–voltage (J–V) curves for TiO₂/MAPbI₃/HTMs/Au fabricated with Py-A, Py-B, Py-C, Sp-A, and Sp-B as HTMs, and without HTM. The efficiency dependence on the overlayer thickness with Py-C and spiro-OMeTAD; inset is dark J–V curves for Py-C and Sp-B devices.

Source: Reprinted with permission from ref 19. © 2013 American Chemical Society.

Several types of novel molecules have been developed through connecting triarylamine moieties through an aromatic linker. As shown in Figure 7.3a, Jeon et al. synthesized pyrene-core arylamine derivatives where a pyrene core was substituted with different numbers of N,N-di-p-methoxyphenylamine moieties.[19] The adjustment of different ratios of N,N-di-p-methoxyphenylamine to pyrene gave rise to the different optical and electronic structures of the pyrene derivatives, where the HOMO levels of these derivatives and the alignment with perovskite conduction band (CB) are shown in Figure 7.3b. The best PCE of 12.4% achieved from the PSC under AM1.5G illuminations, as compared to the 12.7% efficiency when spiro-OMeTAD was used as HTM (Figure 7.3c and d). Hence, it suggests that small-molecule materials with very simple structures can be effective HTMs.

Later, the same group further developed more spiro-OMeTAD isomers and reported a higher PCE of 16.7% with the novel small-molecule HTMs.[20] The molecular structures of this further developed HTM material are shown in Figure 7.4a. The position of one of the two para (p)-OMe substituents in each of the quadrants of spiro-OMeTAD was replaced by meta (m)- or ortho (o)-OMe substituent. The corresponding energy level alignment for these new HTM HOMO levels and the perovskite material CB value is presented in Figure 7.4b. This higher efficiency of po–spiro-OMeTAD as compared to the other derivatives is due to a larger fill factor (FF) which arises from a low series resistance and a high shunt resistance. Thus, in this work, it is demonstrated that the cell performance is strongly dependent on the position of the methoxy group as illustrated by the J–V curves in Figure 7.4c. It should be noted that these efficiency values are generally lower than the latest high efficiency record as device efficiency is also influenced by other factors, especially the quality of perovskite thin films fabricated by improved and novel deposition methods in various laboratories. On the other hands, studies of developing new HTMs bring more possibilities in reducing the cost of producing high-efficiency PSCs.

Li et al. synthesized a new HTM (H101) with an ethylenedioxythio-phene core substituted with triphenylamine (TPA) moieties and achieved PCE up to 13.8% through optimization of the concentration of cobalt dopant FK102, which was comparable to devices based on spiro-OMeTAD (13.7%), Figure 7.5a.[16] The molecular structure of this new HTM is rather simple and easy to synthesize. The HOMO level is at −5.16 eV. However, upon being added with the cobalt dopant tris(2-(1H-pyrazol-1-yl)pyridine) cobalt(III) (FK102), the HOMO level can reach to lower levels to achieve higher voltage output. The effect of oxidizing the HTM can be seen in the UV-Vis spectra in Figure 7.5b where the absorption in 500–700 nm range by the oxidized species increases gradually with increasing doping content. The optimized device structure of the PSCs and the corresponding J–V curves of devices with different doping content in the HTM are shown in Figures 7.5c and d. Efficiency of 13.8% was achieved with H101, doped with 15% FK102. Further development from this resulted in two TPA-based compounds containing tetrasubstituted thiophene and tetrasubstituted bithiophene core units, which have lower HOMO level as compared to the previous molecule.[21] Thus, these two materials help to increase the V_{oc} of the device as it is influenced by the difference between the HOMO level of HTM and the quasi-state energy levels of the VB of the perovskite material. The final device efficiency after optimization for both materials reaches beyond 15%, surpassing the cells fabricated in the same conditions with spiro-OMeTAD as HTM.

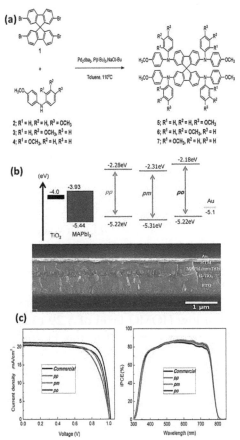

FIGURE 7.4 (a) Synthetic routes for spiro-OMeTAD derivatives. (b) Energy level diagram of the corresponding materials used in our devices and cross-sectional structure of the representative device. (c) Current density–voltage (J–V) curves and the corresponding IPCE spectra for TiO$_2$/MAPbI$_3$/HTMs/Au fabricated with pp-, pm-, po- and commercial pp-spiro-OMeTAD as HTMs.

Source: Reprinted with permission from ref 20. © 2014 American Chemical Society.

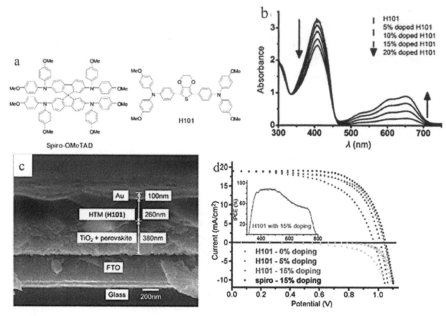

FIGURE 7.5 (a) Chemical structures of spiro-OMeTAD and H101. (b) Absorption spectra of H101 with different doping ratios in chlorobenzene. (c) Cross-sectional SEM picture of a typical perovskite solar-cell device and current–voltage curves of solar cells based on spiro-OMeTAD and H101 as HTM. (d) The IPCE spectra of the perovskite solar cell (PSC) with 15% doped H101 as HTM.

Source: Reprinted with permission from ref 16. © 2014 John Wiley & Sons, Inc. Publishing Group.

Krishnamoorthy et al. reported a swivel-cruciform 3,30-bithiophene-based HTM (KTM3) (chemical structure in Figure 7.6a and optical properties in Figure 7.6b) with a low HOMO level of 5.29 eV and demonstrated an optimized PCE of 11%,[22] as compared to the 11.4% efficiency of the standard perovskite device based on spiro-OMeTAD. The same research group also developed a series of other small-molecule HTMs based on core triptycene in short synthetic routes with high yields, exhibiting promising conversion efficiencies of 12.38%.[23]

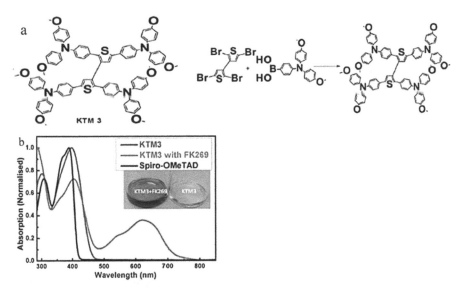

FIGURE 7.6 (a) Chemical structure and synthetic scheme of KTM3. (b) Normalized absorption spectra of KTM3, spiro-OMeTAD, and doped KTM3. The inset shows the KTM3 solution.
Source: Reprinted with permission from ref 22. © 2014 Royal Society of Chemistry.

In another work, a new HTM (V886) based on methoxydiphenylamine-substituted carbazole was reported by Gratia and coworkers with very comparable performance to the state-of-art efficiency of Spiro-OMeTAD.[24] This new HTM possesses high solubility in organic solvents (> 1000 mg/ mL in chlorobenzene) and can be very easy to synthesize through a simple two-step synthesis, which makes it appealing to the commercial interests for PSCs. In the pristine form, V886 absorbs light in the UV region below 450 nm, whereas in the chemically oxidized form, it shows visible absorption bands at 628 and 814 nm. The cross-section scanning electron microscopy of the best V886 device shows a typical thin-film perovskite capping layer of around 150 nm over the 200-nm-thick mesoscopic TiO_2 underlayer. The performance of V886 was tested in $CH_3NH_3PbI_3$-based solar cells and a maximum PCE of 16.9% was obtained under AM1.5G illumination with the average PCE values exceeding 14% observed. The measured J_{sc} is 21.4 mA/ cm^2, V_{oc} is 1.09 V, and FF is 0.73. The best device with Spiro-OMeTAD as the HTL prepared in the same device fabrication procedure displayed a PCE of 18.4% due to higher FF values. On the other hand, Arora et al. have done research into four different HTMs based on fluorene–dithiophene derivative (SO7–10), synthesized through a facile route and successfully employed into the fabrication of PSCs based on formamidinium lead bromide $CH(NH_2)_2PbBr_3$.[25] The purpose was to achieve high photovoltage output from the deeper HOMO level values of the materials. Under AM1.5G illumination, the mesoscopic PSC with SO7 as the HTM achieved a large photovoltage (V_{oc}) of 1.5 V and an efficiency (η) of 7.1%. In comparison, the device using the state-of-the-art Spiro-OMeTAD as HTM has a V_{oc} of 1.47 V and a maximum η of 6.9%. It was also supported via a density functional theory approach with GW (it is an approximation of green's function [G] and screening coulombic interaction[W]) simulations including spin–orbit coupling and electrochemical measurements that deeper HOMO levels of the new HTM, making them promising HTMs for PSCs for high photovoltage output.

Other types of TPA-based molecular HTMs have also been attempted and shown reasonable performances, including butadiene derivatives, diacetylide-TPA compounds, N,N,N′,N′-tetraphenylbenzidine (TPB) derivatives and π-conjugated linear modified TPA-based compounds.[26–30] Lee and coworkers synthesized three carbazole-based molecules with two-arm- and three-arm-type structures, which are linked through phenylene, diphenylene or TPA core units. PCEs of over 13% were obtained from these three carbazole-based HTMs with the highest value reaching 14.79% for HTM.[31] Grätzel and coworkers introduced two acceptor–donor–acceptor (A–D–A) oligothiophenes with low bandgap values as HTMs consisting of electron-rich thiophene-pyrrole-based S,N-heteropentacene central units attached with terminal dicyanovinylene acceptor groups.[32] $MAPbI_3$-based devices fabricated from these two HTMs without additives have overall efficiencies of 9.5–10.5%. Both HTMs have high absorption in the range of 600–800 nm, complementary to the perovskite $MAPbI_3$, as the absorption of $MAPbI_3$ becomes weaker after 600 nm. The IPCE spectra of the devices made with these two HTMs show enhancements over the whole region between 400 and 800 nm due to more effective charge extraction and/or light harvesting. An obvious additional band between 680 and 800 nm can be seen for HTM, which is correlated well with the absorption maximum in the film.[32] Therefore, these two oligomers not only act as HTMs but also contribute to the light absorption in the low energy region of the solar spectrum forming a dual light-harvesting system with the perovskite. The same research group later reported another molecularly engineered hole-transport material (FDT) with a dissymmetric fluorene–dithiophene (FDT) core substituted by N,N-di-p-methoxyphenylamine donor groups. When employed on state-of-the-art devices with mixed perovskite film fabricated through anti-solvent method, the material showed comparable PCEs of 20.2% with the efficiency of 19.7% from the control devices with spiro-OMeTAD material.[14] Thus, it has been demonstrated that alternative new HTMs show promising potential to replace the commonly used spiro-OMeTAD in high-performance PSCs.

In perovskite devices, closer match of the HOMO levels of HTMs to the valence bands (VBs) of the perovskite materials usually result in an improved device performance, particularly V_{oc}. Thus, it normally requires the HTMs to be doped with additives for both lower HOMO levels and higher conductivities. Most of the previous organic HTMs require additives such as Li^+ salt and cobalt complex dopant to improve their conductivities and lower down the HOMO levels. However, these additives and cobalt dopants are generally hydroscopic in nature, which tend to absorb moisture in the atmosphere, thus creating challenges for maintaining the long-term device stability in ambient working conditions. Meanwhile, such requirements of additives in the HTM solution for higher efficiency also put constraints on the potential of the material for real-life applications as it would require large-scale solution homogeneity. Thus, some research interests and efforts have also been put in dopant-free HTMs for high-performance PSC devices.

For example, one A–D–A structure, phenoxazine (POZ)-based small-molecule HTM was reported by Sun and coworkers, containing an electron-rich benzodithiophene (BDT) unit as a central building block, flanked by POZ units, and with electron-withdrawing 3-ethylrodanine.[33] Planar devices based on this pristine HTM yield an overall efficiency of 13.2% under AM1.5G illumination, significantly higher than that of dopant-free spiro-OMeTAD (8.9%). The better performance based on this new HTM is mainly ascribed to a higher FF, which is strongly related to a much higher hole conductivity in its pristine form as compared to spiro-OMeTAD.[33] Another dopant-free small-molecule HTM based on a tetrathiafulvalene derivative was also introduced in mesoscopic $MAPbI_3$-based PSCs, showing a similar efficiency (11.03%) in comparison to doped spiro-OMeTAD with dopants (11.4%).[34] Moreover, the stability of devices based on the pristine form of this HTM is also better than its counterpart spiro-OMeTAD with dopants with slower degradation under ambient conditions. The absence of the hydroscopic additives is suggested to help to improve the device stability as it is less moisture sensitive. In addition, it also supports that the developments of HTMs with high hole mobility in their pristine forms are important ways to improve the perovskite devices' long-term stability.

Similarly, Li et al. reported another electron donor–acceptor (D–A)-substituted dipolar chromophore (BTPA-TCNE) material to serve as an efficient HTM without dopants for PSCs.[35] It is synthesized through reacting a triphenylamine-based Michler's base and tetracyanoethylene. This chromophore also possesses a zwitterionic resonance structure in its ground state. Moreover, BTPA-TCNE shows an antiparallel molecular packing in its crystalline state, which reduces its overall molecular dipole moment to facilitate charge transport. As a result, BTPA-TCNE can be employed as an effective dopant-free HTM to achieve a high efficiency of 17.0% in PSCs with the conventional n-i-p planar configuration, higher than the doped spiro-OMeTAD HTM.

With the continuous efforts in developing new novel HTMs for PSCs, we can expect the efficiency to continue to increase and reduction in the cost of fabrication by replacing the spiro-OMeTAD as the HTM. New dopant-free HTMs would also eliminate the need of introducing hydroscopic additives and dopant materials, thus helping to enhance the device's air stability in ambient working environment.

Another important class of charge carrier conducting materials is conducting polymers, which have been widely used in the process of organic photovoltaics. Development of new generation of hole or electron-conducting polymers with high charge carrier mobility and long diffusion length has been an important research area for the advancement in organic photovoltaic field. Hence, as the research in HTM for PSCs progresses, researchers started to use conducting polymers to replace small molecule HTMs as the hole mobilities are generally much higher than small-molecule materials.[36–38]

Several polymer HTMs, such as poly(triarylamine) (PTAA) and poly(3-hexylthiophene-2,5-diyl) (P3HT), with higher hole mobilities and good film-forming properties, have been tested by Heo et al., PTAA demonstrated the highest efficiency of up to 12% in early studies compared to 6.7% for P3HT;[19] later, it has been improved to a certified PCE of 16.2% ($J_{SC} = 19.6$ mA/cm^2, $V_{oc} = 1.11$ V and FF = 0.74) with a mixed halide perovskite and PTAA.[39] Based on pristine P3HT in conjunction with MAPbI$_{3-x}$Cl$_x$ in planar PSCs, an overall efficiency of exceeding 10% was reached through the optimization of deposition parameters and precursor concentrations.[40] Chen et al. reported a P3HT/multiwalled carbon nanotube (MWNT) composite material as the HTM which showed higher PCE (6.45%) than the pristine P3HT (4.1%).[41] In this study, MWNTs function as efficient nanostructured charge transport channels to significantly enhance the conductivity. In another report, doped with Li-TFSI and a pyridine-based additive (2,6-di-tert-butylpyridine, D-TBP), the efficiency of pristine P3HT is significantly increased from

9.2 to 12.4% due to an increase of hole conductivity by 2 orders of magnitude.[42] Johansson et al. used transient photovoltage decay measurements to investigate the electron lifetime difference of devices based on P3HT and spiro-OMeTAD and found out that the recombination rate in device with P3HT is more than 10 times faster.[43] It was highlighted that the design of the molecular structure of HTM should aim to avoid close contact between the HTM and the perovskite to reduce the recombination rate. Park et al. prepared Li-doped poly(3-hexylthiopehene) (P3HT) nanofibrils (LN-P3HT) by cooling down a P3HT solution and used it as an HTL.[44] The nanofibrils employed in the devices resulted in improved conductivity and air stability, yielding efficiency of 13.12% with active area of 1 cm^2, potentially for high-performance, flexible, and air-stable PSCs. More importantly, after 500 bending cycles, the high performance and mechanical stability are retained; also, it shows good resistance to moisture at a relative humidity of 30% with 87% retention of the initial device performance. Therefore, this efficient and air-stable organic HTL is a promising method for high-performance large-area flexible PSCs.

On the other hand, Snaith and coworkers introduced C_{60}-substituted benzoic acid self-assembled monolayer as an interlayer between mesoporous titania and perovskite in P3HT-based devices[45] as shown in Figure 7.7a. The corresponding energy level diagram in Figure 7.7b shows that C_{60} self-assembled monolayer (SAM) interlayer facilitates the charge transfer between the perovskite and TiO_2 layer. As shown in the device performance in Figure 7.7c, the presence of this interlayer significantly enhances the photovoltage and FF by a reduction of energy loss due to better interfacial passivation layer. The role of this C_{60} SAM interlayer in passivating interfacial charge recombination was further investigated and the result is shown in Figure 7.8. The density of states for TiO_2 devices incorporating the C_{60} SAM interlayer measured through photoinduced absorption spectroscopy (Fig. 7.8a) shows a marked reduction of sub-bandgap states. This is consistent with charge "stored" in the fullerene molecules, which have a comparatively narrower density of states than the exponential tail to the density of sub-bandgap states in the TiO_2. It was proposed that the C_{60} SAM interfacial layer effectively acts as an "electron reservoir," where electrons transferred from both the perovskite and P3HT to this fullerene interfacial layer are "trapped" inside. The electrons are mostly mediated through the C_{60} SAM, regardless of whether they originate from light-absorbed perovskite or P3HT.

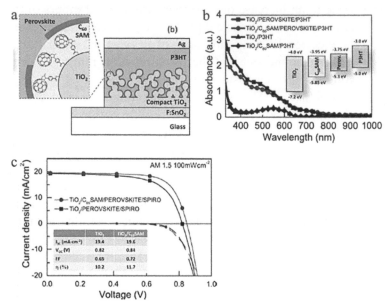

FIGURE 7.7 (a) Absorption spectra of P3HT and perovskite films with and without C_{60} SAM (Self assembled monolayer) fullerene functionalization. The materials were cast on glass substrates which had been coated in a 600-nm-thick mesoporous titania layer. Inset: Energy levels of the system components relative to vacuum. (b) Schematic of the device structure. (c) $J–V$ characteristics taken under AM1.5 simulated sunlight (100 mW/cm irradiance) and in the dark for optimized Spiro-OMeTAD devices with and without C_{60} SAM fullerene functionalization.

Source: Reprinted with permission from ref [45]. © 201x American Chemical Society.

In another study, Habisreutinger et al. employed P3HT-functionalized single-walled carbon nanotubes (SWCNTs) to substitute the commonly used organic HTMs and achieved a high PCE of up to 15.3% with an average efficiency of 10%.[46] The schematic structure of the device is illustrated in Figure 7.9a. In this work, two different types of carbon nanotubes with different metallic to semiconducting nanotube ratios were evaluated to compare the performance of the nanotubes in transporting the charge carriers away. It can be observed in Figures 7.9b and c that higher content of more conducting carbon nanotubes facilitates faster charge transport and thus the devices made with CG200 carbon nanotube have higher photocurrent and efficiencies. poly(methylmethacrylate) (PMMA) layer was incorporated to further promote better device performance. Besides good efficiency achieved, more notably, the moisture sensitivity and thermal instability of PSC were also addressed. As shown in Figure 7.10a, perovskite films coated with different HTMs are subjected to a temperature of 80°C on a hot plate in ambient air. After 96 h, all perovskite films coated with the organic hole transporters have become predominantly yellow. However, the insulating polymer PMMA shows as an effective protective layer as compared to the standard HTMs because of its hydrophobic nature and capability of inhibiting the intrusion of moisture into the perovskite structure while also inhibiting the evaporation of the methylammonium iodide. Hence, even after 96 h at 80°C in air, no significant degradation in the perovskite covered by PMMA layer was observed. For the complete device, the thermal stability of the system was also much enhanced through using this polymer-functionalized SWCNTs embedded in an insulating polymer matrix to replace the organic hole transport material. As shown in Figure 7.10b, the efficiency values remained in similar range after the devices were subjected to 80°C for 96 h, while the cells with organic HTMs in this work suffered severe degradation and showed no efficiencies after the same heat treatment process.

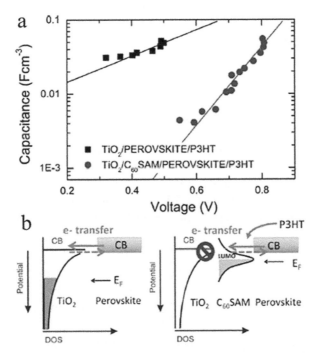

FIGURE 7.8 (a) Differential capacitance as a function of open-circuit voltage for devices with and without C_{60} SAM functionalization. Solid lines are exponential fits to the data. (b) Energy diagram of the proposed mechanisms for electron transfer from perovskite. The shaded regions in the density of states of the TiO_2 and fullerene represent filled states. *Source:* Reprinted with permission from ref 45.© 201x American Chemical Society.

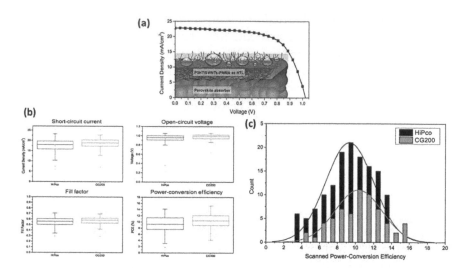

FIGURE 7.9 (a) Schematic illustration of the PSCs with P3HT/ single-walled carbon nanotubes-PMMA as hole-transporting layer (HTL). (b) and (c) Device performance comparison between perovskite devices with the two different nanotube types HiPCO and CG200 (more metallic carbon nanotube content) nanotubes in the HTL.
Source: Reprinted with permission from ref 46. © 201x American Chemical Society.

Organic polymeric HTMs also have the flexibility in tuning the oxidization potentials and the surface properties, which enables them to match close to the energy levels of various perovskite materials and to achieve high voltage output. For example, Cai et al. demonstrated the use of a diketopyrrolopyrrole (DPP)–carbazole-based polymer HTM (PCBTDPP) with HOMO level of 5.4 eV to achieve V_{oc} of 1.16 V in $CH_3NH_3PbBr_3$ devices.[47] Another even higher V_{oc} value of 1.5 V was reported by Edri and coworkers with 4,40-bis(N-carbazolyl)-1,10-biphenyl (CBP) in chlorine-doped $CH_3NH_3PbBr_3$ devices.[48]

Other types of thiophene-based conducting polymers have also been incorporated as HTMs in PSCs. Qiu et al. introduced polymer PCBTDPP as an HTM in both MAPbBr$_3$- and MAPbI$_3$-based solar cells.[47] With MAPbBr$_3$ perovskite as the absorber, this HTM demonstrates a much higher V_{oc} of 1.16 V as compared to P3HT as a result of several factors including a deeper HOMO level (–5.4 eV), a high hole mobility of 0.02 cm^2/V·s and a large difference between the conduction band of MAPbBr$_3$ and the quasi-Fermi level of TiO_2. When substituting Br with I in the perovskite structure, the photovoltaic performance is further enhanced to 5.55% due to the broader light absorption wavelength range. Another DPP-containing thiophene-based polymer PDPPDBTE was also studied in mesoscopic MAPbI$_3$-based PSCs,[49] which displays a PCE of 9.2%, higher than that of spiro-OMeTAD (7.6%), attributable to a deeper HOMO level (5.4 eV) and high hole mobility of such polymer (estimated 10.3 cm^2/V · s, estimated from a transient mobility spectroscopy method) associated with a lower series resistance and thus a high FF. Devices using this polymeric HTM also exhibited an excellent long-term durability with 90% of their initial efficiencies remained after 1000 h under a 20% relative humidity. However, as a comparison, spiro-OMeTAD-based devices only remained 70%. This excellent long-term durability of this PDPPDBTE HTM can be probably attributed to the more hydrophobic properties of such polymer, inhibiting moisture from penetrating into the perovskite surface.

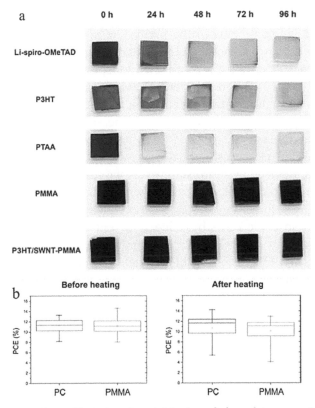

FIGURE 7.10 (a) Photos illustrating the visible degradation of the perovskite layer. The color shifts from almost black to yellow for all organic HTLs except for the films covered with PMMA only or a composite of carbon nanotubes and PMMA. (b) Temperature stressing of PSCs. Device performances before and after being exposed to 80°C in air for 96 h. *Source:* Reprinted with permission from ref 46. © 201x American Chemical Society.

Some other types of thiophene-based conducting polymers have also been studied as HTMs in PSC devices, giving moderate overall efficiencies ranging between 4.2 and 8.7%.[50] Seok et al. compared three types of thiophene-based polymeric HTMs with PTAA where PTAA demonstrates the best performance up to 12% under standard AM1.5G illumination condition.[19] The reason for its superior performance could be related to its higher hole mobility and a better film-forming ability due to the stronger interaction between the perovskite with PTAA as compared to other polymers.

However, further studies are required to confirm the specific chemical interaction between PTAA and the perovskite. In this study, a pillared structure was used, consisting of three-dimensional composites of $TiO_2/$ $MAPbI_3$ where an overlayer of $MAPbI_3$ coexist on top of the mesoscopic $TiO_2/MAPbI_3$ composite film, and polymer PTAA penetrates the scaffold to a much smaller degree as elaborated in Figure 7.11a. The corresponding energy level alignment between the perovskite materials and the polymer PTAA is also shown in Figure 7.11b. Under the same conditions, a lower overall efficiency of only 8.4% is obtained for spiro-OMeTAD as the result of a lower V_{oc} and FF. Based on PTAA HTM, Seok et al. further fabricated colorful PSCs through adjusting the chemical composition of $MAPbI_{3-x}Br_x$.[51] As illustrated in Figures 7.11c and d, with the substitution of iodide with bromide, a maximum PCE of 12.3% was achieved under 1 sun illumination (AM1.5G). More interestingly, devices with $MAPbI_{3-x}Br_x$ (x=0.2) also have improved device long-term stability as the result of their compact and stable structure.[51] In the following study, with PTAA as HTM together with $MAPbI_{3-x}Br_x$ as absorber (x=0.1–0.15), a certified PCE of 16.2% has been achieved in mesoscopic PSCs by using a solvent-engineering technique to deposit high-quality perovskite films.[39] Furthermore, Seok and coworkers incorporated $MAPbBr_3$ into $FAPbI_3$ to stabilize the perovskite phase of $FAPbI_3$, which further improved the PCE to more than 18% with PTAA as the HTM.[52]

In other works, two triarylamine polymer derivatives with deeper HOMO levels were studied in comparison to PTAA in mesoscopic PSCs based on $MAPbI_3$ and $MAPbBr_3$.[53] As discussed previously, the photovoltage output is influenced by both the VB of the perovskite absorbers and the HOMO levels of HTMs. As the VB of $MAPbBr_3$ is at −5.68 eV, deeper HOMO levels of the triarylamine polymer derivatives (−5.44 eV and −5.51 eV respectively), as compared to PTAA (−5.14 eV), result in increase of photovoltages by 70–110 mV, leading to V_{oc} up to ≈1.40 V. However, the J_{sc} of these devices are relatively lower as the result of lower driving forces for hole injections.

However, the larger V_{oc} increase of the polymeric HTMs as compared to the decrease of J_{sc} still gives rise to better overall efficiencies with the highest value of 6.7%. Yang et al. introduced polyflorene-derivative polymers consisting of fluorine and arylamine groups as HTMs in MAPbI$_3$ absorber-based mesoscopic PSCs, giving rise to promising efficiencies (10.9–12.8%) compared to the corresponding values of devices made from spiro-OMeTAD (9.8–13.6%).[54]

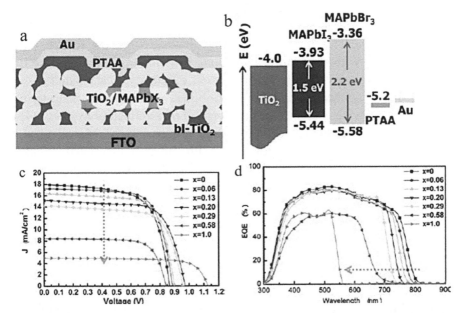

FIGURE 7.11 (a) Schematic illustration of the device architecture of mesoscopic PSCs with PTAA as the HTP and (b) the corresponding energy level diagram. (c) and (d) Photocurrent density–voltage (J–V) characteristics and corresponding external quantum efficiency spectra of the heterojunction solar cells based on MAPb(I$_{1-x}$Br$_x$)$_3$ (x = 0, 0.06, 0.13, 0.20, 0.29, 0.58, 1.0).
Source: Reprinted with permission from ref 51. © 2014 American Chemical Society.

Meantime, there are also other works to develop new dopant-free polymeric HTMs for eliminating the hydroscopic additives added into the HTM solutions and increasing the device air stability. Liao et al. synthesized and implemented a new polymeric HTM series based on semiconducting 4,8-dithien-2-yl-benzo[1,2-d;4,5-d'] bistriazole-alt-benzo[1,2-b:4,5-b'] dithiophenes (pBBTa-BDTs), yielding high PCEs and environmentally stable perovskite cells.[55] These dopant-free HTMs achieved a stabilized

PCE of 12.3% with the thin-film planar heterojunction PSCs. This high performance is attributed to efficient hole extraction/collection and balanced electron/hole transport. The smooth, conformal coatings achieved with these polymers suppress the underneath perovskite film degradation; thus, it significantly enhances the solar cell stability compared to other typical molecular HTMs. Meanwhile, Kim et al. also reported a dopant-free polymeric HTM, based on benzo[1,2-b:4,5: b']dithiophene and 2,1,3-benzothiadiazole, which had resulted in very high efficiency of 17.3% and stable PSCs with for over 1400 h at 75% humidity as compared to device failure after 900 h for HTMs with additives.[56] This HTM actually consists of a random copolymer (RCP). This highest efficiency obtained from the RCP-based PSCs is 17.3% without additives [lithium bis(trifluoromethanesulfonyl) imide and tert-butylpyridine]. The observed efficiency is attributed to a deep HOMO energy level and high hole mobility. In addition, the long-term stability of the device is dramatically improved by eliminating hygroscopic dopants and by introducing a hydrophobic polymer layer. Matsui et al. synthetized triarylamine-based polymers with different functional groups as hole-transport materials (V873) for PSCs.[57] Without adding in additives, devices employing poly(triarylamine) with methylphenylethenyl functional groups showed a PCE of 12.3%, whereas widely used additive-free poly[bis (4-phenyl)(2,4,6-trimethylphenyl)amine] (PTAA) demonstrated 10.8%. Notably, the devices with V873 also achieved stable performance at maximum power point under 1 sun illumination for approximately 40 h at room temperature, and in the dark under elevated temperature (85°C) for more than 140 h. However, the additive-containing devices degraded significantly within the same time frame. Thus, these results also show the progress and potential of these dopant-free polymeric HTMs towards stable PSC under real working conditions and industrial stress tests.

KEYWORDS

- sensitized solar cells
- hole-transporting materials
- power-conversion efficiency
- conducting polymers
- bandgap values

REFERENCES

1. Kim, H-S.; et al. Lead Iodide Perovskite Sensitized All-Solid-State Submicron Thin Film Mesoscopic Solar Cell with Efficiency Exceeding 9%. *Sci. Rep.* **2012,** *2*(591), 1–7.
2. Lee, M. M.; et al. Efficient Hybrid Solar Cells Based on Meso-Superstructured Organometal Halide Perovskites. *Science* **2012,** *338*(6107), 643–647.
3. Etgar, L.; et al. Mesoscopic CH3NH3PbI3/TiO2 Heterojunction Solar Cells. *J. Am. Chem. Soc.* **2012,** *134*(42), 17396–17399.
4. Ball, J. M.; et al. Low-temperature Processed Meso-Superstructured to Thin-Film Perovskite Solar Cells. *Energy Environ. Sci.* **2013,** *6*(6), 1739–1743.
5. Jeng, J. Y.; et al. CH3NH3PbI3 Perovskite/Fullerene Planar-Heterojunction Hybrid Solar Cells. *Adv. mater.* **2013,** *25*(27), 3727–3732.
6. Yang, L.; et al. Comparing Spiro-OMeTAD and P3HT Hole Conductors in Efficient Solid State Dye-Sensitized Solar Cells. *Phys. Chem. Chem. Phys.: PCCP* **2012,** *14*(2), 779–789.
7. Hsu, C. Y.; et al. Solid-State Dye-Sensitized Solar Cells Based on Spirofluorene (Spiro-OMeTAD) and Arylamines as Hole Transporting Materials. *Phys. Chem. Chem. Phys.: PCCP* **2012,** *14*(41), 14099–14109.
8. Zhang, W.; et al. Ultra Smooth Organic–Inorganic Perovskite Thin-Film Formation and Crystallization for Efficient Planar Heterojunction Solar Cells. *Nat. Commun.* **2015,** *6*.
9. Chen, Q.; et al. Planar Heterojunction Perovskite Solar Cells Via Vapor-Assisted Solution Process. *J. Am. Chem. Soc.* **2014,** *136*(2), 622–625.
10. Chen, C. W.; et al. Efficient and Uniform Planar-type Perovskite Solar Cells by Simple Sequential Vacuum Deposition. *Adv. mater.* **2014,** *26*(38), 6647–6652.
11. Bi, D.; et al. Efficient Luminescent Solar Cells Based on Tailored Mixed-Cation Perovskites. *Sci. Adv.* **2016,** *2*(1).
12. Li, X.; et al. A Vacuum Flash–Assisted Solution Process for High-Efficiency Large-Area Perovskite Solar Cells. *Sci.* **2016,** *353*(6294), 58–62.
13. Shi, D.; et al. Spiro-OMeTAD Single Crystals: Remarkably Enhanced Charge-Carrier Transport Via Mesoscale Ordering. *Sci. Adv.* **2016,** *2*(4).
14. Saliba, M.; et al. A Molecularly Engineered Hole-Transporting Material for Efficient Perovskite Solar Cells. *Nat. Energy.* **2016,** *1*, 15017.
15. Di Giacomo, F.; et al. High efficiency CH3NH3PbI(3–x)Clx Perovskite Solar Cells with Poly(3-hexylthiophene) Hole Transport Layer. *J. Power Sources.* **2014,** *251,* 152–156.
16. Li, H.; et al. A Simple 3,4-Ethylenedioxythiophene Based Hole-Transporting Material for Perovskite Solar Cells. *Angew. Chem.* **2014,** *53*(16), 4085–4088.
17. Qin, P.; et al. Perovskite Solar Cells with 12.8% Efficiency by Using Conjugated Quinolizino Acridine Based Hole Transporting Material. *J. Am. Chem. Soc.* **2014,** *136*(24), 8516–8519.
18. Qin, P.; et al. A Novel Oligomer as a Hole Transporting Material for Efficient Perovskite Solar Cells. *Adv. Energy Mater.* **2015,** *2,* 1400980.
19. Jeon, N.J.; et al. Efficient Inorganic-Organic Hybrid Perovskite Solar Cells Based on Pyrene Arylamine Derivatives as Hole-Transporting Materials. *J. Am. Chem. Soc.* **2013,** *135*(51), 19087–19090.
20. Jeon, N.J.; et al. o-Methoxy Substituents in Spiro-OMeTAD for Efficient Inorganic–Organic Hybrid Perovskite Solar Cells. *J. Am. Chem. Soc.* **2014,** *136*(22), 7837–7840.

21. Li, H.; et al. Hole-Transporting Small Molecules Based on Thiophene Cores for High Efficiency Perovskite Solar Cells. *ChemSusChem.* **2014,** *7*(12), 3420–3425.

22. Krishnamoorthy, T.; et al. A Swivel-Cruciform Thiophene Based Hole-Transporting Material for Efficient Perovskite Solar Cells. *J. Mater. Chem. A.* **2014,** *2*(18), 6305.

23. Krishna, A.; et al. Novel Hole Transporting Materials Based on Triptycene Core for High Efficiency Mesoscopic Perovskite Solar Cells. *Chem. Sci.* **2014,** *5*(7), 2702–2709.

24. Gratia, P.; et al. A Methoxydiphenylamine-Substituted Carbazole Twin Derivative: An Efficient Hole-Transporting Material for Perovskite Solar Cells. *Angew. Chem. Int. Ed.* **2015,** *54*(39), 11409–11413.

25. Arora, N.; et al. High Open-Circuit Voltage: Fabrication of Formamidinium Lead Bromide Perovskite Solar Cells Using Fluorene–Dithiophene Derivatives as Hole-Transporting Materials. *ACS Energy Letters.* **2016,** *1*(1), 107–112.

26. Qin, P.; et al. A Novel Oligomer as a Hole Transporting Material for Efficient Perovskite Solar Cells. *Adv. Energy Mater.* **2014,** *5*(2), p1400980

27. Qin, P.; et al. Inorganic Hole Conductor-Based Lead Halide Perovskite Solar Cells with 12.4% Conversion Efficiency. *Nat. Commun.* **2014,** *5,* 3834.

28. Lv, S.; et al. Mesoscopic TiO2/ CH3NH3PbI3 Perovskite Solar Cells with New Hole-Transporting Materials Containing Butadiene Derivatives. *Chem. Commun. (Camb)* **2014,** *50*(52), 6931–6934.

29. Wang, J.; et al. Novel Hole Transporting Materials with a Linear Π-Conjugated Structure for Highly Efficient Perovskite Solar Cells. *Chem. Commun. (Camb)* **2014,** *50*(44), 5829–5832.

30. Song, Y.; et al. Energy Level Tuning of TPB-Based Hole-Transporting Materials for Highly Efficient Perovskite Solar Cells. *Chem. Commun. (Camb)* **2014,** *50*(96), 15239–15242.

31. Sung, S.D.; et al. 14.8% Perovskite Solar Cells Employing Carbazole Derivatives as Hole Transporting Materials. *Chem. Commun. (Camb)* **2014,** *50*(91), 14161–14163.

32. Qin, P.; et al. Low Band Gap S, N-Heteroacene-Based Oligothiophenes as Hole-Transporting and Light Absorbing Materials for Efficient Perovskite-Based Solar Cells. *Energy Environ. Sci.* **2014,** *7*(9), 2981–2985.

33. Cheng, M.; et al. Phenoxazine-Based Small Molecule Material for Efficient Perovskite Solar Cells and Bulk Heterojunction Organic Solar Cells. *Adv. Energy Mater.* **2015,** *5*(8), 1401720.

34. Liu, J.; et al. A Dopant-Free Hole-Transporting Material for Efficient and Stable Perovskite Solar Cells. *Energy Environ. Sci.* **2014,** *7*(9), 2963–2967.

35. Li, Z. A.; et al. Rational Design of Dipolar Chromophore as an Efficient Dopant-Free Hole-Transporting Material for Perovskite Solar Cells. *J. Am. Chem. Soc.* **2016,** *138*(36), 11833–11839.

36. Malinkiewicz, O.; et al. Perovskite Solar Cells Employing Organic Charge-Transport Layers. *Nat. Photon.* **2014,** *8*(2), 128–132.

37. Docampo, P.; et al. Efficient Organometal Trihalide Perovskite Planar-Heterojunction Solar Cells on Flexible Polymer Substrates. *Nat. Commun.* **2013,** *4,* 2761.

38. Heo, J. H.; et al. Efficient Inorganic–Organic Hybrid Heterojunction Solar Cells Containing Perovskite Compound and Polymeric Hole Conductors. *Nat. Photonics.* **2013,** *7*(6), 486–491.

39. Jeon, N.J.; et al. Solvent Engineering for High-Performance Inorganic-Organic Hybrid Perovskite Solar Cells. *Nat. Mater.* **2014,** *13*(9), 897–903.

40. Conings, B.; et al. Perovskite-Based Hybrid Solar Cells Exceeding 10% Efficiency with High Reproducibility Using a Thin Film Sandwich Approach. Adv Mater, **2014**, *26*(13), 2041–2046.

41. Chen, H.; et al. Efficient Panchromatic Inorganic-Organic Heterojunction Solar Cells with Consecutive Charge Transport Tunnels In Hole Transport Material. *Chem. Commun. (Camb)* **2013**, *49*(66), 7277–7279,

42. Guo, Y.; et al. Enhancement in the Efficiency of an Organic-Inorganic Hybrid Solar Cell with a Doped P3HT Hole-Transporting Layer on a Void-Free Perovskite Active Layer. *J. Mater. Chem. A.* **2014**, *2*(34), 13827–13830.

43. Bi, D.; et al. Effect of Different Hole Transport Materials on Recombination in CH3NH3PbI3 Perovskite-Sensitized Mesoscopic Solar Cells. *J. Phys. Chem. Lett.* **2013**, *4*(9), 1532–1536.

44. Park, M.; et al. High-Performance Flexible and Air-Stable Perovskite Solar Cells with a Large Active Area Based on Poly(3-Hexylthiophene) Nanofibrils. *J. Mater. Chem. A.* **2016**, *4*(29), 11307–11316.

45. Abrusci, A.; et al. High-Performance Perovskite-Polymer Hybrid Solar Cells Via Electronic Coupling with Fullerene Monolayers. *Nano Lett.* **2013**, *13*(7), 3124–3128.

46. Habisreutinger, S.N.; et al. Carbon Nanotube/Polymer Composites as a Highly Stable Hole Collection Layer in Perovskite Solar Cells. *Nano Lett.* **2014**, *14*(10), 5561–5568.

47. Cai, B.; et al. High Performance Hybrid Solar Cells Sensitized by Organolead Halide Perovskites. *Energy Environ. Sci.* **2013**, *6*(5), 1480.

48. Edri, E.; et al. Chloride Inclusion and Hole Transport Material Doping to Improve Methyl Ammonium Lead Bromide Perovskite-Based High Open-Circuit Voltage Solar Cells. *J. Phys. Chem. Lett.* **2014**, *5*(3), 429–433.

49. Kwon, Y.S.; et al. A Diketopyrrolopyrrole-Containing Hole Transporting Conjugated Polymer for use in Efficient Stable Organic-Inorganic Hybrid Solar Cells Based on a Perovskite. *Energy Environ. Sci.* **2014**, *7*(4), 1454–1460.

50. Lee, J.-W.; et al. Enhancement of the Photovoltaic Performance of CH3NH3PbI3 Perovskite Solar Cells through a Dichlorobenzene-Functionalized Hole-Transporting Material. *ChemPhysChem.* **2014**, *15*(12), 2595–2603.

51. Noh, J. H.; et al. Chemical Management for Colorful, Efficient, and Stable Inorganic–Organic Hybrid Nanostructured Solar Cells. *Nano lett.* **2013**, 13, 4, 1764–1769.

52. Jeon, N. J.; et al. Compositional Engineering of Perovskite Materials for High-Performance Solar Cells. *Nat.* **2015**, *517*(7535), 476–480.

53. Ryu, S.; et al. Voltage Output of Efficient Perovskite Solar Cells with High Open-Circuit Voltage and Fill Factor. *Energy Environ. Sci.* **2014**, *7*(8), 2614–2618.

54. Zhu, Z.; et al. Polyfluorene Derivatives are High-Performance Organic Hole-Transporting Materials for Inorganic–Organic Hybrid Perovskite Solar Cells. *Adv. Funct. Mater.* **2014**, *24*(46), 7357–7365.

55. Liao, H-C.; et al. Dopant-Free Hole Transporting Polymers for High Efficiency, Environmentally Stable Perovskite Solar Cells. *Adv. Energy Mater.* **2016**, *6*(16), 1600502-n/a.

56. Kim, G-W.; et al. Dopant-Free Polymeric Hole Transport Materials for Highly Efficient and Stable Perovskite Solar Cells. *Energy Environ. Sci.* **2016**, *9*(7), 2326–2333

57. Matsui, T.; et al. Additive-Free Transparent Triarylamine-Based Polymeric Hole-Transport Materials for Stable Perovskite Solar Cells. *ChemSusChem.* **2016**, *9*(18), 2567–2571.

CHAPTER 8

INORGANIC HOLE-TRANSPORTING MATERIALS

Inorganic p-type materials are often considered as potential candidates in hole-transporting layer (HTL) in perovskite solar cells (PSCs) because they possess higher hole mobility compared to their organic counterparts, as well as high chemical stability, flexibility, and inexpensive deposition methods.[1,2]

The employment of inorganic p-type materials for HTL in PSC follows two possibilities for device architecture: the "normal" structure and the "inverted" structure. The normal structure has the FTO substrate as photo-anode, which will collect electrons, and in the inverted architecture, the FTO substrate acts as photocathode-collecting holes.

A typical PSC with a normal architecture contains TiO_2 as the electron-transporting layer (ETL) and spiro-OMeTAD as the HTL. Spiro-OMeTAD is a small molecule and exhibits low hole transporting which can be replaced by several inorganic materials such as copper thiocyanite (CuSCN) and copper iodide (CuI). In an inverted architecture, a p-type inorganic meso-porous layer is deposited on the FTO layer, onto which the perovskite is formed. Besides the abovementioned materials, this configuration normally employs nickel oxide (NiO) nanoparticles made into a mesoporous layer—using the structure developed for p-type dye-sensitized solar cells (DSSCs) (Figure 8.1).

CuSCN is an intrinsic p-type semiconductor with large bandgap (> 3.5 eV) and exhibits transparent characteristics in the visible and near-infrared region, making it an increasing interesting candidate for many optoelectronic devices.[3,4] CuSCN is commercially readily available at low cost which can easily be purchased from many international sources. It can also be processed by solution methods such as doctor blading, spin-coating or electrodeposition. Just before the "solid-state perovskite solar cell era," Anthopolous et al. have reported the optical and electrical properties of CuSCN thin film deposited from saturated solution in dipropyl sulfide with several techniques, mainly drop casting and spin-coating. By measuring field effect of a top-gate bottom-contact thin-film transistor fabricated by sequential spin-coating of the CuSCN layer and the CYTOP dielectric to determine charge transport, they have found the maximum hole mobility of solution-processed CuSCN thin film is around 0.01–0.1 $cm^2/V \cdot s$, much higher than other organic hole-transporting material (HTM).[5]

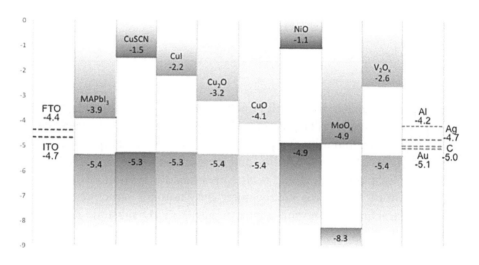

FIGURE 8.1 Energy level of commonly used inorganic p-type semiconductors as hole-transporting layer (HTL) in perovskite solar cells (PSCs).
Source: Adapted from refs 1, 16, 24, 30, 41.

Before being introduced in PSC, CuSCN has long been investigated as an HTM for solid-state DSSC and other thin-film photovoltaics.[6,7] The research group of Professor Seigo Ito (University of Hyogo) was first to publish a paper on using CuSCN as HTM in PSC.[8] Although the PCE obtained by Ito et al. was only 4.85%, lower than cells fabricated with organic HTM such as spiro-OMeTAD at that time (about 12%), they were the first to use carbon-double-bond-free HTM in PSC, which also has exhibited good stability and coverage of the perovskite layer. Around the same time, Ivan Mora-Sero et al. also published an article with n-i-p structure with CuSCN as HTM and a mix chloride–iodide perovskite ($CH_3NH_3PbI_{3-x}Cl_x$) achieving 6.4%.[9]

Following up on this direction, Seigo Ito and his collaborators have published several other articles addressing issues related to the optimization of perovskite layer in the PSC with CuSCN as HTM. It was proven that TiO_2 mesoporous layer was necessary when employing CuSCN to reduce shunting between the HTM and the transparent conducting oxide (TCO)[10] as the planar configuration show surprisingly low J_{sc} and PCE. A significant improvement in this configuration and material is the achievement of PCE as high as 12.4% with 600–700 nm thickness of CuSCN layer from the same group.[2] The major difference reported in this article is the utilization of sequential deposition for perovskite in which the first step of depositing PbI_2 was done twice. It was explained that one deposition of PbI_2 resulting in a porous $CH_3NH_3PbI_3$ layer and the CuSCN has high chance to be shunted with the anode below. Hence, increasing the perovskite thickness was crucial to prohibit shunting and increased the performance significantly. Nonetheless, the authors suspected that the interdiffusion process—in which $CH_3NH_3PbI_3$ partially dissolved in the CuSCN—is the reason why in their previous study, they were not successful to fabricate planar structured CuSCN-based PSC without the TiO_2 mesoporous layer.[10,11] The interdiffusion phenomenon was later carefully studied by X-ray diffraction. The crystal structure of perovskite was studied after an HTM (inorganic or organic) was deposited on it.[12,13]

Just a year later, another group has shown that it is possible to fabricate a planar structure device of PSC with CuSCN by inverting its configuration to p-i-n. The CuSCN layer is deposited directly on the TCO substrate—in this case, ITO.[1] Unlike previous groups, the CuSCN film was potentiostatically electrodeposited on the ITO and from an aqueous solution containing $CuSO_4$ and KSCN. The thickness was controlled by varying depositing time from 7 to 100 s. The ITO substrate can almost be covered with CuSCN with an optimized thickness of 57 nm from 50 s deposition. The authors have achieved an efficiency of 16.6% with one-step deposition of the perovskite layer, the highest efficiency of planar p-i-n perovskite with inorganic HTM at that time and proved that the limitation proposed by Ito et al. earlier can be partially solved by inverting the device configuration, opening new potential for this material. This work was soon followed up by Amassian et al.[3] who although could not achieve as high efficiency, they have proposed a more facile method to deposit a smooth layer of CuSCN by spin-coating a solution of CuSCN in dipropyl sulfide. The work highlights CuSCN's superior performance compared to PEDOT:PSS—a typical option for bottom HTL in p-i-n PSC (PEDOT:PSS stands for poly(3,4-ethylenedioxythiophene)-poly(styrenesulfonate)—a conducting polymer) (Figure 8.2). The enhanced performance was attributed to the better alignment of energy levels at the interface between the perovskite and CuSCN.

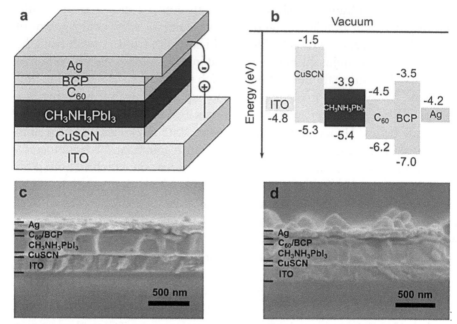

FIGURE 8.2 Configuration, energy level, and cross-section scanning electron microscope (SEM) images of p-i-n planar PSC with CuSCN as the HTL.
Source: Reprinted with permission from ref 1. © 2015 American Chemical Society.

Overall, high-efficiency PSC can be achieved by solution-processed CuSCN, either from a solution of the commercially available compound itself by spin-coating or doctor blading, or from electrodeposition of precursors containing $CuSO_4$ and KSCN. Two configurations were applied with n-i-p and inverted p-i-n structures. Several authors have also noticed out the effect of interdiffusion phenomenon between CuSCN layer and perovskite, in which perovskite could partially dissolve into CuSCN leading to shunting or shorting of the device. This effect was later confirmed by the study pseudohalide-induced perovskite $CH_3NH_3Pb(SCN)_2I$ in which two iodide (I⁻) ions in $CH_3NH_3PbI_3$ were replaced by two pseudohalide ions SCN⁻.[14] Nonetheless, CuSCN has been proven to be a promising candidate to replace expensive organic HTMs and it has also exhibited better stability.

Copper iodide is another wide bandgap p-type material with a close energy level to CuSCN and has also appeared to be a good candidate for HTL in PSC thanks to its high hole mobility, that is, 0.5–2cm²/V · s.[15]—over 20 times higher than CuSCN. In fact, although it has not gained as much attention as CuSCN due to the less popular employment in DSSC earlier, the first published work on inorganic p-type material applied in PSC was with CuI with an efficiency of 6.0%.[16] The major drawback in this work by Kamat et al. was the low V_{oc} (around 0.5–0.6V)—much lower than expected. It is worth noting that in this research, CuI-based PSC has shown good stability during short continuous illumination and extended storage in ambient, nonetheless, only the J_{sc} maintained its value, while V_{oc} decreased significantly. There was no clear evidence for the reason of V_{oc} reduction but it was suspected to be due to thermal effects and charge transport issues since the V_{oc} could recover to its original state after the reduction. Another group from Monash University has also reported their fabrication of n-i-p PSC with CuI as the HTL achieving 7.5% efficiency.[17] Handling a CuI solution can be a crucial step in the fabrication of thin-film HTL. Dipropyl sulfide is the common solvent used for most copper-based inorganic p-type semiconductors including CuI. This solvent evaporates fast but the dissolution of compounds is relatively poor. Despite the best efficiency with CuI was obtained by spin-coating CuI in dipropyl sulfide, Spiccia et al. have indicated that doctor blading was found to be more effective than spin-coating in a planar n-i-p configuration. The reason behind this finding was that the authors could not obtain uniform CuI film with spin-coating, while they found no clear evidence for the influence of film thickness ranged from over 200 nm to 1 μm. Although the CuI layer deposited by doctor blading is rough and show many large particles with gaps in between, the interesting finding in this work is the employment of graphite as an electrode, showing better performance which can efficiently replace Au.[17]

A more recent study recorded a physical method to deposit CuI by two-step gas–solid deposition. In this method, a thin film of Cu was first thermally evaporated on the substrate, followed by reaction at room temperature of the Cu layer with iodine vapor. This method forms larger grains, which resulted in better uniformity and compactness during a shorter period of deposition compared to direct thermal evaporation of CuI (Figure 8.3).[18] Without the use of any solvent, the iodization of Cu is also expected to have little impact on the perovskite layer underneath. The main advantage of this method is the high quality of CuI layer, which principally has much higher hole mobility compared to other organic HTL counterparts, resulting in a remarkably high J_{sc} of 32.72 mA/cm². Although the fill factor (FF) remains low (31%), this is a very significant method to produce high-efficiency PSC.

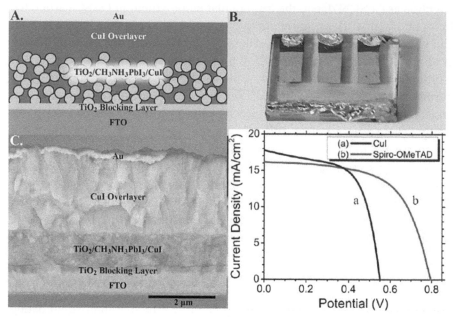

FIGURE 8.3 Device structure, photo, SEM cross-section image and I–V characteristic curve of PSC made with CuI as HTM.
Source: Reprinted with permission from ref 19. © 2014 American Chemical Society.

A planar inverted structure was later published by researchers at Xiamen University with high efficiency of 13.55% with p-i-n configuration: FTO/HTL/perovskite/PCBM/Al.[20] In this work, the authors have compared spin-coated CuI thin film to PEDOT:PSS film and found a slight improvement in device performance with the former material. A remarkable higher J_{sc} is attributed to the higher hole mobility of inorganic p-type semiconductor versus its organic counterpart. A reduction of FF with CuI film was thought to come from the poorer coverage of CuI on the FTO surface, leading to some leakage path and some lower shunt resistance. Although both CuI and PEDOT:PSS-based devices exhibited low hysteresis effect, CuI-based solar cell maintained 90% stability after 14 days storage in air, while the latter reduced to only 27%. This result is crucial for the development of inverted planar device with low-cost and highly stable inorganic HTL. This inverted structure continually shows superior performance in a later article published by Chunhui Huang et al.[21] By using single-step deposition of perovskite to reduce the surface roughness and adding a bathocuproine (BCP) layer to improve the V_{oc}, the authors have achieved a significantly high efficiency of 16.8%. Similar to the previous study, in this research, CuI was also shown to have better stability and maintained 93% of its initial efficiency after almost 300 h of storage in air.

Copper oxides (CuO_x) have previously been employed in solar cells with p-n junction device architecture. Cuprous oxide (Cu_2O) has a high hole mobility of around 100 –200 $cm^2/V \cdot s$,[22] and has broad absorption spectra with a bandgap of 2.2–2.8 eV. Although Cu_2O's bandgap is not ideal for single-junction solar cell, its energy level is well aligned with perovskite material; hence, it can work well as HTM to extract holes from the perovskite layer and improve the efficiency. CuO is an intrinsic p-type semiconductor with copper vacancies as acceptors to conduct holes[23] and has a calculated bandgap of 1 eV. In a theoretical study of HTMs for PSC, researchers from Qatar have used a wxAMPS and SCAPS software to simulate the photovoltaic performance of PSC with various HTMs including spiro-OMeTAD, NiO, CuI, CuSCN, and Cu_2O and found that the best device can be obtained with Cu_2O with calculated efficiency of 24%.[22] However, the simulation was based on the ideal crystallinity of perovskite on TiO_2 layer and energy level obtained from the literature, which gives rise to the explanation of better band alignment at the Cu_2O/perovskite interface compared to other materials.

Methods to deposit Cu_2O and CuO are often more complicated than CuSCN or CuI, such as electrodeposition, thermal oxidation, sputter and metal-organic chemical vapor deposition. Therefore, they have not yet been widely employed for low-cost PSC. Liming Ding et al. were the first to use these oxides in PSC using solution method by spin-coating CuI layers and later convert to Cu_2O by reaction with NaOH or further oxidize in O_2 with heat treatment to form CuO.[24] The perovskite $CH_3NH_3PbI_3$ layer fabricated on Cu_2O surface had better crystallinity compared with the one on CuO, attributing to the better quality and performance of PSCs made with Cu_2O. Both these two oxides can quench the photoluminescence of $CH_3NH_3PbI_3$ proving that they are suitable for HTM. Their deep valence bands that match well with the highest occupied molecular orbital of perovskite lead to a significant high V_{oc} of over 1 V.

Cu_2O-based PSC with the configuration: FTO/TiO_2 blocking layer/$CH_3NH_3PbI_{3-x}Cl_x$/CuO_x/Au fabricated with the CuO_x layer deposited by rotating angular sputtering method in a study by Shahverdi et al. showing lower efficiency compared to the inverted structure presented earlier with just 8.93%.[25] This is in agreement with other cases of Cu-based p-type semiconductors that the inverted structure works better than normal structure, especially in planar architecture. There has not been a clear investigation regarding this issue; nonetheless, from these findings, one can speculate that the crystallinity and coverage of perovskite layer are probably better on these p-type materials than on TiO_2 compact layer. In Shahverdi's study, they have also optimized the thickness of Cu_2O, which affects the coverage of the perovskite layer—in this case, it is highly rough—and the conductivity of the HTL to be around 300 nm. This finding can be a guideline for further works and optimization of Cu_2O in PSC in general.

Among inorganic p-type semiconductors investigated for PSC, NiO is probably the most studied material for DSSC,[26] quantum dot-based solar cell (CdS, CdSe)[27,28] and later on PSC. To characterize the charge separation mechanism at the NiO/perovskite interface, several techniques have been employed including photoluminescence (with femtosecond photoluminescence [PL] up-conversion spectroscopy) and photoinduced absorption. It has been clearly shown that NiO can quench the PL of $CH_3NH_3PbI_3$ effectively in comparison to high bandgap Al_2O_3. From a PL transient study by Diau group[29] in combination with steady-state PL by Wang et al.,[30] the surface state of perovskite relaxation process occurs in both NiO and Al_2O_3 but much more significant in the former than the latter. The hole extraction time from perovskite to NiO is calculated to be around 5 ns with the maximum PL at 770 nm. The Taiwanese groups have pioneered the application of NiO in PSC starting with several publications dated back in 2014 with NiO_x mesoporous layer and $PC_{61}BM$ and C_{60} as ETL, starting with an efficiency of 7.8% with planar device,[31] then 9.51% in a mesoporous configuration[30] and 11.6%[32] within the same year by improving the quality of NiO_x bottom electron blocking layer using sputtering method. Also, with p-i-n planar configuration PSC, by treating the NiO_x layer with a thin coating of PEDOT:PSS, Park et al.[33] have increased the PCE, mainly thanks to higher FF of the PSC to 15.1%. The higher FF is attributed to the better quality of the HTL and improved charge injection at the perovskite/NiO_x interface. Another chemical approach to deposit better NiO_x layer is to change the precursor to nickel nitride, which has been proven to grow better crystals. In combination with ZnO for an all-metal-oxide-transporting layers device, Yang et al. have reported an efficiency of 16.1%.[34]

Doping NiO with copper has also been investigated to increase the V_{oc} thanks to the larger grain size of Cu-doped NiO_x that offers better coverage on the electrode and prevent shunting and more specially, better energy alignment with higher E_g perovskite—in this case, $[CH_3NH_3Pb(I_{1-x}Br_x)_3]$ (Figure 8.4). The doping strategy has also been featured in *Science*, published paper by Grätzel and Han et al. who developed heavily doped both p- and n-type materials (doping $Ni_xMg_{1-x}O$ with Li and doping TiO_2 with Nb) to obtain a remarkably large area and highly stable device with efficiency of over 16%.[35]

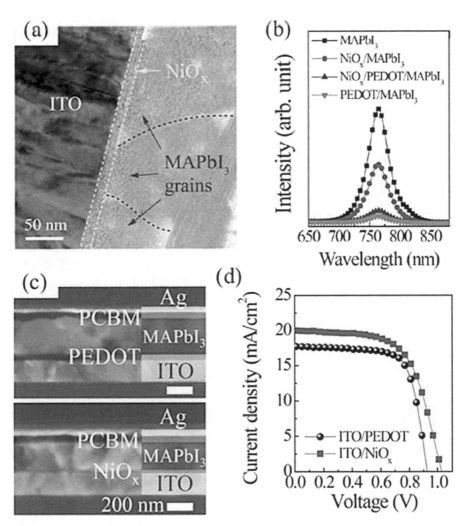

FIGURE 8.4 NiO_x employed as HTL in PSCs. (a) ITO/NiO_x interface showing $MAPbI_3$ grains; (b) photoluminescence quenching of $MAPbI_3$ with the presence of NiO_x and PEDOT; (c) cross-section SEM image; and (d) device performance of PSC fabricated with PEDOT and NiO_x.

Source: Reprinted with permission from ref 33. © 2015 American Chemical Society.

It is crucial to improve the FF in PSC and one way is to enhance the quality of the NiO blocking layer for better charge separation and hole transport like the strategies developed earlier with PEDOT:PSS treatment or sputtered thin film. Seok et al.[36] demonstrated pulsed laser deposition of NiO film in another report with a similar configuration as previous groups, with the nanostructure and transparency of NiO film being controlled by the pressure of oxygen. By optimizing the fabrication of NiO nanostructured layer, with deposition condition of 200 mTorr partial oxygen pressure, the authors have achieved an efficiency of 17.8%—the highest reported with NiO as HTL in a planar configuration.

Beside the mentioned materials, a range of other possibilities has been investigated to function as an HTL in PSC. Among these, perovskite itself can also act as an HTM thanks to its ambipolar semiconductor characteristic. In the early stage of PSC development, Etgar et al.[37] has proven the hole-transporting property of perovskite in a remarkable paper that changed the previously made assumption about this material—that it is not only a light absorber which replaces the dye in DSSC but it can also conduct both electron and hole. Although the efficiency that Etgar et al. achieved was relatively low of only 5.5%, this finding opened many other research directions and in a later development with a better crystallinity of perovskite, the efficiency of this configuration without additional HTL has been improved to over 10%(Figure 8.5).[38]

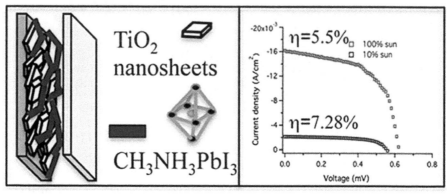

FIGURE 8.5 Perovskite can act as HTL in device configuration FTO/TiO$_2$ blocking layer/TiO$_2$ nanosheets/CH$_3$NH$_3$PbI$_3$/Au.

Source: Reprinted with permission from ref 37. © 2012 American Chemical Society.

Molybdenum oxide (MoO_x) can also be used as the HTL in PSC despite it being an n-type semiconductor. MoO_3 has very deep conduction band of -4.7 eV.[39] With optimization, an efficiency of 14.87% has been achieved by Li et al.[39] MoOx thin film can be successfully applied in tandem solar cells because it is transparent and can be deposited by thermal evaporation, which can prevent the damage of the layer underneath. The tandem device with silicon solar cell has been demonstrated by Baliff et al.[40] Vanadium oxide (V_2O_5 or VO_x) is another wide bandgap option to be applied in PSC. Among the studies employing this material, it is worth noting the solution process proposed by Huang et al.[41] who achieved 14.8% with pristine VO_x and 17.5% efficiency when using VO_x in addition to PEDOT:PSS in a p-i-n planar device.

To reduce the evaporation cost of the metal electrode such as Au or Ag, carbon-based electrode with materials such as carbon nanotubes, graphene oxide, and graphene have been deposited using low-cost, low-temperature processes such as thermal press, inkjet printings,[42] among others. The promising performance of devices with carbon-based electrode is a step forward the industrial orientation development of PSC[43]; hence, stability of these devices is critical. Most of the reported researches have confirmed the efficiency degradation of PSC with carbon-based electrode to be lower than 20% after over 10 days (Figure 8.6). Table 8.1 has the details of the inorganic p-type materials which have been employed and their deposition methodology.

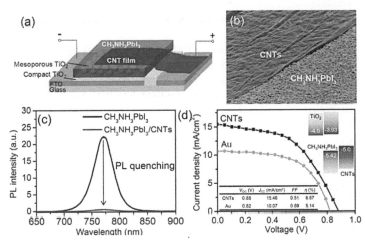

FIGURE 8.6   PSC with carbon nanotubes as electrode, replacing both the HTL and the metal electrode.
Source: Reprinted with permission from ref 44. © 2014 American Chemical Society.

TABLE 8.1 PSC with Inorganic p-Type Materials as Hole-Transporting Layer (HTL).

HTL	Device architecture	Deposition method of deposition	PV performance				Ref.
			V_{oc} (V)	J_{sc} (mA/ cm²)	FF (%)	PCE (%)	
CuSCN	FTO/dense TiO₂/ nanocrystalline TiO₂/ CH₃NH₃PbI₃/CuSCN/ Au	Doctor-blade solution in dipropyl sulfide	0.63	14.5	53	4.85	[8]
CuSCN	FTO/TiO₂/ CH₃NH₃PbI₃₋ₓClₓ/ CuSCN/Au	Drop-cast CuSCN solution in dipropyl sulfide	0.727	14.4	61.7	6.4	[9]
CuSCN	ITO/CuSCN/ CH₃NH₃PbI₃ (one-step)/C₆₀/BCP/ Ag	Potentiostatically electrodeposition of CuSO4, KSCN	1.00	21.9	75.8	16.6	[1]
CuSCN	TiO₂/ CH₃NH3PbI₃(2xPbI₂)/ CuSCN/Au	Doctor-blade solution in dipropyl sulfide	1.02	19.2	58	11.4	[2]
CuSCN	ITO/CuSCN/ perovskite/PCBM/ LiF/Ag	Spin-coat solution in dipropyl sulfide	1.06	15.76	63.2	10.5	[3]
CuI	FTO/TiO₂/ CH₃NH₃PbI₃/CuI/Au	Drop-cast CuI solution in dipropyl sulfide	0.55	17.8	62	6.0	[16]
CuI	FTO/TiO₂/ CH₃NH₃PbI₃/CuI/ graphite	Doctor-blading CuI solution in dipropyl sulfide	0.78	16.7	57	7.5	[17]
CuI	FTO/TiO₂/ CH₃NH₃PbI₃/CuI/Au	Gas–solid–vapor deposition	0.73	32.72	31	7.40	[45]
CuI	FTO/CuI/CH₃NH₃PbI₃/ PCBM/Al	Spin-coat CuI solution in acetonitrile	1.04	21.06	62	13.58	[20]
CuI	ITO/CuI/ CH₃NH₃PbI₃/C₆₀/ BCP/Ag	Spin-coat CuI solution in acetonitrile	0.99	22.6	71.3	16.8	[21]
Cu₂O	ITO/Cu₂O/ CH₃NH₃PbI₃/ PC₆₁BM/Ca/Al	Spin-coat CuI, then later react with NaOH	1.07	16.52	75.5	13.35	[24]
CuO	ITO/CuO/ CH₃NH₃PbI₃/ PC₆₁BM/Ca/Al	Spin-coat CuI, then later react oxidize in air	1.06	15.82	72.5	12.16	[24]
Cu₂O	FTO/TiO₂/ CH₃NH₃PbI₃₋ₓClₓ/ Cu₂O/Au	Angular sputtering	0.96	15.8	59	8.93	[25]

TABLE 8.1   *(Continued)*

HTL	Device architecture	Deposition method of deposition	PV performance				Ref.
			V_{oc} (V)	J_{sc} (mA/ cm^2)	FF (%)	PCE (%)	
NiO	ITO/NiO/ CH$_3$NH$_3$PbI$_3$/ PC$_{61}$BM/BCP/Al	NiO spin-coated from nickel formate dehydrate	0.92	12.43	68	7.8	[31]
NiO	FTO/blocking layer NiO/mesoscopic NiOnc/perovskite/ PC$_{71}$BM	Blocking layer NiO spin-coated from nickel formate dehydrate, mesoporous NiO spin-coated from nanopowder slurry	1.04	13.24	69	9.51	[30]
NiO	ITO/blocking layer sputtered NiO$_x$/ mesoporous NiO/ CH$_3$NH$_3$PbI$_3$/ PC$_{61}$BM/BCP/Al	Low-temperature sputtering of NiO$_x$ for blocking layer, spin-coating slurry of NiO nanopowder for mesoporous layer	0.96	19.8	61	11.6	[32]
NiO	ITO/NiO/ CH$_3$NH$_3$PbI$_3$/ZnO/Al	NiO spin-coated from nickel nitride	1.01	21.0	76	16.1	[34]
NiMg(Li) O	FTO/NiMg (Li)O/ CH$_3$NH$_3$PbI$_3$/PCBM/ Ti(Nb)O$_2$/Ag	Spray pyrolysis of solution containing nickel(II) acetylace-tonate with doping cations (magnesium acetate tetrahydrate and lithium acetate) in super-dehydrated acetonitrile/ethanol mixture	1.072	20.62	74.8	16.2	[35]
NiO	ITO/NiO/ CH$_3$NH$_3$PbI$_3$/ PC$_{61}$BM/LiF/Al	Pulsed laser deposi-tion of NiO	1.06	20.2	81.3	17.3	[36]
MoO$_x$	ITO/MoO$_3$/ PEDOT:PSS/ CH$_3$NH$_3$PbI$_3$/C$_{60}$/ Bphen/Ag	Spin-coat of (NH$_4$)$_6$Mo$_7$O$_{24}$– 4H$_2$O followed by drying at 120°C	1.00	21.49	69	14.87	[39]
V$_2$O$_x$	ITO/PEDOT:PSS/ V$_2$O$_x$/CH$_3$NH$_3$PbI$_3$/ C$_{60}$/BCP/Ag	Spin-coat precursor solution at 5000 rpm for 30 s, annealed at 150°C	1.05	21.2	78.5	17.5	[41]

TABLE 8.1 *(Continued)*

HTL	Device architecture	Deposition method of deposition	PV performance				Ref.
			V_{oc} (V)	J_{sc} (mA/ cm²)	FF (%)	PCE (%)	
Carbon nanotube	FTO/TiO$_2$/ CH$_3$NH$_3$PbI$_3$/CNT	Carbon nanotube aerosol synthesized by chemical vapor deposition, assembled into a freestanding film then transferred on top of perovskite layer	0.88	15.46	51	6.87	[43]
Carbon	FTO/TiO$_2$/ CH$_3$NH$_3$PbI$_3$/C	Doctor-blading paste of C and ZrO$_2$	0.95	17.2	71	11.6	[44]

KEYWORDS

- **intrinsic p-type semiconductor**
- **fill factor**
- **bathocuproine**
- **p-type semiconductors**
- **photoluminescence**

REFERENCES

1. Ye, S.; Sun, W.; Li, Y.; Yan, W.; Peng, H.; Bian, Z.; Liu, Z.; Huang, C. CuSCN-Based Inverted Planar Perovskite Solar Cell with an Average PCE of 15.6%. *Nano Lett.* **2015,** 15(6), 3723–3728.
2. Qin, P.; Tanaka S; Ito, S.; Tetreault, N.; Manabe, K.; Nishino, H.; Nazeeruddin, M. K.; Grätzel M, Inorganic Hole Conductor-Based Lead Halide Perovskite Solar Cells with 12.4% Conversion Efficiency. *Nat. Commun.* **2014,** 5, 3834.
3. Zhao, K.; Munir, R.; Yan, B.; Yang, Y.; Kim, T-S.; Amassian, A. Solution-Processed Inorganic Copper(I) Thiocyanate (CuSCN) Hole Transporting Layers for Efficient p–i–n Perovskite Solar Cells. *J. Mater. Chem. A* **2015,** 3, 20554–20559.
4. Wijeyasinghe, N.; Anthopoulos, T. D. Copper(I) Thiocyanate (CuSCN) as a Hole-Transport Material for Large-Area Opto/Electronics. *Semicond. Sci. Technol.* **2015,** 30(10), 104002.
5. Pattanasattayavong, P.; Ndjawa, G. O. N.; Zhao, K.; Chou, K. W.; Yaacobi-Gross, N.; O'Regan, B. C.; Amassian, A.; Anthopoulos, T. D. Electric Field-Induced Hole

Transport in Copper(I) Thiocyanate (CuSCN) Thin-Films Processed from Solution At Room Temperature. *Chem. Commun. (Camb).* **2013**, *49*(39), 4154–4156.

6. O'Regan, B.; Lenzmann, F.; Muis, R.; Wienke, J. A Solid-State Dye-Sensitized Solar Cell Fabricated with Pressure-Treated P25-TiO$_2$ and CuSCN: Analysis of Pore Filling and IV Characteristics. *Chem. Mater.* **2002**, *14*(20), 5023–5029.

7. Moon, S-J.; Itzhaik, Y.; Yum, J.; Zakeeruddin, S. M.; Hodes, G.; Grätzel ,M. Sb$_2$S$_3$-Based Mesoscopic Solar Cell Using an Organic Hole Conductor. *J. Phys. Chem. Lett.* **2010**, *1*(10), 1524–1527.

8. Ito, S.; Tanaka, S.; Vahlman, H.; Nishino, H.; Manabe, K.; Lund, P. Carbon-Double-Bond-Free Printed Solar Cells from TiO$_2$/CH$_3$NH$_3$PBI$_3$/CuSCN/Au: Structural Control and Photoaging Effects. *ChemPhysChem.* **2014**, *15*(6), 1194–1200.

9. Chavhan, S.; Miguel, O.; Grande, H-J.; Gonzalez-Pedro, V.; Sanchez, R. S.; Barea, E. M.; Mora-Sero, I.; Tena-Zaera, R. Organo-Metal Halide Perovskite-Based Solar Cells with CuSCN as Inorganic Hole Selective Contact. *J. Mater. Chem. A.* **2014**, *2*, 12754–12760.

10. Murugadoss, G.; Mizuta, G.; Tanaka, S.; Nishino, H.; Umeyama, T.; Imahori, H.; Ito, S. Double Functions of Porous TiO$_2$ Electrodes on CH$_3$NH$_3$PbI$_3$ Perovskite Solar Cells: Enhancement of Perovskite Crystal Transformation and Prohibition of Short Circuiting. *APL Mater.* **2014**, *2*(8), 10–16.

11. Ito, S. Inorganic Hole-Transporting Materials for Perovskite Solar Cell. **In** *Organic-Inorganic Halide Perovskite Photovoltaics*; Park, N. G., Grätzel, M., Miyasaka, T., Eds; Springer: Cham, 2016; pp 343–366.

12. Xiao, M.; Huang, F.; Huang, W.; Dkhissi Y; Zhu, Y.; Etheridge, J.; Gray-Weale, A.; Bach, U.; Cheng, Y-B.; Spiccia, L. A Fast Deposition-Crystallization Procedure for Highly Efficient Lead(Supporting). *Angew. Chemie.* **2014**, *126*(37), 10056–10061.

13. Ito, S.; Kanaya, S.; Nishino, H.; Umeyama, T.; Imahori, H. Material Exchange Property of Organo Lead Halide Perovskite with Hole-Transporting Materials. *Photonics.* **2015**, *2*(4), 1043–1053.

14. Jiang, Q.; Rebollar, D.; Gong, J.; Piacentino, E. L.; Zheng, C.; Xu, T. Pseudohalide-Induced Moisture Tolerance in Perovskite CH$_3$NH$_3$Pb(SCN)$_2$I Thin Films. *Angew. Chemie - Int. Ed.* **2015**, *54*(26), 7617–7620.

15. Inudo, S.; Miyake, M.; Hirato, T. Electrical Properties of CuI Films Prepared by Spin Coating. *Phys. Status Solidi Appl. Mater. Sci.* **2013**, *210*(11), 2395–2398.

16. Christians, J. A.; Fung, R. C. M.; Kamat, P. V. An Inorganic Hole Conductor for Organo-Lead Halide Perovskite Solar Cells. Improved Hole Conductivity with Copper Iodide. *J. Am. Chem. Soc.* **2013**, *136*(2), 758–764.

17. Sepalage, G. A.; Meyer, S.; Pascoe, A.; Scully, A. D.; Huang, F.; Bach, U.; Cheng, Y-B.; Spiccia, L. Copper(I) Iodide as Hole-Conductor in Planar Perovskite Solar Cells: Probing the Origin of J-V Hysteresis. *Adv. Funct. Mater.* **2015**, *25*(35), 5650–5661.

18. Nejand, B. A.; Gharibzadeh, S.; Ahmadi, V.; Shahverdi, H. R. Novel Solvent-free Perovskite Deposition in Fabrication of Normal and Inverted Architectures of Perovskite Solar Cells. *Sci. Rep.* **2016**, *vol. 6*, 33649.

19. Christians, J. A.; Fung, R. C. M. Kamat, P. V. An Inorganic Hole Conductor for Organo-Lead Halide Perovskite Solar Cells. Improved Hole Conductivity with Copper Iodide. *J. Am. Chem. Soc.*, 2014, 136(2), 758–764.

20. Chen, W-Y.; Deng, L-L.; Dai, S-M.; Wang, X.; Tian, C-B.; Zhan, X-X.; Xie, S-Y.; Huang, R-B.; Zheng, L. Low-Cost Solution-Processed Copper Iodide as an Alternative

to PEDOT:PSS Hole Transport Layer for Efficient and Stable Inverted Planar Heterojunction Perovskite Solar Cells. *J. Mater. Chem. A.* **2015,** *3*(38), 19353–19359.

21. Sun, W.; Ye, S.; Rao, H.; Li, Y.; Liu, Z.; Xiao, L.; Chen, Z.; Bian, Z.; Huang, C. Room-Temperature and Solution-Processed Copper Iodide as Hole Transport Layer for Inverted Planar Perovskite Solar Cells. *Nanoscale* **2016,** *8,* 15954–15960.

22. Hossain, M. I.; Alharbi, F. H.; Tabet, N. Copper Oxide as Inorganic Hole Transport Material for Lead Halide Perovskite Based Solar Cells. *Sol. Energy* **2015,** *120,* 370–380.

23. Meyer, B. K.; Polity, A.; Reppin, D.; Becker, M.; Hering, P.; Klar, P. J.; Sander, T.; Reindl, C.; Benz, J.; Eickhoff, M.; Heiliger, C.; Heinemann, M.; Bläsing, J.; Krost, A.; Shokovets, S.; Müller, C.; Ronning, C. Binary Copper Oxide Semiconductors: from Materials Towards Devices. *Phys. Status Solidi.* **2012,** *249*(8), 1487–1509.

24. Zuo, C.; Ding, L. Solution-Processed Cu_2O and CuO as Hole Transport Materials for Efficient Perovskite Solar Cells. *Small,* **2015,** *11*(41), 5528–5532.

25. Nejand, B. A.; Ahmadi, V.; Gharibzadeh, S.; Shahverdi, H. R. Cuprous Oxide as a Potential Low-Cost Hole-Transport Material for Stable Perovskite Solar Cells. *ChemSusChem.* **2016,** *9*(3), 302–313.

26. Odobel, F.; Pellegrin, Y.; Gibson, E. A.; Hagfeldt A; Smeigh, A. L.; Hammarström L, Recent Advances and Future Directions to Optimize the Performances of P-Type Dye-Sensitized Solar Cells. *Coord. Chem. Rev.* **2012,** *256,* 21–22, 2414–2423.

27. Kang, S. H.; Zhu, K.; Neale, N. R.; Frank, A. J. Hole Transport in Sensitized CdS – NiO Nanoparticle Photocathodes. *Chem. Commun.* **2011,** *47,* 10419–10421.

28. Chan, X-H.; Robert Jennings, J.; Anower Hossain, M.; Koh Zhen, K. Yu; Wang, Q. Characteristics of p-NiO Thin Films Prepared by Spray Pyrolysis and Their Application in CdS-Sensitized Photocathodes. *J. Electrochem. Soc.* **2011,** *158*(7), H733–H740.

29. Hsu, H. Y.; Wang, C. Y.; Fathi, A.; Shiu, J. W.; Chung, C. C.; Shen, P. S.; Guo, T. F.; Chen, P.; Lee, Y. P.; Diau, E. W. G. Femtosecond Excitonic Relaxation Dynamics of Perovskite on Mesoporous Films of Al_2O_3 and NiO Nanoparticles. *Angew. Chemie—Int. Ed.* **2014,** *53*(35), 9339–9342.

30. Wang, K-C.; Jeng, J.; Shen, P.; Chang, Y-C.; Diau, EW-G.; Tsai, C.; Chao, T-Y.; Hsu, H-C.; Lin, P.; Chen, P.; Guo, T-F.; Wen, T-C. P-Type Mesoscopic Nickel Oxide/Organometallic Perovskite Heterojunction Solar Cells. *Sci. Rep.* **2014,** *4,* 4756.

31. Jeng, J. Y.; Chen, K. C.; Chiang, T. Y.; Lin, P. Y.; Da Tsai, T.; Chang, Y. C.; Guo, T. F.; Chen, P.; Wen, T. C.; Hsu, Y. J. Nickel Oxide Electrode Interlayer in $CH_3NH_3PBI_3$ Perovskite/PCBM Planar-Heterojunction Hybrid Solar Cells. *Adv. Mater.* **2014,** *26*(24), 4107–4113.

32. Wang, K.; Shen, P.; Li, M.; Chen, S.; Lin, M.; Chen, P.; Guo, T. Low-Temperature Sputtered Nickel Oxide Compact Thin Film as Effective Electron Blocking Layer for Mesoscopic $NiO/CH_3NH_3PBI_3$ Perovskite Heterojunction Solar Cells. *ACS Appl. Mater. Interfaces* **2014,** *6,* 11851–11858.

33. Park, I. J.; Park, M. A.; Kim, D. H.; Do Park, G.; Kim, B. J.; Son, H. J.; Ko, M. J.; Lee, D. K.; Park, T.; Shin, H.; Park, N. G.; Jung, H. S.; Kim, J. Y. New Hybrid Hole Extraction Layer of Perovskite Solar Cells with a Planar p-i-n Geometry. *J. Phys. Chem. C.* **2015,** *119*(49), 27285–27290.

34. You, J.; Meng, L.; Song, T-B.; Guo, T-F.; (Michael) Yang, Y.; Chang, W-H.; Hong, Z.; Chen, H.; Zhou, H.; Chen, Q.; Liu, Y.; De Marco, N.; Yang, Y. Improved Air Stability of Perovskite Solar Cells Via Solution-Processed Metal Oxide Transport Layers. *Nat. Nanotechnol.* **2016,** *11,* 75-81.

35. Han, L.; Chen, W.; Wu, Y.; Yue, Y.; Liu, J.; Zhang, W.; Yang, X.; Chen, H.; Bi, E.; Ashraful, I.; Grätzel, M. Efficient and Stable Large-Area Perovskite Solar Cells with Inorganic Charge Extraction Layers. *Sci.* **2015**, *350*(6263), 944–948.

36. Park, J. H.; Seo, J.; Park, S.; Shin, S. S.; Kim, Y. C.; Jeon, N. J.; Shin, H-WW.; Ahn, T. K.; Noh, J. H.; Yoon, S. C.; Hwang, C. S.; Seok, S. I., Efficient $CH_3NH_3PBI_3$ Perovskite Solar Cells Employing Nanostructured P-Type NiO Electrode Formed by a Pulsed Laser Deposition. *Adv. Mater.* **2015**, *27*(27), 4013–4019.

37. Etgar, L.; Gao, P.; Xue, Z.; Peng, Q.; Chandiran, A. K.; Liu, B.; Nazeeruddin, M. K.; Grätzel, M. Mesoscopic $CH_3NH_3PBI_3/TiO_2$ Heterojunction Solar Cells. *J. Am. Chem. Soc.* **2012**, *134*(42), 17396–17399.

38. Xiao, Y.; Han, G.; Li, Y.; Li, M.; Wu, J. Electrospun Lead-Doped Titanium Dioxide Nanofibers and the in Situ Preparation of Perovskite-Sensitized Photoanodes for Use in High Performance Perovskite Solar Cells. *J. Mater. Chem. A.* **2014**, *2*(40), 16856–16862.

39. Hou, F.; Su, Z.; Jin, F.; Yan, X.; Wang, L.; Zhao, H.; Zhu, J.; Chu, B.; Li, W. Efficient and Stable Planar Heterojunction Perovskite Solar Cells with an MoO_3/PEDOT:PSS Hole Transporting Layer. *Nanoscale* **2015**, *7*(21), 9427–9432.

40. Löper, P.; Moon, S-J.; Martín de Nicolas, S.; Niesen, B.; Ledinsky, M.; Nicolay, S.; Bailat, J.; Yum, J-H.; De Wolf, S.; Ballif, C. Organic–Inorganic Halide Perovskite/Crystalline Silicon Four-Terminal Tandem Solar Cells. *Phys. Chem. Chem. Phys.* **2015**, *17*(3), 1619–1629.

41. Peng, H.; Sun, W.; Li, Y.; Ye, S.; Rao, H.; Yan, W.; Zhou, H. Solution Processed Inorganic V_2O_x as Interfacial Function Materials for Inverted Planar-Heterojunction Perovskite Solar cells with Enhanced Efficiency. *Nano Res.* **2016**, *9*(10), 2960–2971.

42. Liu, Z.; Shi, T.; Tang, Z.; Sun, B.; Liao, G. Using Low-Temperature Carbon Electrode for Preparing Hole- Conductor-Free Perovskite Heterojunction Solar Cells Under High Relative Humidity. *Nanoscale* **2016**, *8*(13), 7017–7023.

43. Zhou, H.; Shi, Y.; Wang, K.; Dong, Q.; Bai, X.; Xing, Y.; Du, Y.; Ma, T., Low-Temperature Processed and Carbon-Based ZnO/ $CH_3NH_3PBI_3$/C Planar Heterojunction Perovskite Solar Cells. *J. Phys. Chem. C.* **2015**, *119*(9), 4600–4605.

44. Li, Z.; Kulkarni, S. A.; Boix, P. P.; Shi, E.; Cao, A.; Fu K; Batabyal, S. K.; Zhang, J.; Xiong, Q.; Wong, L. H.; Mathews, N.; Mhaisalkar, S. G. Laminated Carbon Nanotube Networks for Metal Electrode-Free Efficient Perovskite Solar Cells. *ACS Nano.* **2014**, *8*(7), 6797–6804.

45. Gharibzadeh, S.; Nejand, B. A.; Moshaii, A.; Mohammadian, N.; Alizadeh, A. H.; Mohammadpour, R.; Ahmadi, V.; Alizadeh, A. Two-Step Physical Deposition of a Compact CuI Hole-Transport Layer and the Formation of an Interfacial Species in Perovskite Solar Cells. *ChemSusChem.* **2016**, *9*(15), 1929–1937.

CHAPTER 9

HOLE-TRANSPORTING-FREE PEROVSKITE SOLAR CELLS

As elaborated previously, perovskite material $CH_3NH_3PbI_3$ has also been demonstrated to be capable to transport charge carriers in meso-super-structured and thin-film planar structure device configuration, triggering researchers to investigate on its hole-transporting properties concurrently and resulting in other types of perovskite solar cell (PSC) structures without hole-transporting layer.[1-4] Thus, besides the typical p-i-n structure PSCs with organic or inorganic hole-transporting materials (HTMs) which have achieved high efficiencies, there have also been attempts to construct perovskite-based solar cells completely free of HTMs. In addition, the removal of HTM layer can significantly reduce the fabrication cost as the conventionally used spiro-OMeTAD material is more expensive than other components of the solar cell.

The first demonstration was realized by Lioz et al. by simply constructing the device with direct contact between gold electrode and perovskite layer without the commonly used spiro-OMeTAD layer.[2] In this configuration, as shown in Figure 9.1a, electron–hole pair separation mostly occurs at the TiO_2 and perovskite interface and holes are transported through the perovskite layer directly to the gold cathode as illustrated in the energy diagram in Figure 9.1b. Efficiency of 5.5% was obtained from the device. With this demonstration of hole carrier transport within perovskite layer, further research of high-efficiency HTM-free PSC have been pursued by many other research groups.

Further optimization on this configuration by depositing thicker perovskite layer to have more homogeneous coverage over the underneath TiO_2 layer increased the device efficiency to 8% with a current density of 18.8 mA/cm².[5] The thicker perovskite layer with large crystal sizes formed function simultaneously as light harvesters and hole transport materials also. Furthermore, when the two-step deposition technique was introduced to optimize the perovskite thin film layer, the solar cell efficiency reached

10.85%, with fill factor (FF) of 68%, V_{oc} of 0.84 V, and J_{sc} of 19 mA/cm², the highest efficiency to date of a hole-conductor-free PSC.[1] More analysis of the capacity voltage measurement helped to reveal that the depletion region formed in the TiO_2–perovskite interface assists the charge separation and suppresses charge recombination processes, both contributing to higher performance. The performance of the hole-conductor-free PSC is strongly dependent on the depletion layer width, created at the TiO_2–$CH_3NH_3PbI_3$ junction. In this work, when the total depletion width fraction value is at the maximum of 0.49, the device efficiency reaches highest. It indicates that the depletion region can assist in the charge separation and suppress the back-recombination reaction, and consequently contributes to the increase in the power-conversion efficiency (PCE) of the cells. The stability of this device was also investigated and that X-ray diffraction measurements show no changes in the crystallographic structure of the $CH_3NH_3PbI_3$ perovskite over time, indicating the high stability of these hole-conductor-free PSCs.

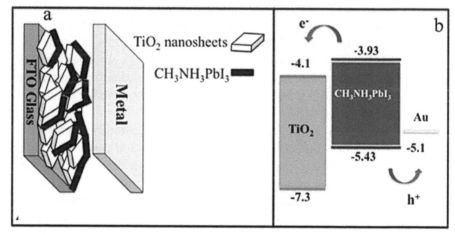

FIGURE 9.1   (a) Scheme of the device structure. (b) Energy level diagram of the $CH_3NH_3PbI_3$/TiO_2 heterojunction (HJ) solar cell.
Source: Reprinted with permission from ref 2. © 2012 American Chemical Society.

Regarding this HTM-free structure PSCs, this interface between the perovskite film and the metallic cathode electrode is very critical due to the lack of hole conductor. Cohen et al. demonstrated in their work the effect of antisolvent treatment on organometal halide perovskite film in improving the interface between perovskite and the Au cathode in hole-conductor-free perovskite-based solar cell in Figure 9.2a, achieving impressive PCE of

11.2% for hole-conductor-free cells with gold contact.[6] The I–V curve of the best performing cell is shown in Figure 9.2b, where the open-circuit voltage (V_{oc}) of 0.91 V, FF of 0.65, and current density (J_{sc}) of 19 mA/cm². It was found out in this work that antisolvent (toluene) surface treatment has improving effects on both the morphology of the perovskite layer and the electronic properties of the perovskite. By conductive atomic force microscopy (cAFM) and surface photovoltage measurement, it was shown that the perovskite film after antisolvent treatment becomes more conductive. It can be observed in Figures 9.2c and d that the slope of the I–V plot is smaller for antisolvent-treated perovskite than for nontreated film, which suggests different carrier densities in the two samples. After toluene treatment, the perovskite film becomes more intrinsic than without toluene treatment.

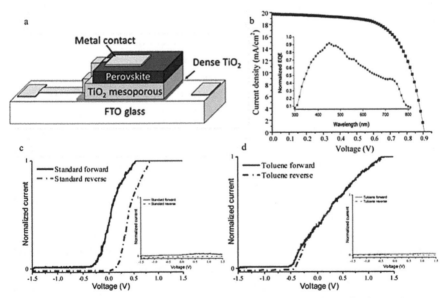

FIGURE 9.2 (a) Structure of HTM-free perovskite-based solar cell. (b) I–V curve of toluene-treated HTM-free perovskite-based solar cell. (c) and (d) I–V plots measured on a single perovskite grain by cAFM (c) without and (d) with toluene treatment. Inset: I–V plot of single nonconductive grain.
Source: Reprinted with permission from ref 6. © 2016 American Chemical Society.

Moreover, this treatment also helps to reduce the hysteresis effect commonly found in perovskite-based solar cells. On bare perovskite film, I–V plot of a single perovskite grain measured by cAFM shows that hysteresis disappears after toluene treatment. The reason for this reduction of hysteresis effect is believed to be the removal of excess halide and methylammonium (MA) ions from the perovskite film surface, leading to a new positive charge on the Pb atoms, resulting in a more conductive perovskite surface and beneficial for the hole-conductor-free solar cell structure. Further elucidation on the influence of toluene treatment on the electronic properties of perovskite through the surface photovoltage technique revealed that the contact potential difference (ΔCPD) for nontreated perovskite film is higher than that of antisolvent-treated film by 100 mV, while the bandgap structures of the two remain the same. Hence, it can be concluded that subsequent to toluene treatment, the perovskite film becomes slightly more intrinsic.

Shi et al. also achieved similar high-level efficiency of 10.49% on hole-conductor-free organic lead iodide thin film solar cells fabricated by a sequential deposition method.[3] In addition, the ideal current–voltage model for a single heterojunction (HJ) solar cell was applied to clarify the junction property of the cell and the results confirmed that the $TiO_2/CH_3NH_3PbI_3/Au$ cell is a typical HJ cell and the intrinsic parameters of the cell are comparable to that of the high-efficiency thin-film solar cells. From the plots of $-dV/dJ$ versus $(J_{sc}-J)^{-1}$ and the linear fitting curves as shown in the paper by Shi et al., it can be found that there is a good linear relationship between $-dV/dJ$ and $(J_{sc}-J)^{-1}$ both at illumination and in dark, which indicates that the cell we fabricated is a well-behaved HJ solar cell. Furthermore, the ideality factor and series resistance were derived from the slope and intercept of the linear fitting results. The results of ideality factors larger than 1.3 and small difference of 0.05 between ideality factors in dark and light difference indicate that the cell agrees well with the HJ solar cell model. Other methods to derive the ideality factor are by $\ln(J_{sc}-J)$ versus $(V+R_SJ)$ plot and R_{CT} versus bias voltage plot, which also reveal the similar ideality factor results.

Meanwhile, in another work, Cohen et al. focused on the impact of perovskite deposition process to achieve uniform coverage and optimal film thickness on the device's high efficiency.[7] Detailed study of individual deposition process parameters in hole-conductor-free PSCs, such as spin velocity, annealing temperature, dipping time, and methylammonium iodide (MAI) concentration on the photovoltaic performance was carried out and efficiency of 9.4% was obtained. Thus, it provides better understanding and control over the perovskite deposition through highly efficient, low-cost perovskite-based solar cells.

Regarding the deposition method of perovskite layer, Gamliel et al. adopted spray pyrolysis process to deposit uniform perovskite films with good coverage over bottom layer, whose scanning electron microscope

(SEM) photos are shown in Figure 9.3.[8] The deposited perovskite film was made into planar configuration hole-conductor (HTM)-free perovskite-based solar cells. The $CH_3NH_3PbI_3$ perovskite crystals deposited using spray technique have sizes in micrometer range and the film thickness is controlled by the number of passes during the process. As illustrated in Figure 9.3, 10 passes during the spray process resulted in around 3.4-μm thickness film. The efficiency of 6.9% was demonstrated for this simple PSC structure without HTM layer. Charge accumulation at the $Au/CH_3NH_3PbI_3$ interface was observed through capacitance–voltage measurements, while the compact $TiO_2/CH_3NH_3PbI_3$ junction showed a space charge region, which inhibits the recombination. The realization of this simple planar HTM-free PSC shows the potential to make large-scale solar cells while maintaining a low cost.[8]

To explore more into different types of perovskite materials for HTM-free solar cells, Aharon et al. used $TiO_2/CH_3NH_3PbI_nBr_{3-n}$ (where $0 \leq n \leq 3$) as hole conductor and light harvester in the solar cell.[9] Adjusting the relative concentrations of MAI and MABr shows that all the different compositions of the hybrid $CH_3NH_3PbI_nBr_{3-n}$ can conduct holes. The energy diagram of the hybrid perovskite material in the solar cell structure is illustrated in Figure 9.4, both $MAPbI_3$ and $MAPbBr_3$ as the light harvester and hole conductor. To adjust the relative compositions, the hybrid perovskite was deposited in two steps, enabling better control of the perovskite composition and effective tuning of the bandgap. The X-ray diffraction reveals the trend of change in the lattice parameter by introducing the Br⁻ ions. The hybrid iodide/bromide perovskite hole-conductor-free solar cells show PCE at 8.54% under 1 sun illumination with J_{sc} of 16.2 mA/cm² and very good stability. The results of this work open the possibility for graded structure of PSCs without the need for hole conductor.[9] Ma et al. have demonstrated for the first time a hole-transport-material-free planar solar cell with cesium lead mixed halide perovskite ($CsPbIBr_2$), deposited by dual-source thermal evaporation, achieving an efficiency of 4.7%. The addition of iodine into the bromide lowers the bandgap for wider solar spectrum absorption. Compared to the hybrid halide perovskites, $CsPbIBr_2$ demonstrates better thermal stability.[10] Dymshits et al. demonstrated high open-circuit voltage of 1.35 V using $Al_2O_3/CH_3NH_3PbBr_3$ HTM-free PSCs. Surface photovoltage spectroscopy was applied to reveal that contact potential difference under light measured of $CH_3NH_3PbBr_3$ was more than twice that of $CH_3NH_3PbI_3$. This results in smaller surface potential for the $Al_2O_3/CH_3NH_3PbBr_3$ cells. A longer recombination lifetime for the $Al_2O_3/CH_3NH_3PbBr_3$ cells than for the

$TiO_2/CH_3NH_3PbI_3$ cells or for the $TiO_2/CH_3NH_3PbBr_3$ cells was also found, matching with the high open-circuit voltage. Hence, it indicated that the perovskite/metal oxide interface has a major effect on influencing the open-circuit voltage in perovskite-based solar cells.[11]

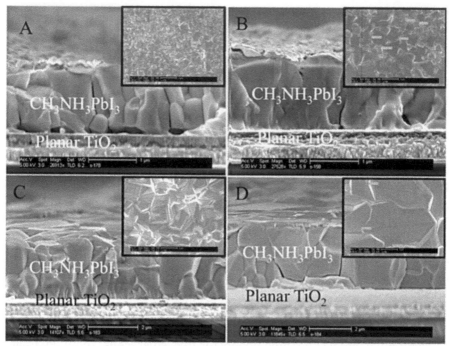

FIGURE 9.3 Cross-sectional view and (inset) top-surface view of perovskite layer deposited by spray pyrolysis process with (A) Four passes of perovskite spray, (B) six passes of perovskite spray; (C) eight passes of perovskite spray, (D) 10 passes of perovskite spray. Scale bars in the insets are 2 μm.

Source: Reprinted with permission from ref 8. © 2015 American Chemical Society.

Other attempts to construct HTM-free PSCs to lower down the cost include replacing the gold contact with carbon-based material, such as graphite and carbon nanotubes (CNTs).[4,12–14]

Zhang et al. developed a low-temperature-processed (100°C) carbon paste and applied it as counter electrode (CE) material by doctor-blading technique to substitute noble metallic materials in hole-conductor-free perovskite/TiO_2 HJ solar cells. Under optimized conditions, PCE value of 8.31% has been achieved with this carbon counter electrode (CCE), with good stability over 800 h.[15] Electrochemical impedance spectroscopy study demonstrated good charge transport characteristics of low-temperature-processed CCE. Zhou and coworkers also successfully prepared full-solution-processed $TiO_2/CH_3NH_3PbI_3$ HJ solar cells based on a low-temperature carbon electrode, avoiding the high-temperature processes at which organ lead halide is unstable.[16] PCE of mesoporous (M-) $TiO_2/CH_3NH_3PbI_3/$ Carbon HJ solar cells based on a low-temperature-processed carbon electrode achieved 9%. The M-$TiO_2/CH_3NH_3PbI_3$/carbon HJ solar cells devices without encapsulation exhibited advantageous stability, that is, over 2000 h in the air in the dark. The processing of low-cost carbon electrodes at low temperature on top of the $CH_3NH_3PbI_3$ layer without destroying its structure reduces the cost and simplifies the fabrication process of perovskite HJ solar cells. This also provides higher flexibility to choose and optimize the device, as well as investigate the underlying active layers.

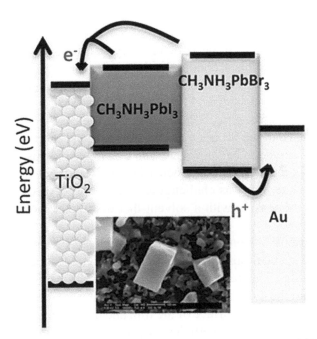

FIGURE 9.4 Energy level diagram of the pure $MAPbI_3$ and the pure $MAPbBr_3$ perovskites used hybrid perovskite HTM-free solar cells.
Source: Reprinted with permission from ref 9. © 2014 American Chemical Society.

Ku et al. constructed a mesoscopic MA lead iodide perovskite/TiO_2 HJ solar cell which is developed with low-cost CCE by a fully screen-printable process.[17] The CCE is printed on top of an insulating ZrO_2 layer and a TiO_2

nanoparticle (NP) layer. Perovskite material is deposited by dripping and infiltrating into the mesoscopic structure of the device. In the initial state, this mesoscopic HJ solar cell achieved a good PCE of 6.64%, higher than that of the flaky graphite-based device and comparable to the HTM-free PSCs with Au electrode, as well as good stability. The same research group further optimized the device and replaced TiO_2 NP with TiO_2 nanosheet (NS) containing high levels of exposed (001) facets in the device.[4] Figure 9.5a illustrates the architecture of such CCE PSCs. It was claimed that the high reactivity of (001) facets in TiO_2 NSs improves the interfacial properties between the perovskite and the electron collector. Thus, as the result, device efficiency of up to 10.64% was obtained with the hole-conductor-free fully printable mesoscopic $TiO_2/CH_3NH_3PbI_3$ HJ solar cell, while in comparison, device with TiO_2 NP shows lower efficiency of 7.36% as shown in Figure 9.5b. This high-efficiency and low-cost carbon materials CE gives this HTM-free PSC a promising prospect toward low-cost photovoltaic devices.

Similarly, in the work by Yi et al. PSCs with $ITO/ZnO/CH_3NH_3PbI_3/$ graphite/carbon black electrode structure were fabricated by spin-coating, which was done at ambient conditions.[18] The formation of the perovskite layer was also through two-step conversion process, where spin-coated PbI_2 thin films reacted with CH_3NH_3I solution and were converted into $CH_3NH_3PbI_3$ perovskite films. Electrochemically exfoliated graphite film was incorporated and the FF, open-circuit potential and short-circuit current density were observed to be improved. The best device in this work yielded 10.2% PCE. The incorporation of electrochemically exfoliated graphite film onto the perovskite film was challenging because of the necessity to form well-dispersed suspension with a solvent that does not dissolve perovskite film.

Among various solvents tried, only 4-methyl-2-pentanone produced stable graphite suspensions after the centrifugation and does not dissolve the perovskite film. The treatment on the graphite film with $NaBH_4$ prior to dispersing it into the 4-methyl-2-pentanone solvent was also done to reduce the oxidizing group on the surface of the film and hence increase the stability of the perovskite film underneath. The work also compared the difference between device with and without graphite layer as the hole extraction layer by electrochemical impedance spectroscopy. Upon incorporation of the graphite layer, the overall electron transfer resistance increased by more than an order of magnitude at zero bias but it dropped back to the previous values when 0.550 V bias was applied. The increase of the resistance was likely due to the initial formation of a poor electrical contact between the perovskite and graphite layers, which was later substantially improved when the bias was applied.

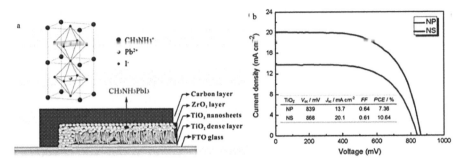

FIGURE 9.5 (a) Schematic of a hole-conductor-free mesoscopic TiO$_2$/CH$_3$NH$_3$PbI$_3$ HJ solar cell based on anatase nanosheets (NSs) and carbon counter electrons. (b) J–V curves under AM1.5 simulated sunlight of 100 mW/cm and of the mesoscopic TiO$_2$/CH$_3$NH$_3$PbI$_3$ HJ solar cells based on P25 TiO$_2$ nanoparticles and TiO$_2$ NSs.
Source: Reprinted with permission from ref 4. © 2014 American Chemical Society.

Zhang et al. reported two-step conversion process to make the perovskite layer in the mesoporous graphite/carbon black CEs with different size flaky graphite material in hole-conductor-free mesoscopic PSCs.[19] With investigation study through conductivity measurements, current–voltage characteristics, and impedance spectroscopy measurements, it is shown that as the conductor in CCEs graphite significantly affect the square resistance of CCEs and FF and PCE of the devices. The optimized CCE thickness is 9 μm, with which the device efficiency exceeds 11% for the fully printable hole-conductor-free mesoscopic PSCs due to the low square resistance and large pore size of graphite-based CCEs. Mei et al. fabricated a higher efficiency HTM-free mesoscopic TiO$_2$/CH$_3$NH$_3$PbI$_3$ HJs with the same double-layer TiO$_2$ and ZrO$_2$ scaffold structure and carbon as a back contact and obtained a certified PCE of 12.8% and exhibit excellent stability under long-term light soaking.[13] In this work, the perovskite was introduced by drop-casting a solution of PbI$_2$, MA) iodide, and 5-ammonium valeric acid (5-AVA) iodide through a porous carbon film. The better infiltration of perovskite material into the bottom TiO$_2$ layer was realized by the addition of the 5-AVA, which also created mixed-cation perovskite (5-AVA)$_x$(MA)$_{1-x}$PbI$_3$ crystals with lower defect concentration as well as more complete contact with the TiO$_2$ scaffold. Thus, it resulted in a longer exciton lifetime and a higher quantum yield for photoinduced charge separation. A record-high certified PCE of 12.8% for HTM-free PSCs and stability for >1000 h in ambient air under full sunlight were achieved. Yang et al. further studied the size effect of the TiO_2 photoanode on the hole-conductor-free fully printable mesoscopic PSCs based on the CCE and $(5-AVA)_x(MA)_{1-x}PbI_3$ perovskite.[20] With TiO_2 NPs with an optimized diameter of 25 nm, a champion device exhibits an efficiency of 13.41%. Jiang et al. further simplified the fabrication process by removing the underlying TiO_2 compact layer.[21] This compact-layer-free, hole-conductor-free, fully printable mesoscopic PSC has a PCE of over 13%, which is comparable to that of the device with a TiO_2 compact layer. This result shows a promising future in printable solar cells with further simpler fabrication process and lower costs. Hu et al. designed and constructed high-efficiency, low-cost, highly stable, hole-conductor-free, solid-state PSCs, with TiO_2 as the electron transport layer and carbon as the hole collection layer in ambient air. Atomic layer deposition (ALD) process was used to deposit uniform, pinhole-free TiO_2 films with various thicknesses on FTO electrodes. The effect of TiO_2 compact film thickness on the device perfor-mance was investigated in hole-conductor-free PSCs with carbon as the CE fabricated in ambient air. The best performance devices on optimized TiO_2 compact film (by 2000 cycles ALD) can achieve efficiency of as high as 7.82%. Furthermore, long-term stability was also demonstrated in the cells which maintained over 96% of initial PCE after 651 h (about 1 month) storage in ambient air.[22]

Though higher efficiency values have been achieved in different types of HTM-free PSCs, the lack of proven stability has become a major obstacle on the path of PSCs toward commercial viability due to the sensitivity of $MAPbI_3$ toward moisture in the ambient environment. Li et al. performed extensive stability tests to demonstrate the durability of carbon electron HTM-free PSCs based on a triple-layer architecture employing carbon as a back contact, including outdoor tests in the hot desert climate and indoor long-term light soaking as well as heat exposure during 3 months at 80–85°C.[23] No evidence for device degradation under the test conditions was observed; thus, it confirmed that the carbon–electrode HTM-free perovskite device architecture provides a promising path toward efficient and stable perovskite photovoltaics. Xu et al., on the other hand, used highly ordered mesoporous carbon (OMC) in mesoscopic $TiO_2/CH_3NH_3PbI_3$ HJ solar cells as CE.[24] The OMCs were synthesized by a template method and mixed with flaky graphite. The PSC devices based on OMC material have an FF of 0.63 and a PCE (η) of 7.02%, a good improvement compared with the carbon-black-based devices. The higher FF and η were attributed to the decrease of charge transfer resistance at the interface because of the uniform mesopores and interconnected structures in the CCE.

Besides graphite film as the hole-conducting layer, other types of carbon-based materials have also been under investigation to partially or fully replace commonly used spiro-OMeTAD for low-cost, high-efficiency HTM-free PSCs. Li et al. reported a semitransparent perovskite/CNTs solar cells where the $CH_3NH_3PbI_3$ PSCs were fabricated by directly laminating CNT network films onto a $CH_3NH_3PbI_3$ substrate as a hole collector. Figure 9.6a illustrates the architecture of such perovskite device with CNT films as the CE. The colored surface view SEM image in Figure 9.6b shows the CNT film is laminated on top of the perovskite film, which acts as the hole conductor. Intimate contact between the flexible CNT film and $CH_3NH_3PbI_3$ perovskite film can be observed. In the absence of an organic HTM and metal contact, the CNT CE PSCs showed an efficiency of up to 6.87%. The semi-transparency nature of such perovskite cell shows photovoltaic output with dual-side illuminations. The current density–voltage curve in Figure 9.6c shows the performance of the perovskite cells illuminated from both sides. Further efficiency enhancement was achieved by adding spiro-OMeTAD layer to the CNT network as a composite electrode that improved the efficiency to 9.90% due to the enhanced hole extraction and reduced recombination in solar cells. Moreover, the flexible and transparent CNT network film shows potential for realizing flexible and semitransparent PSCs.[12] On the other hand, Aitola et al. also demonstrated a hybrid HTM CE based on a thin single-walled carbon nanotube (SWCNT) film and a drop-cast spiro-OMeTAD HTM layer, with better efficiency obtained.[25] This hybrid CE devices showed high efficiency with the average of 13.6% and the highest record cell yielding 15.5% efficiency as compared to the average 17.7% for control cell with spin-coated spiro-OMeTAD HTMs and an Au CE (highest 18.8%). When no spiro-OMeTAD was drop-cast, the HTM-free perovskite showed highest efficiency of 11%.

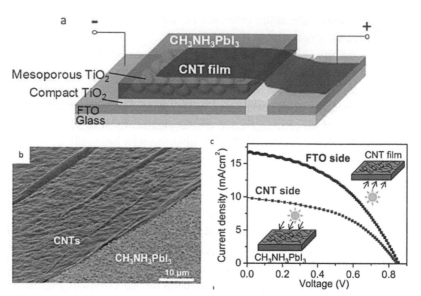

FIGURE 9.6 (a) Schematic of $CH_3NH_3PbI_3$ perovskite solar cell (PSC) with carbon nanotube (CNT) film electrode. (b) Tilted scanning electron microscope (SEM) image of $CH_3NH_3PbI_3$ perovskite substrate (blue) partly covered by CNT film (purple). (c) Light J–V curves of $CH_3NH_3PbI_3$ perovskite/CNTs solar cell with illumination from FTO and from CNT side under the condition of AM1.5 100 mW/cm2.
Source: Reprinted with permission from ref 12. © 2014 American Chemical Society.

Gopi et al. demonstrated cost-effective novel CNT paste that was applied to FTO substrate by the facile doctor blade method and processed at low temperature (100°C). In this work, a new method was reported for cost-efficient PSCs with the use of conventional hole transporters by simply directly clamping a selective hole extraction electrode made of CNT and a TiO_2/perovskite photoanode. Most importantly, $CH_3NH_3PbI_3$ and CNTs formed a solar cell with an efficiency of up to 7.83% under optimized conditions, fabricated in air without high-vacuum deposition which simplifies the fabrication process. On the other hand, the perovskite devices also exhibited good stability over 50 h.[26] The abundance, low cost, and excellent properties of the CNT material provide good prospects for future applications in PSCs.

In addition, as mentioned previously, Habisreutinger et al. employed P3HT-functionalized single-walled carbon nanotubes (SWCNTs) to substitute the commonly used organic HTMs and achieved a high PCE of up to 15.3% with an average efficiency of 10%.[14] Two different types of CNTs with different metallic to semiconducting nanotube ratios were evaluated to compare the performance of the nanotubes in transporting the charge carriers away. Higher content of more conducting CNTs facilitates faster charge transport and thus the devices made with CG200 CNT have higher photocurrent and efficiencies. PMMA layer was incorporated to further promote better device performance and stability. Besides good efficiency achieved, more notably, the moisture sensitivity and thermal instability of PSC were also addressed. The insulating polymer poly(methylmethacrylate) (PMMA) turned out to be a very effective protective layer as compared to the standard HTMs because of its hydrophobic nature and capability of inhibiting the intrusion of moisture into the perovskite structure as well as inhibiting the evaporation of the MA iodide. Thus, in the complete device, the thermal stability of the system was also much enhanced by using these polymer-functionalized SWCNTs embedded in an insulating polymer matrix to replace the organic hole transport material; hence, the efficiency values remained in the similar range after the devices were subjected to 80°C for 96 h, while other cells with organic HTMs suffered severe degradation and showed no efficiencies.

The same research group continued to explore the use of polymer-decorated CNTs in conducting hole carriers in PSCs. In another recent work, they demonstrated a planar n–i–p PSC design with a steady-state efficiency of up to 18.8% in the absence of any electronic dopants.[27] SnO_2 is used as an electron-accepting n-type layer to achieve faster electron carrier extraction. The perovskite film later is modified to the mixture-type perovskite with the composition of $(FA_{0.83}MA_{0.17}Pb(I_{0.83}Br_{0.17})_3)$, as this composition recently has produced the highest efficiency record. Polymer-wrapped single-walled CNTs are mixed with pristine spiro-OMeTAD to act as the double-layer structure of hole-transporting p-type layer. The structure of the device is shown in the SEM image in Figure 9.7a. The SWCNTs are clearly visible and protruding from the interface between the perovskite absorber and the undoped spiro-OMeTAD layer. In Figure 9.7b, the device efficiency at maximum power output point for the SWCNTs hole-transporting layer is highest as compared to those of neat spiro and Li^+ salt-added spiro layers. The addition of SWCNTs enhances the hole conductivity of this pristine spiro-OMeTAD, which is commonly known for its relatively low conductivity in undoped form. The effect of adding this SWCNTs to conduct the hole carriers can be clearly seen in the transient photoluminescence decay in Figure 9.7c. Compared to both neat spiro- and Li-doped spiro, spiro with SWCNT shows the fastest charge extraction; thus, it leads to better device performance. Thus, it can demonstrate that the adoption of this polymer-decorated SWCNTs charge transport channels within the pristine spiro layer can help to improve the charge transport within this layer, leading to higher device performance. The thermal stability of the devices with polymer-decorated SWCNTs in the HTM layer is also superior than the devices based on the only doped spiro-OMeTAD layer. It is shown in Figure 9.7d, when exposed to 85°C in the dark at a relative humidity of ~45%, the devices with polymer-decorated SWCNTs in the HTM layer decrease 30% of the initial performance in the first 48 h and remain almost constant thereafter. However, the devices with doped spiro-OMeTAD show a significant decrease in performance in the first 24 h and continuous drop furthermore. Thus, polymer-decorated SWCNTs HTM provides a pathway for dopant-free PSCs which is crucial for long-term stability and the demonstrated better thermal stability and has a higher potential toward industrial production and applications.

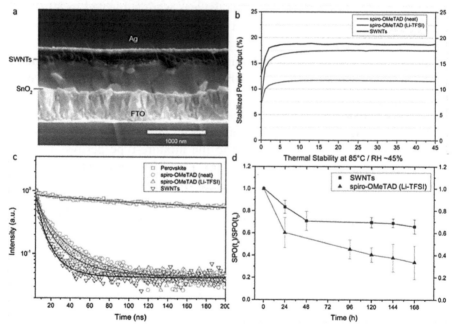

FIGURE 9.7 (a) SEM image of a device cross section, including the interlayer of SWCNTs, (b) SPO obtained from photocurrent tracking at the maximum power point voltage (corresponding to the transient current-voltage sweeps. (c) Time-resolved photoluminescence spectra in the first 200 ns. The decay was fitted with a stretched exponential function. (d) Thermal stress test of devices with doped spiro-OMeTAD (triangles) and the SWCNT interlayer without dopant (squares). The devices were exposed to 85°C in the dark at a relative humidity of ~45%.

Source: Reprinted with permission from ref 27. © 2017 American Chemical Society.

In other attempts to remove the organic hole-transporting layer, all inorganic PSCs have been introduced. This is due to the main instability against moisture and high temperature comes from the organic–inorganic hybrid perovskite and the organic hole-conducting layers. To overcome this instability issue, Liang et al. reported a successful fabrication of all-inorganic PSCs without organic components,[28] in which fabrication process can be completed in an ambient environment without humidity control. The PSC structure consists of functional layers of fluorine-doped tin oxide (FTO)/compact TiO_2 (c-TiO_2)/mesoporous TiO_2 (m-TiO_2)/inorganic perovskite $CsPbBr_3$/carbon as illustrated in Figure 9.8a. The energy level alignment of each individual layers in the structure is shown in Figure 9.8b where the carbon layer extracts the hole carriers from the conductor $CsPbBr_3$, similar to other carbon HTM structure PSCs. The highest PCE achieved from initial batch of all-inorganic PSCs was 6.7%, as shown in Figure 9.8c, which is considerably still lower than other hybrid perovskite-based solar cells. The benefits of all inorganic components in PSCs include its enhanced stability and lower fabrication cost. In the elaboration of Figure 9.8d, all-inorganic PSCs present no performance degradation in humid air (90–95% relative humidity) for over 2640 h without encapsulation. The devices also withstand more extreme temperature conditions (100 and −22°C). Moreover, by the elimination of expensive HTMs and noble-metal electrodes, the cost was significantly reduced. This study opens the door for next-generation PSCs with long-term stability under harsh conditions, making practical application of PSCs a real possibility.

In summary, we have seen various research works from worldwide research groups on the elimination of organic or inorganic hole-transporting layer in PSC. Without the conventionally used spiro-OMeTAD layer, fast charge extraction requires good interfacial contact between the perovskite layer and the metal cathode layer or the carbon layer. Till date, carbon-based materials have shown good performance and potential in replacing spiro and other organic HTMs. Thus, from these ongoing research activities in HTM-free perovskite device development, we can anticipate further reduction of the fabrication cost and wider commercial interests into this research area.

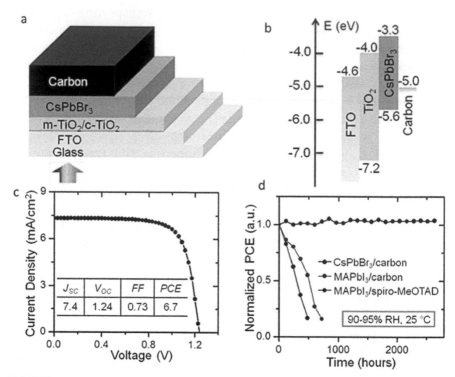

FIGURE 9.8   (a) Schematic cross-sectional view of CsPbBr$_3$/carbon-based all-inorganic PSCs with the configuration of FTO/c-TiO$_2$/m-TiO$_2$/CsPbBr$_3$/carbon. (b) Energy level diagram of the all-inorganic PSCs, showing smooth electron injection and hole extraction. (c) J–V plot of CsPbBr$_3$/carbon-based all-inorganic PSCs. The inset shows the corresponding photovoltaic parameters. (d) Normalized PCEs of CsPbBr$_3$/carbon-based all-inorganic PSCs and MAPbI$_3$/carbon-based and MAPbI$_3$/spiro-MeOTAD-based hybrid PSCs as a function of storage time in the humid air (90–95% RH, 25°C) without encapsulation.
Source: Reprinted with permission from ref 28. © 2016 American Chemical Society.

KEYWORDS

- **hole-transporting layer**
- **fill factor**
- **power-conversion efficiency**
- **conductive atomic force microscopy**
- **heterojunction**

REFERENCES

1. Aharon, S.; et al. Depletion Region Effect of Highly Efficient Hole Conductor Free $CH_3NH_3PBI_3$ Perovskite Solar Cells. *Phys. Chem. Chem. Phys.* **2014**, *16*(22), 10512–10518.

2. Etgar, L.; et al. Mesoscopic $CH_3NH_3PBI_3$/Tio$_2$ Heterojunction Solar Cells. *J. Am. Chem. Soc.* **2012**, *134*(42), 17396–17399.

3. Shi, J.; et al. Hole-Conductor-Free Perovskite Organic Lead Iodide Heterojunction Thin-Film Solar Cells: High Efficiency and Junction Property. *Appl. Phys. Lett.* **2014**, *104*(6), 063901.

4. Rong, Y.; et al. Hole-Conductor-Free Mesoscopic TiO_2/$CH_3NH_3PBI_3$ Heterojunction Solar Cells Based on Anatase Nanosheets and Carbon Counter Electrodes. *J. Phys. Chem. Lett.* **2014**, *5*(12), 2160–2164.

5. Laban, W. A.; Etgar, L. Depleted Hole Conductor-Free Lead Halide Iodide Heterojunction Solar cells. *Energy Environ. Sci.* **2013**, *6*(11), 3249–3253.

6. Cohen, B-E.; et al. Impact of Antisolvent Treatment on Carrier Density in Efficient Hole-Conductor-free Perovskite-Based Solar Cells. *J. Phys. Chem. C.* **2016**, *120*(1), 142–147.

7. Cohen, B-E.; Gamliel, S.; Etgar, L. Parameters Influencing the Deposition of Methylammonium Lead Halide Iodide in Hole Conductor Free Perovskite-Based Solar cells. APL Materials. **2014**, *2*(8), 081502.

8. Gamliel, S.; et al. Micrometer Sized Perovskite Crystals in Planar Hole Conductor Free Solar Cells. *J. Phys. Chem. C.* **2015**, *119*(34), 19722–19728.

9. Aharon, S.; Cohen, B. E.; Etgar, L. Hybrid Lead Halide Iodide and Lead Halide Bromide in Efficient Hole Conductor Free Perovskite Solar Cell. *J. Phys. Chem. C.* **2014**, *118*(30), 17160–17165.

10. Ma, Q.; et al. Hole Transport Layer Free Inorganic $CsPbIBr_2$ Perovskite Solar Cell by Dual Source Thermal Evaporation. *Adv. Energy Mater. 2016*, *6*(7), 1502202-n/a.

11. Dymshits, A.; Rotem, A.; Etgar, L. High Voltage in Hole Conductor Free Organo Metal Halide Perovskite Solar Cells. *J. Mater. Chem. A.* **2014**, *2*(48), 20776–20781.

12. Li, Z.; et al. Laminated Carbon Nanotube Networks for Metal Electrode-Free Efficient Perovskite Solar Cells. *ACS Nano.* **2014**, *8*(7), 6797–6804.

13. Mei, A.; et al. A Hole-Conductor-Free, Fully Printable Mesoscopic Perovskite Solar Cell with High Stability. *Sci.* **2014**, *345*(6194), 295–298.

14. Habisreutinger, S. N.; et al. Carbon Nanotube/Polymer Composites as a Highly Stable Hole Collection Layer in Perovskite Solar Cells. *Nano Lett.* **2014**, *14*(10), 5561–5568.

15. Zhang, F.; et al. Structure Engineering of Hole–Conductor Free Perovskite-Based Solar Cells with Low-Temperature-Processed Commercial Carbon Paste As Cathode. *ACS Appl. Mater. Interfaces* **2014**, *6*(18), 16140–16146.

16. Zhou, H.; et al. Hole-Conductor-Free, Metal-Electrode-Free TiO_2/$CH_3NH_3PBI_3$ Heterojunction Solar Cells Based on a Low-Temperature Carbon Electrode. *J. Phys. Chem. Lett.* **2014**, *5*(18), 3241–3246.

17. Ku, Z.; et al. Full Printable Processed Mesoscopic $CH_3NH_3PBI_3$/TiO_2 Heterojunction Solar Cells with Carbon Counter Electrode. *Sci. Rep.* **2013**, *3*, 3132.

18. Jin, Y.; Chumanov, G. Solution-Processed Planar Perovskite Solar Cell without a Hole Transport Layer. *ACS Appl. Mater. Interfaces.* **2015**, *7*(22), 12015–12021.

19. Zhang, L.; et al. The effect of carbon counter electrodes on fully printable mesoscopic perovskite solar cells. *J. Mater. Chem. A.* **2015,** *3*(17), 9165–9170.
20. Yang, Y.; et al. The Size Effect of TiO$_2$ Nanoparticles on a Printable Mesoscopic Perovskite Solar Cell. *J. Mater. Chem. A.* **2015,** *3*(17), 9103–9107.
21. Jiang, X.; et al. Efficient Compact-Layer-Free, Hole-Conductor-Free, Fully Printable Mesoscopic Perovskite Solar Cell. *J. Phys. Chem. Lett.* **2016,** *7*(20), 4142–4146.
22. Hu, H.; et al. Atomic Layer Deposition of TiO$_2$ for a High-Efficiency Hole-Blocking Layer in Hole-Conductor-Free Perovskite Solar Cells Processed in Ambient Air. *ACS Appl. Mater. Interfaces.* **2016,** *8*(28), 17999–18007.
23. Li, X.; et al. Outdoor Performance and Stability Under Elevated Temperatures and Long-Term Light Soaking of Triple-Layer Mesoporous Perovskite Photovoltaics. *Energy Technol.* **2015,** *3*(6), 551–555.
24. Xu, M.; et al. Highly Ordered Mesoporous Carbon for Mesoscopic CH$_3$NH$_3$PBI$_3$/TiO$_2$ Heterojunction Solar Cell. *J. Mater. Chem. A.* **2014,** *2*(23), 8607–8611.
25. Aitola, K.; et al. Carbon Nanotube-Based Hybrid Hole-Transporting Material and Selective Contact for High Efficiency Perovskite Solar Cells. *Energy Environ. Sci.* **2016,** *9*(2), 461–466.
26. Gopi, C. V. V. M.; et al. Low-Temperature Easy-Processed Carbon Nanotube Contact for High-Performance Metal and Hole-Transporting Layer-Free Perovskite Solar Cells. *J. Photochem. Photobiol. A.* **2017,** *332,* 265–272.
27. Habisreutinger, S. N.; et al. Dopant-Free Planar n–i–p Perovskite Solar Cells with Steady-State Efficiencies Exceeding 18%. *ACS Energy Lett.* **2017,** *2*(3), 622–628.
28. Liang, J.; et al. All-Inorganic Perovskite Solar Cells. *J. Am. Chem. Soc.* **2016,** *138*(49), 15829–15832.

PART IV
LEAD-FREE PEROVSKITE SOLAR CELLS

HEMANT KUMAR MULMUDI and ANITA WING YI HO-BAILLIE

CHAPTER 10

TIN-BASED PEROVSKITES

Methyl-ammonium lead perovskites have been identified as materials which are "defect tolerant," which means that the fundamental electronic properties more or less remain the same, irrespective of the processing methodology.[1] The intrinsic and extrinsic defects have minimal effect on the mobility and lifetime of the materials. Low exciton binding energies, low effective mass and high mobilities are figure of merit for a material to have exceptional optoelectronic properties. The demerits of using Pb in the system are that it is toxic and can be ecologically hazardous. The maximum level of lead ion content set by the United States Environmental Protection Agency is at the level of 0.15 and 15 μgL^{-1} in air and water, respectively.[2] The solubility product of Pb^{2+} in PbI_2 (10^{-8}) is much higher than the conventional Cd^{2+} in CdTe (10^{-22}),[3] hence there is a very high chance that the degradation of perovskites can cause ecological damage if they seep into groundwater. Also, the degradation of conventional $MAPbI_3$ under light, ambient conditions[4-6] drive researchers to find material systems which are more stable. Finding replacements for Pb become a priority. An obvious choice is to look at elements from the same periodic table as Pb, such as Sn, Ge, which are less toxic. They are expected to have similar electronic and optical properties as the lead-based perovskites. First principle calculations have shown that Bi^{3+} and Sb^{3+} can form $B_2X_9^{3-}$ anions consisting of edge or face-sharing octahedra and can bind with alkali metals resulting in several crystal structures and optical bandgap.[7] A substitution of lead with nontoxic and environmentally benign elements forming lead-free metal halide perovskites can be generally achieved via two approaches:

i. By homovalent substitution of lead with isovalent cations, such as group-14 elements (Ge, Sn), alkaline-earth metals (Mg, Ca, Sr, Ba), transitions metals (Mn, Fe, Ni, Pd, Cu, Cd), and lanthanides (Eu, Tm, Yb).[8]

ii. By heterovalent substitution with aliovalent metal cations, such as transition metals (Au), main group elements (Tl, Sb, Bi, Te), lanthanides (La, Ce, Pr, Nd, Sm, Eu, Gd, Dy, Er, Tm, Lu), and actinides (Pu, Am, Bk). However, a successful replacement of the divalent lead cation can be accomplished via a mixed valence approach, that is, an equal proportion of mono- and trivalent metal cations to give an overall divalent state in average to balance the total charge and valence. In addition, double halide perovskites ($A_2B^IB^{II}X_6$), which are based on the mixture of different mono- and trivalent metal cations, are a further approach towards heterovalent substitution. Another possible avenue is based on the mixture of higher valent metal cations and vacancies to accommodate the total charge neutrality, which is accompanied by a considerable change in structure leading to $A_3B_2X_9$-type perovskites (B = Sb, Bi).[8]

Homovalent and heterovalent substitution approaches lead to a wide range of lead-free metal halide perovskite semiconductors based on various elements in the periodic table (Figure 10.1).[8] This chapter will focus on the developments associated with these material systems, device performance, and current challenges of lead-free alternatives (Figure 10.2).

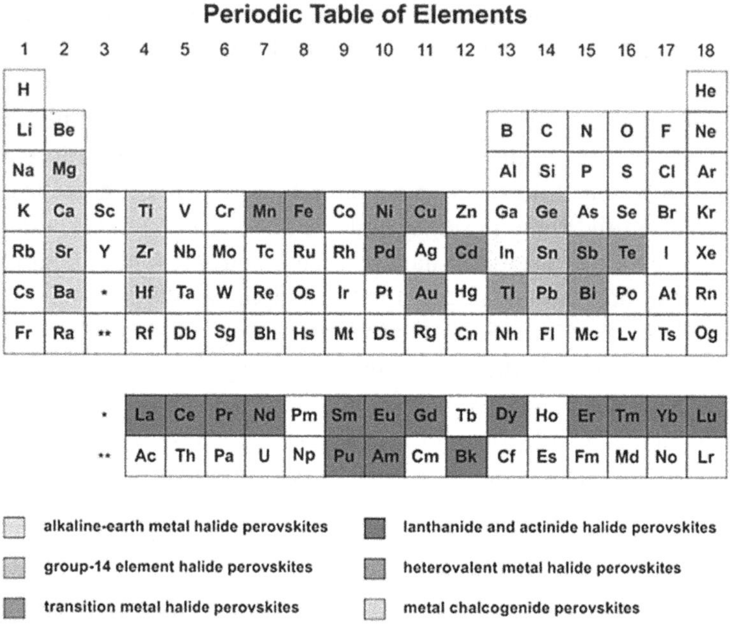

Periodic Table of Elements

FIGURE 10.1   Lead replacement options available from the periodic table.
Source: Adapted from ref 8.

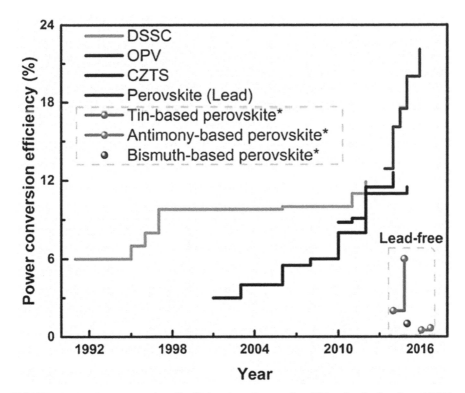

FIGURE 10.2 Best research cell efficiencies of emerging PV technologies from NREL website.
Source: Redrawn from ref. 9 (*also included are uncertified device efficiencies of lead-free perovskite solar devices from literature).

The last few years have seen a concerted effort to find lead-free tin-based perovskites for photovoltaic devices. It is both cheap and nontoxic, and being a group 14 metal, it is expected to show much of the same coordination chemistry and electronic properties found in the lead analogues. Sn^{2+} metal cations are the most obvious substitute for Pb^{2+} in the perovskite structure because of the similar s^2 valence electronic configuration to Pb^{2+} and the similar ionic radius (Pb^{2+}: 119 pm, Sn^{2+}: 110 pm).[10] Consequently, materials containing tin have long been sought after as a route to an earth-abundant and nontoxic solar absorber. Initially, partial substitution of Pb with Sn to form $CH_3NH_3Sn_xPb_{1-x}I_3$ enabled a reduction in bandgap relative to $MAPbI_3$, as seen in Figure 10.3. The absorption onset shifted from 1000 to 1200 nm when x increased from 0.3 to 1.0. It was also reported that an anomalous trend in the bandgap of compositions lying between $CH_3NH_3PbI_3$

and $CH_3NH_3SnI_3$ may have the same origin as the similar anomalous trend observed in $Pb_{1-x}Sn_xTe$.[11] In these chalcogenide systems, a band inversion occurs with varying x which is associated with a systematic change in the atomic orbital composition of the conduction and valence bands. This is because the conduction band minimum in SnTe has a similar orbital composition as the valence band maximum in PbTe and vice versa.[11] The device structure of the solar cells consisted of fluorine-doped tin oxide (FTO)/ compact TiO_2/mesoporous (mp)TiO_2/perovskite/P3HT/Au. The device performance of the solar cells using varying compositions of Sn: Pb can be found in Figure 10.4a.

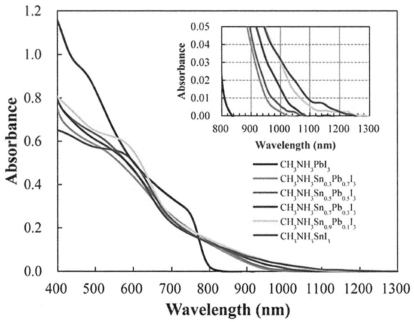

FIGURE 10.3   Absorption spectra of $CH_3NH_3Pb_{(1-x)}Sn_xI_3$ films deposited on porous TiO_2.
Source: Adapted with permission from ref 12. © 2014 American Chemical Society.

As the amount of Pb increased in the perovskite mixture, the conductivity of the Sn-Pb mixture reduced and lead to a more diode-like behavior as is seen from the J–V curves. The pure Sn-based perovskite was found to show a metal-like behavior. The incident photon-to-current efficiency (IPCE) of the devices is shown in Figure 10.4b. The mixtures of the Sn-Pb show an anomalous behavior with the onset of IPCE in the infrared region up to 1050 nm. This was the first demonstration of a working solar cell based on Sn-Pb mixture.

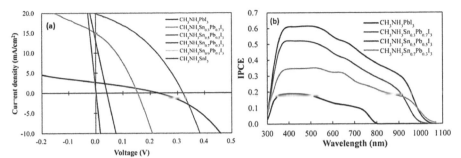

FIGURE 10.4   (a) J–V curves for $CH_3NH_3Sn_xPb_{(1-x)}I_3$ perovskite solar cells (PSCs) and (b) incident photon-to-current efficiency (IPCE) spectrum of the corresponding cells. *Source:* Adapted with permission from ref 12. © 2014 American Chemical Society.

The most studied tin halide perovskites are $CH_3NH_3SnI_3$ and $CH(NH_2)_2SnI_3$. In analogy to the lead halide perovskites, the structural properties of the tin-based perovskites, that is, dimensionality and connectivity of the perovskite lattice can be greatly affected by the size and functionality of the A-site cation as well as by the halide used. Small monovalent A-site cations (e.g. $CH_3NH_3^+$, $CH(NH_2)_2^+$, Cs^+) lead to the formation of three-dimensional structures, whereas larger ones (e.g. cyclobutylammonium, tropylium) cause a reduced dimensionality such as two-dimensional layered, one-dimensional chain-like, or zero-dimensional structures. The structural changes lead to variations in optoelectronic properties of perovskite systems. The first study on an entirely lead-free tin halide perovskite semiconductor used as absorber material, namely methylammonium tin iodide ($CH_3NH_3SnI_3$), was reported by Noel et al.[13] Simultaneously, there were two other reports on tin perovskite by Hao et al. and Mulmudi et al.[14,15] In the work by Noel et al., solar cells yielding power conversion efficiency (PCE) values over 6% were prepared in the device architecture glass/FTO/c-TiO_2/mp-TiO_2/$CH_3NH_3SnI_3$/Spiro-OMeTAD/Au (FTO, c: compact, mp). The reproducibility of devices was an issue which had to be addressed. Using mp TiO_2 has been beneficial due to the shorter charge carrier diffusion lengths of the tin halide perovskite compared to the lead-based analogue. Because of the challenging stability of tin halide perovskites, solar cell preparation had to be performed entirely in inert atmosphere starting from highly pure precursor materials. It is also remarkable that an open-circuit voltage (V_{OC}) of 0.88 V was obtained using an absorber material which has a relatively low bandgap of 1.23 eV. The obtained short-circuit current density (J_{SC}) was 16.8 mA cm^{-2} and the fill factor (FF) was 42%. Hao et al., who substituted the I$^-$ anion with other halides, produced a range of different tin halide perovskite analogues $CH_3NH_3SnX_3$ (X=Cl, Br) with a calculated band 1.7–3.0 eV.[14] By variation of the I: Br ratio, the optical bandgap can be engineered between 1.3 eV ($CH_3NH_3SnI_3$) and 2.15 eV ($CH_3NH_3SnBr_3$). Based on this approach, Hao et al. reported a mixed iodide–bromide tin perovskite semiconductor ($CH_3NH_3SnIBr_2$) with an optical bandgap of 1.75 eV yielding a PCE of 5.73% in meso-structured perovskite solar cells (PSCs).[14] Mulmudi et al. used inorganic $CsSnI_3$ as an absorber in solid-state PSCs. The high short-circuit photocurrents (22 mAcm^{-2}) stem from the fact that the absorption has been extended to 950 nm (Figure 10.5). Addition of SnF_2 into $CsSnI_3$ was found to be critical in attaining photovoltaic performance through the reduction of Sn vacancies, thereby rendering the $CsSnI_3$ less conducting in nature. The addition of 20% SnF_2 to $CsSnI_3$ was found to positively influence the solar cell performance in meso-structured PSCs and a PCE of 2.02% was obtained. IPCE was measured which shows the extension of the absorption to the near infrared region and an integrated photocurrent density of 20 mAcm^{-2} (Figure 10.5b). In another work that followed, controlled grain-coarsening of $CsSnI_3$ films based on heat treatment and using a planar device architecture (NiO_x as hole transport layer (HTL), PCBM as electron transport layer (ETL)) solar cells with a PCE of 3.31% have been reported by Wang et al.[16]

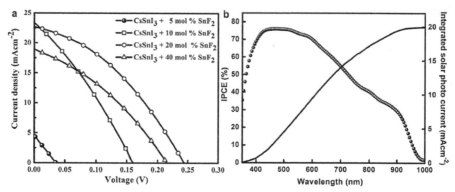

FIGURE 10.5 (a) J–V curves of photovoltaic devices fabricated with different amounts of SnF_2 addition, indicating how SnF_2 is necessary for functional devices; (b) representative IPCE spectrum of a 20% SnF_2 -$CsSnI_3$ device showing a photocurrent response extending to the infrared region ~950 nm. The integrated solar photocurrent density sums up to ~20 mAcm^{-2} in good agreement with the current densities obtained from the particular solar cell (J_{sc}=19.6 mAcm^{-}, V_{oc}=0.188 V, FF=0.27, η=1.02%).
Source: Adapted from ref 15.

CsSnBr$_3$ possesses a direct bandgap of 1.75 eV and solar cells with efficiencies of up to 2.1% have been reported using this material as an absorber layer.[17] However, CsSnBr$_3$-based solar cells currently suffer from a low V$_{OC}$ (up to 420 mV) stemming most likely from a mismatch of the energy levels of the materials (TiO$_2$, CsSnBr$_3$, Spiro-OMeTAD) used in these devices (Figure 10.6), which gives space for further optimization by investigating better suited ETLs and HTLs. Mixed chloride/bromide cesium tin halide perovskites reveal PCE values of up to 3.2% as well as good thermal and device stabilities.[18] The increase in open circuit voltage of tin-based perovskite is obtained by incorporation of Br anion in CsSnI$_{3-x}$Br$_x$.[18] The crystal structure transforms from orthorhombic to cubic for CsSnI$_3$ to CsSnBr$_3$. The optical bandgap of the perovskite was found to increases by 38.5% with a blue shift in the absorption spectra when I− is progressively replaced by Br−. Supplementing the increment in the bandgap, low charge carrier densities (6.32 × 1015 cm^{-3}) and increased charge recombination resistance aided in achieving high open circuit voltages of more than 400 mV.[18] Addition of SnF$_2$ complemented by 100°C post-annealing of perovskite was found to be beneficial, contributing to the attainment of high photocurrent densities. The transition of nearly linear (for iodine rich) to perfectly linear (for bromine-rich) dependence of short circuit current densities with respect to varying light intensities reflects the reduction of Sn cation vacancies by Br− addition, which are the dominant defects in CsSnI$_3$.

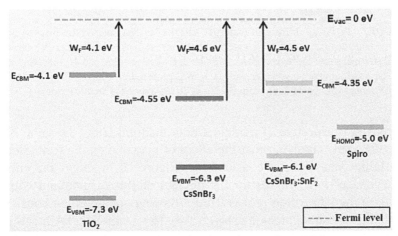

FIGURE 10.6 Energies of the valence band maximum (from UPS measurements) and conduction band minimum (calculated from the UPS data and the optical bandgap, i.e., neglecting the exciton binding energy) and the Fermi level (from UPS data) of dense TiO$_2$, pristine CsSnBr$_3$, and CsSnBr$_3$ (with 20 mol % SnF$_2$).

Source: Adapted from ref 17.

By introducing the CH $(NH_2)_2^+$ ion into tin iodide perovskites forming CH $(NH_2)_2SnI_3$, the bandgap is widened to 1.41 eV. CH $(NH_2)_2SnI_3$ has, in contrast to CH $(NH_2)_2PbI_3$, only one thermally accessible phase, which is stable up to 200°C. By adding SnF_2, which increases the stability of Sn^{2+}, a PCE of 2.1% could be obtained.[19] Further optimization using the SnF_2–pyrazine complex causing a more homogeneous distribution of SnF_2 in the perovskite led to PCE values of 4.8%.[20] A single phase of $FASnI_3$ with a black color was obtained upon annealing, which has a well-matched X-ray diffraction (XRD) pattern with the previously reported orthorhombic crystal structure of $FASnI_3$ (Figure 10.7a), and there is no significant peak originating from SnF_2 in the film. Unlike $FAPbI_3$, this phase of $FASnI_3$ is kept at room temperature for several days without any phase transition.

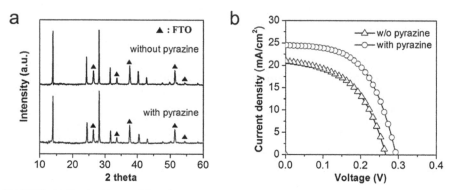

FIGURE 10.7   (a) X-ray diffraction patterns (XRD) of FASnI3 perovskite films with and without pyrazine deposited on FTO glass; (b) Photocurrent density–voltage (J–V) curves of FASnI3 PSCs measured by reverse scan with 10 mV voltage steps and 10 ms delay times under AM 1.5 G illumination when $FASnI_3$ layer was prepared.
Source: Reprinted with permission from ref 20. © 2016 American Chemical Society.

Moreover, irrespective of pyrazine, both the films show the same XRD pattern of $FASnI_3$, implying that the effect of pyrazine on the formation of the crystalline phase is not considerable. Figure 10.7b shows the current density–voltage (J–V) curves for the devices fabricated with and without pyrazine. In the fabrication process, the following configuration consisting of the FTO substrate/blocking layer (bl)-TiO_2/ mp-TiO_2/$FASnI_3$/Spiro-OMeTAD/Au was used. A cross-sectional scanning electron microscopic image of the whole device indicates the deposition of the dense and flat $FASnI_3$ absorbing layer on the mp-TiO_2 scaffold; the $FASnI_3$ perovskite forms a thick capping layer (300 nm) with full infiltration into mp-TiO_2

(400 nm), and a thin Spiro-OMeTAD layer (200 nm) exists as an overlayer. In the figure, we see that the PCE of FASnI$_3$ PSC fabricated without pyrazine is 2.8% with short-circuit current density (J$_{sc}$) of 20.9 mA/cm^2, an open-circuit voltage (V$_{oc}$) of 0.26 V, and an FF of 50%. In contrast, the PCE of a device fabricated in the presence of pyrazine in the mixture solution was improved to 4.0% (J$_{sc}$ = 21.5 mA/cm^2, V$_{oc}$ – 0.29 V, and FF – 55%).

The efficiency shows good agreement with the steady-state power output, measured at a maximum power voltage of 0.202 V. Moreover, no hysteresis behavior in the J–V curves was observed in both reverse (from V$_{oc}$ to J$_{sc}$) and forward (from J$_{sc}$ to V$_{oc}$) scans. Recently, the efficiency of this material could be further increased to 6.22%.[21] In this study, poly polystyrene sulfonate was used as hole selective contact and fullerene (C60) was used as ETL in contrast to both the aforementioned reports, in which Spiro-OMeTAD and TiO$_2$ have been used. Moreover, this study points out that similar to the lead-based perovskites, solvent treatment during spin-coating is crucial for the performance of tin-based PSCs and diethyl ether dripping was found to give the best results in terms of PCE and reproducibility. Song et al. presented an elegant and effective process that involves a reducing vapor atmosphere during the preparation of Sn-based halide PSCs to solve this problem of Sn^{2+} oxidation to Sn^{4+}, using MASnI$_3$, CsSnI$_3$, and CsSnBr$_3$ as the representative absorbers.[22] This process enables the fabrication of remarkably improved solar cells with PCEs of 3.89, 1.83, and 3.04% for MASnI$_3$, CsSnI$_3$, and CsSnBr$_3$, respectively. In this work, hydrazine which is known as a strong reducing agent and also a strong base was used to prevent or suppress the high-oxidation Sn^{4+} formation. The schematic for reducing the vapor pressure is shown in Figure 10.8. The respective J–V curves are also shown. All devices show significant PCE improvement when hydrazine vapor is employed.

In previous reports for MASnI$_3$ and CsSnI$_3$, several factors have been suggested to account for the excess p-doping, even shorting behavior in devices, including the impurity of the SnI$_2$ starting material and unexpected defect formation as well as oxidized Sn (Sn^{4+}) impurity.[23,24] Without the hydrazine vapor reaction, we obtain similar results with MASnI$_3$ and CsSnI$_3$ devices showing near shorting behavior, even with the SnF$_2$ additive. However, with proper amounts of hydrazine vapor, the PCEs of MASnI$_3$ and CsSnI$_3$ devices were improved dramatically, from an average of ~0.02–3.40% and ~0.16–1.50%, respectively. This trend is also observed in CsSnBr$_3$ devices, increasing from ~2.36 to 2.82%, mainly due to the improved V$_{oc}$. CsSnBr$_3$ was found to be more stable than the other two perovskites, and it was not required to encapsulate the $CsSnBr_3$ devices for measurements in air for a short time'.

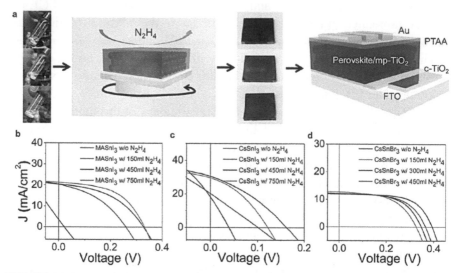

FIGURE 10.8 Top: Scheme of reducing vapor atmosphere process of device fabrication and photovoltaic performances; (a) reducing vapor atmosphere procedure for preparing $MASnI_3$ (top), $CsSnI_3$ (middle), and $CsSnBr_3$ (bottom) perovskite solar cell devices and device structure (glass/FTO/c-TiO$_2$/mp-TiO$_2$-perovskite/perovskite capping layer/PTAA/Au). Bottom: Representative J−V curves for (b) $MASnI_3$ solar cells, (c) $CsSnI_3$ solar cells, and (d) $CsSnBr_3$ solar cells without and with various hydrazine vapor concentrations.
Source: Reprinted with permission from ref 22. © 2017 American Chemical Society.

On the basis of chemical interactions between hydrazine and Sn-based halide perovskites, a possible reaction mechanism responsible for the observed performance change with schematic diagrams was proposed as shown in Figure 10.9. Hydrazine is well recognized as a strong reagent to reduce metal salts and metal oxides to pure metals, as well as an oxygen scavenger ($N_2H_4 + O_2 \rightarrow N_2 + 2H_2O$) to decrease surrounding oxygen levels. With these points taken into consideration, hydrazine in its vapor form was used, instead of directly adding it into the perovskite precursor solutions, to prevent the full reduction of Sn halide salts to Sn metal. Therefore, by generating the more diluted (relative to the form in solution) hydrazine atmosphere during the device fabrication process, not only could the Sn^{4+} impurities be reduced but also the unfavorable reduction of Sn^{2+} could be suppressed.

Reduction

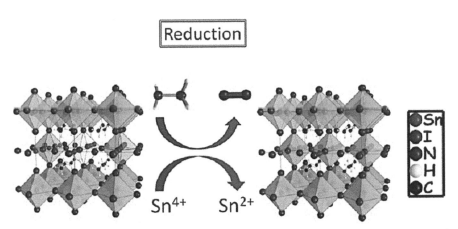

FIGURE 10.9 Proposed possible mechanism of hydrazine vapor reaction with Sn-based perovskite materials.
Source: Reprinted with permission from ref 22. © 2017 American Chemical Society.

The possible reaction path during the thin-film formation is given as $2SnI_6^{2-} + N_2H_4 \rightarrow 2SnI_4^{2-} + N_2 + 4HI$. The reduction of Sn^{4+} to Sn^{2+} decreases the amount of Sn^{2+} vacancies (V_{Sn}), thus suppressing the undesirable p-type electrical conductivity and reverting the perovskites back to the desirable semiconducting behavior. The smaller amount of V_{Sn} would lead to decreased cation antisite (A_{Sn}) and anion antisite (X_{Sn}) formation, further contributing to a lower level of p-type doping. This mechanism could account for the observed device performance improvements from shorting diodes to working diodes, especially in $MASnI_3$ and $CsSnI_3$, and better device performances in $CsSnBr_3$. New strategies and innovations in material design and device engineering are sought to pave pathways for tin-based solar devices with enhanced stability and high power conversion efficiencies.

KEYWORDS

- lead free perovskite
- tin based perovskite
- antimony based perovskite
- bismuth based perovskite
- double perovskite

REFERENCES

1. Brandt, R. E.; et al. Identifying Defect-Tolerant Semiconductors with High Minority-Carrier Lifetimes: Beyond Hybrid Lead Halide Perovskites. *MRS Commun.* **2015,** *5*(2), 265–275.
2. http://www2.epa.gov/lead/lead-laws-and-regulations.
3. Patnaik, P. *Handbook of Inorganic Chemicals;* McGraw-Hill: New York, 2003; Vol. 529.
4. Bryant, D.; et al. Light and Oxygen Induced Degradation Limits the Operational Stability of Methylammonium Lead Triiodide Perovskite Solar Cells. *Energy Environ. Sci.* **2016,** *9*(5), 1655–1660.
5. Divitini, G.; et al. In Situ Observation of Heat-Induced Degradation of Perovskite Solar Cells. *Nat. Energy* **2016,** *1*, 15012.
6. Song, Z.; et al. In-Situ Observation of Moisture-Induced Degradation of Perovskite Solar Cells Using Laser-Beam Induced Current. *2016 IEEE 43rd Photovoltaic Specialists Conference (PVSC).* 2016.
7. Xiao, Z.; et al. Thermodynamic Stability and Defect Chemistry of Bismuth-Based Lead-Free Double Perovskites. *ChemSusChem* **2016,** *9*(18), 2628–2633.
8. Hoefler, S.F.; Trimmel, G.; Rath, T. Progress on Lead-Free Metal Halide Perovskites for Photovoltaic Applications: A Review. *Monatsh Chem.* **2017,** *148,* 795. https://doi.org/10.1007/s00706-017-1933-9.
9. Best Research-Cell Efficiencies, NREL, http://www.nrel.gov/pv/assets/images/efficiency_chart.jpg.
10. Chen, Q.; et al. Under the Spotlight: The Organic–Inorganic Hybrid Halide Perovskite for Optoelectronic Applications. *Nano Today* **2015,** *10*(3), 355–396.
11. Hao, F.; et al. Anomalous Band Gap Behavior in Mixed Sn and Pb Perovskites Enables Broadening of Absorption Spectrum in Solar Cells. *J. Am. Chem. Soc.* **2014,** *136*(22), 8094–8099.
12. Ogomi, Y.; et al. $CH_3NH_3Sn_xPb_{(1-x)}I_3$ Perovskite Solar Cells Covering up to 1060 nm. *J. Phys. Chem. Lett.* **2014,** *5*(6), 1004–1011.
13. Noel, N. K.; et al. Lead-Free Organic-Inorganic Tin Halide Perovskites for Photovoltaic Applications. *Energy Environ. Sci.* **2014,** *7*(9), 3061–3068.
14. Hao, F.; et al. Lead-Free Solid-State Organic-Inorganic Halide Perovskite Solar Cells. *Nat. Photon.* **2014,** *8*(6), 489–494.
15. Kumar, M. H.; et al. Lead-Free Halide Perovskite Solar Cells with High Photocurrents Realized Through Vacancy Modulation. *Adv. Mater.* **2014,** *26*(41), 7122–7127.
16. Wang, N.; et al. Heterojunction-Depleted Lead-Free Perovskite Solar Cells with Coarse-Grained B-γ-CsSnI3 Thin Films. *Adv. Energy Mater.* **2016,** *6*(24), n/a–n/a.
17. Gupta, S.; et al. CsSnBr3, A Lead-Free Halide Perovskite for Long-Term Solar Cell Application: Insights on SnF2 Addition. *ACS Energy Lett.* **2016,** *1*(5), 1028–1033.
18. Sabba, D.; et al. Impact of Anionic Br – Substitution on Open Circuit Voltage in Lead Free Perovskite (CsSnI3-xBrx) Solar Cells. *J. Phys. Chem. C* **2015,** *119*(4), 1763–1767.
19. Koh, T. M.; et al. Formamidinium Tin-Based Perovskite with Low Eg for Photovoltaic Applications. *J. Mater. Chem. A* **2015,** *3*(29), 14996–15000.
20. Lee, S. J.; et al. Fabrication of Efficient Formamidinium Tin Iodide Perovskite Solar Cells through SnF_2 – Pyrazine Complex. *J. Am. Chem. Soc.* **2016,** *138*(12), 3974–3977.

21. Liao, W.; et al. Lead-Free Inverted Planar Formamidinium Tin Triiodide Perovskite Solar Cells Achieving Power Conversion Efficiencies up to 6.22%. *Adv. Mater.* **2016,** *28*(42), 9333–9340.
22. Song, T.-B.; et al. Importance of Reducing Vapor Atmosphere in the Fabrication of Tin-Based Perovskite Solar Cells. *J. Am. Chem. Soc.* **2017,** *139*(2), 836–842.
23. Takahashi, Y.; et al. Charge-Transport in Tin-Iodide Perovskite $CH_3NH_3SnI_3$: Origin of High Conductivity. Dalton Trans. **2011,** *40*(20), 5563–5568.
24. Takahashi, Y.; et al. Hall Mobility in Tin Iodide Perovskite $CH_3NH_3SnI_3$: Evidence for a Doped Semiconductor. *J. Solid State Chem.* **2013,** *205*, 39–43.

CHAPTER 11

GERMANIUM-BASED PEROVSKITES

Another potential candidate for the substitution of lead in the perovskite structure is the group-14 metalloid germanium. In comparison to Pb^{2+}, the divalent germanium cation (Ge^{2+}) is in the same oxidation state but exhibits a lower electronegativity, a more covalent character and an ionic radius (73 pm) lower than Pb^{2+} (119 pm).[1] Nevertheless, Goldschmidt tolerance factor calculations support the formation of germanium halide perovskites, as shown for $CH_3NH_3GeX_3$ (X=Cl, Br, I) compounds with tolerance factor values of 1.005 ($CH_3NH_3GeCl_3$), 0.988 ($CH=NH_3GeBr_3$), and 0.965 ($CH_3NH_3GeI_3$), which coincide with t values reported for the ideal perovskite structure.

Theoretical calculations using density functional theory methods show that germanium halide perovskites have high absorption coefficients as well as similar absorption spectra and carrier transport properties as the lead analogs.[2,3] First-principle calculations of $CsGeX_3$ (X=Cl, Br, I) perovskites show that the bandgaps depend on the halide ion, that is, $CsGeCl_3$ (3.67 eV) > $CsGeBr_3$ (2.32 eV) > $CsGeI_3$ (1.53 eV).[4] Sun et al. extended the theoretical investigations to hybrid germanium halide perovskites, namely to $CH_3NH_3GeX_3$ (X=Cl, Br, I) compounds.[4] The calculated bandgaps based on Perdew–Burke–Ernzerhof functions were found to show a similar trend as for the cesium-based compounds, that is, $CH_3NH_3GeCl_3$ (3.76 eV) > $CH_3NH_3GeBr_3$ (2.81 eV) > $CH_3NH_3GeI_3$ (1.61 eV).

Germanium halide perovskites, however, have rarely been investigated experimentally, which is presumably due to the chemical instability upon oxidation of the divalent Ge^{2+} cation.[3] Due to the reduced inert electron pair effect, this oxidation stability issue is even more prominent in germanium-based perovskites than in tin-based ones. Stoumpos et al. thoroughly investigated the structural, electronic, and optical properties of germanium halide perovskites with the basic formula $AGeI_3$ incorporating Cs^+ and different organic A-site cations.[3] Depending on the A-site cation, different structures can be formed. Small cations such as Cs^+, $CH_3NH_3^+$, or

CH $(NH_2)_2{}^+$ ions form three-dimensional perovskite frameworks based on $GeI_6{}^{4-}$ corner-sharing octahedra. Bigger A-site cations (e.g. guanidinium, trimethylammonium) lead to distortions of the crystal structure and one-dimensional chain-like hexagonal perovskite structures ($CsCdBr_3$-type) consisting of $GeI_6{}^{4-}$ face-sharing octahedra being formed. Using the n-butylammonium ion as A-site cation, the orthorhombic perovskite $(C_4H_9NH_3)_2GeI_4$ is formed exhibiting a two-dimensional structure in which perovskite sheets consisting of corner-sharing GeI_6 octahedra are separated by bilayers of n-butylammonium cations.[5,6] $CsGeI_3$ and $CH_3NH_3GeI_3$ have already been implemented as absorber materials in meso-structured perovskite solar cells yielding power conversion efficiency (PCE) values of 0.11 and 0.20%, respectively.[2]

This moderate performance might be due to the oxidation of Ge^{2+} to Ge^{4+} Already occurring during the fabrication of the solar cell and the limited open-circuit voltage (V_{OC}), in particular of the $CsGeI_3$ (74 mV), was suggested to originate from the defect chemistry in this material.[2] In a patent by Huang et al., a PCE of 3.2% in a meso-structured perovskite solar cell architecture is claimed, with the following photovoltaic parameters: short-circuit current density (J_{sc}) = 10.49 $mAcm^{-2}$, V_{oc} = 0.57V, fill factor (FF) = 0.53.[7] This value is still much lower compared to the theoretically possible PCE values of 27.9% predicted by Shockley-Queisser limit and further innovations in material development and interfaces have to be made to improve the efficiency of germanium-based perovskites to competitive values.[8]

KEYWORDS

- germanium halide perovskites
- open-circuit voltage
- bandgaps
- power conversion efficiency
- fill factor

REFERENCES

1. Shannon, R.; Shannon, R. D.; Prewitt, C. T. Effective Ionic Radii in Oxides and Fluorides. *Acta Crystallogr. B* **1969**, *25*, 925–46. [Central Research Dept. Experimental

Station, EI Du Pont de Nemours and Co., Wilmington, DE and Dept. Earth and Space Sci., SUNY, Stony Brook, NY].

2. Krishnamoorthy, T.; et al. Lead-Free Germanium Iodide Perovskite Materials for Photovoltaic Applications. *J. Mater. Chem. A* **2015**, *3*(47), 23829–23832.

3. Stoumpos, C. C.; et al. Hybrid Germanium Iodide Perovskite Semiconductors: Active Lone Pairs, Structural Distortions, Direct and Indirect Energy Gaps, and Strong Nonlinear Optical Properties. *J. Am. Chem. Soc.* **2015**, *137*(21), 6804–6819.

4. Sun, P.-P.; et al. Theoretical Insights into a Potential Lead-Free Hybrid Perovskite: Substituting Pb^{2+} with Ge^{2+}. *Nanoscale* **2016**, *8*(3), 1503–1512.

5. Liao, W.-Q.; et al. A Lead-Halide Perovskite Molecular Ferroelectric Semiconductor. *Nat. Commun.* **2015**, *6*, 7338.

6. Dai, Y. D.; et al. The Band Structure Analysis of the Hybrid Compound $(C_4H_9NH_3)_2GeI_4$. In *Solid State Phenomena;* Sun, C.-J., Ding, J., Gupta, M., Chow, G.-M., Kurihara, L., Kabacoff, L., Eds.; Trans Tech Publications: Switzerland, 2006; Vol. 111, pp 51–54.

7. Novel Germanium-Containing Perovskite Material and Solar Cell Comprising Same. Google Patents, 2014.

8. Rau, U.; Werner, J. H. Radiative Efficiency Limits of Solar Cells with Lateral Band-Gap Fluctuations. *Appl. Phys. Lett.* **2004**, *84*(19), 3735–3737.

CHAPTER 12

COPPER-BASED PEROVSKITES

Copper is a nontoxic, low-cost earth abundant transition metal. The divalent Cu^{2+} cation is of particular interest for the incorporation into the perovskite structure as a replacement for Pb^{2+} because of its ambient stability and the high absorption coefficient in the visible region.[1] Cu^{2+} has a $3d^9\ 4s^0$ ($t^6_{2g}\ e^3_g$) electronic configuration different to the group-14 main group metal cations of Sn^{2+} and Pb^{2+}, that is, lone pair electrons, which has a considerable effect on the electronic band structure.[1,2] Due to the smaller ionic radius of Cu^{2+} (73 pm) compared to Pb^{2+} (119 pm) or Sn^{2+} (110 pm), the formation of three-dimensional structures is sterically hindered, and thus, hybrid copper halide perovskites form two-dimensional layered structures, which are isostructural to Ruddlesden–Popper phase compounds.[3] These hybrid perovskites have the general formula $(R\text{-}NH_3)_2CuX_4$ incorporating monovalent ammonium cations (R = alkyl, aryl) and halide counterions.[4] The two-dimensional structures form inorganic layers of corner-sharing BX_6 octahedra separated by monolayers of organo-ammonium cations on either side of the metal halide sheets, which are accommodated within the voids of the inorganic framework.[5] The layered structure is stabilized by hydrogen bonding interactions between the ammonium groups and the halogen atoms and by van der Waals interactions between the interdigitating organic moieties.[5] Each successive inorganic perovskite sheet is shifted to give a "staggered" configuration of the layers. Examples are $(CH_3(CH_2)_3NH_3)_2CuBr_4$ and $(p\text{-}F\text{-}C_6H_5C_2H_4NH_3)_2CuBr_4$.[6] A further way to stabilize the layered structure is the use of organic di-ammonium cations (NH^{3+}, $R\text{-}NH^{3+}$, R = alkyl, aryl) in $(NH_3\text{-}R\text{-}NH_3)CuX_4$ compounds.[7] Di-ammonium-based layered structures feature hydrogen bonding interactions of both functional ammonium groups to halogen atoms of the inorganic sheets resulting in an "eclipsed" arrangement of the layers, which are separated by a single organic layer instead of a double or bilayer. The distance between adjacent inorganic layers can be influenced by the length of the organic spacer R, which eventually affects the compound's dimensionality and physical properties.[7]

Cortecchia et al. reported on two-dimensional copper halide perovskites with the general formula $(CH_3NH_3)_2CuCl_xBr_{4-x}$ with a varying Br: Cl ratio, Figure 12.1. Ligand-to-metal charge transfer transitions and Cu d–d transitions influence the absorption properties of this material. In addition, the optical bandgap was found to be tunable via the Br: Cl ratio within the visible to near-infrared range with a shift for higher bromide content: $(CH_3NH_3)_2CuCl_4$ (2.48 eV), $(CH_3NH_3)_2CuCl_2Br_2$ (2.12 eV), $(CH_3NH_3)_2CuClBr_3$ (1.90 eV), and $(CH_3NH_3)_2CuCl_{0.5}Br_{3.5}$ (1.80 eV). The prepared $(CH_3NH_3)_2CuCl_xBr_{4-x}$ perovskites were investigated in solar cells using thick (5 µm) mesoporous TiO_2 scaffolds giving photon-to-current efficiency values of 0.0017% $((CH_3NH_3)_2CuCl_{0.5}Br_{3.5})$ and 0.017% $((CH_3NH_3)_2CuCl_2Br_2)$. However, the photovoltaic performance of layered copper halide perovskites, in general, is limited by various factors including low absorption coefficients, the high effective mass of holes and the intrinsic low conductivity of two-dimensional perovskite structures.

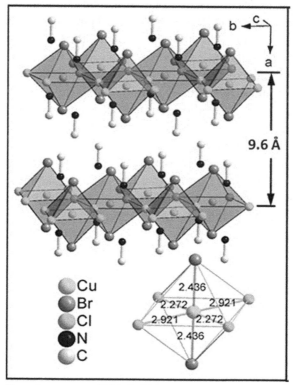

FIGURE 12.1   Crystal structure of $MA_2CuCl_2Br_2$, showing the alternation of organic and inorganic layers and the Cu–X bond lengths in the inorganic framework.
Source: Reprinted with permission from ref. 1. © 2016 American Chemical Society.

The choice of the halide anion plays a key role not only in the engineering of the bandgap but also is essential with regard to the material's stability, film formation properties, and photovoltaic performance. Bromide is responsible for the partial reduction of Cu^{2+} to Cu^+ in the perovskite framework, which is accompanied by the formation of anion vacancies. These crystallographic defects act as electron traps and limit the photovoltaic performance since an additional charge recombination pathway is introduced. This is supported by Cortecchia et al. who found a pronounced photoluminescence with higher bromine contents resulting from the in situ formation of Cu^+ ions and, the corresponding charge carrier recombination at the charge traps, Figure 12.2.

Chloride was found to be essential for the material's stability against the copper reduction and to improve the crystallization of the perovskite accompanied by a shift of the optical bandgap. The presence of the Jahn–Teller active Cu^{2+} metal cation introduces an additional flexibility into the inorganic framework, which also affects hydrogen bonding interactions. This is based on the Jahn–Teller distortion of the CuX_6 octahedra leading to an elongation of two equatorial Cu–X bonds within the octahedral coordination. As a consequence, the layered perovskite adopts a structure in which adjacent Cu^{2+} ions are linked via one short (normal) and one Jahn–Teller elongated bond via a bridging halide ion. The normal bond length is relatively constant, whereas the semi-coordinate bond is considerably elastic, allowing the inorganic layers to adopt a more flexible structure, which enables the interaction with larger organic ammonium cations to be incorporated into the two-dimensional structure. Other layered perovskite analogues with Jahn–Teller active metal cations such as Cr^{2+} show similar structural distortions and ferromagnetic behavior.[8] The absorption spectra of the series $MA_2CuCl_xBr_{4-x}$ show typical features of copper complexes CuX_4^{2-} in square planar coordination (Figure 12.2a), are in agreement with the strong Jahn–Teller distortion. Strong bands for each material with absorption coefficients up to 35,000 cm^{-1} are found below 650 nm, and the corresponding bandgaps determined from Tauc plots (not shown here) are 2.48 eV (500 nm) for MA_2CuCl_4, 2.12 eV (584 nm) for $MA_2CuCl_2Br_2$, 1.90 eV (625 nm) for $MA_2CuClBr_3$, and 1.80 eV (689 nm) for $MA_2CuCl_{0.5}Br_{3.5}$. Upon excitation, the perovskite films showed photoluminescence, which peaked around 515 nm with increasing intensity for higher Br/Cl ratio (Figure 12.2b). The observed green fluorescence can be assigned to the emission of Cu^+ ions and suggests that Cu^{2+} is partially reduced during annealing creating emissive trap states in the material. The reduction process is strongly fostered by the presence of bromine, as suggested by the photoluminescence trend culminating in the stronger emission of $MA_2CuCl_{0.5}Br_{3.5}$, while chlorine helps to stabilize the Cu^{2+} oxidation state. The modulation of the bandgap appears evident from the color of the powders, which changes from yellow to dark brown by increasing the Br/Cl ratio (Figure 12.2c).

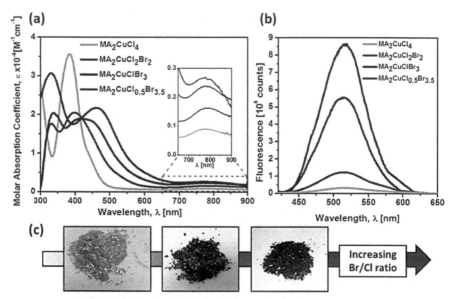

FIGURE 12.2 (a) Absorption coefficient for perovskites of the series $MA_2CuCl_xBr_{4-x}$ showing strong charge-transfer bands below 650 nm and broad d–d transitions between 700 and 900 nm (inset); (b) Photoluminescence of the perovskites $MA_2CuCl_xBr_{4-x}$ with the intensity increasing with higher Br contents; (c) color shift for powders with different Br/Cl ratio: MA_2CuCl_4 (yellow), $MA_2CuCl_2Br_2$ (red), $MA_2CuCl_{0.5}Br_{3.5}$ (dark brown).
Source: Reprinted with permission from ref. 1. © 2016 American Chemical Society.

Although in this work, the TiO_2 mesoporous scaffold was shown to help the electron extraction from the two-dimensional perovskite, the combination of the low absorption coefficient and heavy mass for the holes was found to compromise the solar cell efficiency. The introduction of optoelectronically active cations is seen to overcome these issues stressing the importance of investigation of novel hybrid materials and making two-dimensional copper-based hybrid perovskites an ideal platform to study the further developments.

KEYWORDS

- copper halide perovskites
- hydrogen bonding interactions
- hybrid perovskites
- photovoltaic
- mesoporous TiO_2 scaffolds

REFERENCES

1. Cortecchia, D.; et al. Lead-Free $Ma_2CuCl_xBr_{4-x}$ Hybrid Perovskites. *Inorg. Chem.* **2016**, *55*(3), 1044–1052.
2. Greenwood, N. N.; Earnshaw, A. (Eds.). Germanium, Tin and Lead. In *Chemistry of the Elements;* 2nd ed.; Butterworth-Heinemann: Oxford, 1997; pp 367–405.
3. Cortecchia, D.; et al. Crystal Engineering of a Two-Dimensional Lead-Free Perovskite with Functional Organic Cations by Second-Sphere Coordination. *ChemPlusChem* **2017**, *82*(5) 681–685.
4. Cheng, Z.; Lin, J. Layered Organic-Inorganic Hybrid Perovskites: Structure, Optical Properties, Film Preparation, Patterning and Templating Engineering. *CrystEngComm* **2010**, *12*(10), 2646–2662.
5. Peterson, E. R.; Willett, R. D. Crystal Structure of $(CH_3CH_2CH_2NH_3)_2MnCl_4$. *J. Chem. Phys.* **1972**, *56*(5), 1879–1882.
6. Cui, X.-P.; et al. Cupric Bromide Hybrid Perovskite Heterojunction Solar Cells. *Synth. Met.* **2015**, *209*, 247–250.
7. Mitzi, D. B. Synthesis, Structure, and Properties of Organic-Inorganic Perovskites and Related Materials. In *Progress in Inorganic Chemistry;* Karlin, K. D., Ed.; John Wiley & Sons, Inc., 2007; pp 1–121. Print ISBN:9780471326236, Online ISBN:9780470166499.
8. Guedel, H. U.; Snellgrove, T. R. Jahn-Teller Effect in the 4T2g State of Chromium(III) in Dicesium Sodium Indium(III) Hexachloride. *Inorg. Chem.* **1978**, *17*(6), 1617–1620.

CHAPTER 13

DOUBLE PEROVSKITES

Even though encouraging stability data has already been reported, the main drawback of tin halide perovskites is still the chemical instability of the divalent metal cation, which is due to the oxidation of Sn^{2+} to Sn^{4+} in ambient conditions. As a consequence, the oxidation of Sn^{2+} to the chemically more stable Sn^{4+} analog impedes the charge neutrality of the perovskite and causes the degradation of the perovskite by formation of oxides/hydroxides of tin, and furthermore, Sn^{4+} leads to hole doping of the material.[1,2] To avoid oxidation, inert processing and rigorous encapsulation of the tin-based perovskite devices are necessary. To overcome this oxidation stability issue, double perovskite semiconductors with a basic formula A_2SnX_6 (A=Cs, X=halide) have been explored in Figure 13.1. Double perovskites can also be obtained by transmuting two divalent Pb^{2+} ions into one monovalent ion M^+ and one trivalent ion M^{3+}, that is, $2Pb^{2+} \rightarrow M^+ + M^{3+}$, by keeping the total number of valence electrons unchanged at M^{IV} sites.

Considering that the $6s^2p^0$ configuration of Pb^{2+} is believed to be responsible for the unique optoelectronic properties of $AM^{IV}X^{VII}_3$ perovskites, the available choice of M^{3+} can be isoelectronic Bi^{3+} and Sb^{3+}, and M^+ can be any size-matching monovalent cations. Compatible with the smaller sizes of Bi^{3+}/Sb^{3+} than Pb^{2+}, the small inorganic cations (rather than the large organic cations commonly used in $AM^{IV}X^{VII}_3$) can be adopted at the A site to stabilize the perovskite lattice, opening the avenue for achieving the inorganic halide perovskites with good stabilities. Enormous progress in the development of novel lead-free perovskite semiconductors might arise from the heterovalent substitution approach since further non-divalent cations become amenable. Thallium is a p-block metal with a Tl^+ cation isoelectronic to Pb^{2+} ($6s^2 6p^0$ electronic configuration). The monovalent Tl^+ cation, however, cannot substitute the divalent Pb^{2+} metal cation directly in ABX_3-type perovskites because of the violation of the charge neutrality between cationic and anionic species. Nevertheless, the incorporation of thallium into the perovskite structure is possible via the mixed-valence approach using Tl^+ ($6s^2$) and Tl^{3+} ($6s^0$).[4] Another variation is the mixed thallium–bismuth halide perovskite $CH_3NH_3Tl_{0.5}Bi_{0.5}I_3$, where Pb^{2+} metal cation units of the lead-based analog $CH_3NH_3PbI_3$ are replaced with Tl^+/Bi^{3+} heterovalent ionic pairs. Theoretical investigations of this lead-free hybrid perovskite with regard to its structural and electronic properties via discrete Fourier transform analysis and predicted a direct bandgap of 1.68 eV.[5] According to these calculations, $CH_3NH_3Tl_{0.5}Bi_{0.5}I_3$ is predicted to be a potential alternative solar cell material. However, despite these quite promising considerations and optical properties, thallium-based compounds are presumably no alternative to lead-based perovskites in terms of photovoltaic applications due to the inherent toxicity of thallium.

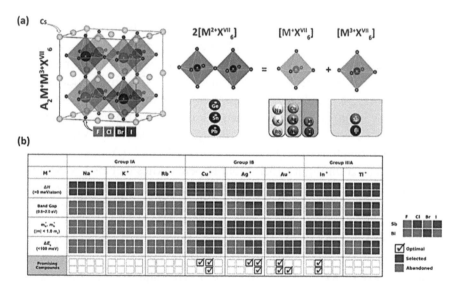

FIGURE 13.1 (a) Space of candidate $A_2M^+M^{3+}X_6$ perovskites for materials screening: left panel shows adopted double-perovskite structure, and the right panel shows the schematic idea of atomic transmutation; (b) Materials screening process by considering gradually the properties relevant to photovoltaic performance.
Source: Reprinted with permission from ref. 3. © 2017 American Chemical Society.

Gold halide perovskites are similar to thallium-based analogs amenable via the mixed-valence approach. Consequently, gold has to be present in a mixture of monovalent Au^+ ($5d^{10}$, t^6_{2g} e^4_g) and trivalent Au^{3+} ($5d^8$, t^6_{2g} e^2_g) metal cations to form ABX_3-type perovskite structures,[2] like in the case of $Cs_2Au^IAu^{III}X_6$ (X=Cl, Br, I) compounds.[6] Due to the presence of mono- and trivalent metal cations, two different coordination spheres are present in mixed-valent gold halide perovskites, that is, linear (twofold) and square planar (fourfold) coordination of Au^+ and Au^{3+}, respectively. In the case of $Cs_2Au^IAu^{III}X_6$ (X=Cl, Br, I), the crystal structure is derived from a distorted ABX_3-type perovskite consisting of BX_2 (linear $[Au^IX_2]^-$ unit) and BX_4 (square-planar $[Au^{III}X_4]^-$ unit) building blocks.[7] The BX_2 and BX_4 units arrange alternately to accomplish the nominal octahedral coordination in the perovskite structure. While linearly coordinated $[Au^IX_2]^-$ units are completed by neighboring $[Au^{III}X_4]^-$ units via four coplanar halide ions forming compressed octahedra, square-planar coordinated $[Au^{III}X_4]^-$ units are completed by apical $[Au^IX_2]^-$ units via two halide ions forming elongated octahedra. The resulting distorted three-dimensional perovskite network can, therefore, be expressed as Cs_2 $[Au^IX_2][Au^{III}X_4]$. The hybrid mixed-valent gold halide perovskites $[NH_3\text{-}R\text{-}NH_3]_2$ $[(Au^{II}_2)(Au^{III}I_4)(I_3)_2]$ (R=heptyl, octyl) feature inorganic metal halide sheets of corner-sharing octahedra which are separated by organic diammonium cations (e.g. $(NH_3(CH_2)_7NH_3)^{2+}$ and $(NH_3(CH_2)_8NH_3)^{2+}$) to give a layered two-dimensional structure. The nominal octahedral coordination of the Au^I center within the $[Au^II_2]^-$ units is accomplished by neighboring $[Au^{III}I_4]^-$ units via four coplanar halide ions forming compressed octahedra, while $[Au^{III}I_4]^-$ units are completed by two asymmetric triiodide ions (I_3^-) in the apical position forming elongated nominal octahedra. Mixed-valent gold halide perovskites such as $Cs_2Au^IAu^{III}X_6$ were predominantly investigated in terms of superconductivity. Further research studies mainly focus on the structural characterization as well as on the electronic and optical behavior. With regard to the optical properties, the choice of the halide counterion plays an essential role in bandgap engineering in mixed-valent systems such as $Cs_2Au^IAu^{III}X_6$ (X=Cl, Br, I). By substitution of chlorine with bromine or iodine, the optical bandgap can be shifted to lower values. Liu et al. determined the optical bandgaps of the corresponding perovskites via optical reflectivity measurements to be 2.04 eV (X=Cl), 1.60 eV (X=Br), and 1.31 eV (X=I).[8] $Cs_2Au^IAu^{III}I_6$, in particular, is a promising absorber material for photovoltaic applications due to the almost ideal bandgap according to the Shockley–Queisser limit, and the three-dimensional distorted ABX_3-type perovskite structure similar to lead-based analogues. However, this class of materials was not characterized with regard to its photovoltaic performance so far.

Three-dimensional perovskite structures containing bismuth have also been obtained in quaternary double perovskites with a basic formula unit of $A_2B^IB^{II}X_6$ by heterovalent substitution of Pb^{2+} by a combination of a monovalent cation such as Ag^+ (B^I) and a trivalent Bi^{3+} (B^{II}) cation. The double perovskite structure (elpasolite) is based on corner-sharing B^IX_6

and $B^{II}X_6$ octahedra alternating along the three crystallographic axes in a rock-salt ordered cubic structure. Cs_2AgBiX_6 (X=Cl, Br) perovskites, for example, can be synthesized via a solution-based or a solid-state reaction, crystallize in the elpasolite structure, and exhibit improved stability in terms of heat and moisture under ambient conditions compared to lead-based halide perovskites.[9,10] The family of pnictogen-noble metal halide double perovskites is especially interesting for photovoltaic applications because of the structural similarity, that is, three-dimensional structures, to lead-based perovskites despite the different valencies of the metal cations incorporated. In addition, a huge variety of material compositions is amenable due to the high number of possible element combinations of monovalent (B^I=Cu^+, Ag^+, Au^+) and trivalent (B^{II}=Sb^{3+}, Bi^{3+}) metal cations together with organic and inorganic cations (A), and halide anions (X).

KEYWORDS

- halide perovskites
- lead-freeperovskite semiconductors
- mixed-valence approach
- photovoltaic applications
- heterovalent substitution

REFERENCES

1. Takahashi, Y.; et al. Charge-Transport in Tin-Iodide Perovskite $CH_3NH_3SnI_3$: Origin of High Conductivity. *Dalton Transactions* **2011,** *40*(20), 5563–5568.
2. Takahashi, Y.; et al. Hall Mobility in Tin Iodide Perovskite $CH_3NH_3SnI_3$: Evidence for a Doped Semiconductor. *J. Solid State Chem.* **2013,** *205*, 39–43.
3. Zhao, X.-G.; et al. Design of Lead-Free Inorganic Halide Perovskites for Solar Cells via Cation-Transmutation. *J. Am. Chem. Soc.* **2017,** *139*(7), 2630–2638.
4. Saparov, B.; Mitzi, D. B. Organic–Inorganic Perovskites: Structural Versatility for Functional Materials Design. *Chem. Rev.* **2016,** *116*(7), 4558–4596.
5. Deng, Z.; et al. Exploring the Properties of Lead-Free Hybrid Double Perovskites Using a Combined Computational-Experimental Approach. *J. Mater. Chem. A* **2016,** *4*(31), 12025–12029.
6. Savory, C. N.; Walsh, A.; Scanlon, D. O. Can Pb-Free Halide Double Perovskites Support High-Efficiency Solar Cells? *ACS Energy Lett.* **2016,** *1*(5), 949–955.

7. Hoefler, S. F.; Trimmel, G.; Rath, T. Progress on Lead-Free Metal Halide Perovskites for Photovoltaic Applications: a Review. *Monatsh Chem.—Chem. Monthly 2017*, *148*, 1–32.

8. Liu, X. J.; et al. Electronic Structure of the Gold Complexes $Cs_2Au_2X_6$ (X=I, Br, and Cl). *Phys. Rev. B* **1999**, *59*(12), 7925–7930.

9. Xiao, Z.; et al. Thermodynamic Stability and Defect Chemistry of Bismuth-Based Lead-Free Double Perovskites. *ChemSusChem* **2016**, *9*(18), 2628–2633.

10. Slavney, A. H.; et al. A Bismuth-Halide Double Perovskite with Long Carrier Recombination Lifetime for Photovoltaic Applications. *J. Am. Chem. Soc.* **2016**, *138*(7), 2138–2141.

CHAPTER 14

BISMUTH-BASED PEROVSKITES

The group-15 metal bismuth is an interesting replacement candidate for lead and tin, which is supported by various aspects: the trivalent Bi^{3+} ion (1) is isoelectronic to Pb^{2+} ($6s^2$ $6p^0$ electronic configuration) featuring the same $6s^2$ lone pair, (2) shows a similar electronegativity (Bi: 2.02, Pb: 2.33, Sn: 1.96), and (3) has an ionic radius (103 pm) comparable to Pb^{2+} (119 pm), and Sn^{2+} (110 pm) metal cations.[1] However, the trivalent Bi^{3+} ion cannot directly replace the divalent Pb^{2+} ion in the perovskite structure due to the different valence states. Bismuth halide perovskites exhibit a huge structural diversity in terms of connectivity (face-, edge- or corner-sharing networks) and dimensionality ranging from zero-dimensional dimer units to one-dimensional chain-like motifs or two-dimensional layered networks up to three-dimensional double perovskite frameworks. Zero-dimensional bismuth halide perovskites with a basic formula unit $A_3Bi_2X_9$ crystallize in the $Cs_3Cr_2Cl_9$ structure type.[2,3] This crystal structure is based on the hexagonal closest packing of A and X atoms forming hexagonally stacked AX_3 layers with trivalent metal cations occupying two-thirds of the emerging octahedral sites, while one-third of the remaining metal sites are vacant. In this way, double octahedral structures are obtained consisting of pairs of face-sharing BiX_6 octahedra to give complex $Bi_2X_9^{3-}$ anionic clusters, which are referred to as isolated metal halide dimer units. The resulting discrete anionic bi-octahedral moieties are surrounded by monovalent cations occupying the A-site of the perovskite structure.

The most intensively studied bismuth halide perovskite in terms of optoelectronic applications is $(CH_3NH_3)_3Bi_2I_9$. Single crystals can be synthesized via a layered-solution crystallization technique,[4,5] while thin films are obtained from solution-based processing (e.g. spin-coating, doctor blading) followed by subsequent thermal annealing at low temperatures or via vapor-assisted methods.[6-11] The $(CH_3NH_3)_3Bi_2I_9$ structure consists of pairs of face-sharing BiI_6 octahedra forming isolated metal halide dimer units of $Bi_2I_9^{3-}$ surrounded by randomly disordered $CH_3NH_3^+$ cations.[5,12]

The bi-octahedral anionic clusters are interconnected via hydrogen bonding interactions. Dipolar ordering of the organic cation and in-plane ordering of the lone pair of the metal upon cooling is accompanied by phase transitions from a hexagonal crystal structure (space group: $P6_3/mmc$) at 300 K to a monoclinic crystal structure (space group: C2/c) at 160 K with an additional first-order phase transition at 143 K (monoclinic, space group: $P2_1$). $(CH_3NH_3)_3Bi_2I_9$ is an environmentally friendly semiconductor with promising stability in ambient atmosphere and under humid conditions.[8] It exhibits a strong absorption band around 500 nm, a pre-edge absorption peak at 2.51 eV indicating the existence of intrinsic excitons, and a high optical absorption coefficient in the order of 105 cm^{-1} comparable to that of lead-based analogs.[5,7,8,11-13]

Hoye et al. explored the potential of methylammonium bismuth iodide as a solar absorber through detailed materials characterization.[8] Vapor processing was shown to be the method that achieved longer photoluminescence (PL) decay times compared to solution processing. The measured characteristic PL decay times were found to be at least 760 ps, with the bulk lifetime possibly closer to 5.6 ns. This work showed that methylammonium bismuth iodide has promising material properties for solar absorbers, and strongly motivates photovoltaic development in this area.

Oez et al. investigated $(CH_3NH_3)_3Bi_2I_9$ in planar heterojunction solar cells in inverted geometry (ITO/PEDOT: PSS/$(CH_3NH_3)_3Bi_2I_9$/PCBM/Ca/Al) and obtained an open-circuit voltage (V_{OC})=0.66 V, fill factor (FF)=49%, and a photon-to-current efficiency (PCE) of about 0.1%.[14] The photovoltaic performance was limited by the low short-circuit current density (J_{SC}) of 0.22 mAcm^{-2}, which is due to the high exciton binding energy and ineffective charge separation in planar configurations.

Singh et al. studied the effect of electron transporting layer on the bismuth-based lead-free perovskite $(CH_3NH_3)_3Bi_2I_9$ (MBI-methyl ammonium bismuth iodide) for photovoltaic applications. Bismuth based halide perovskite have been synthesized at low temperatures (<150°C) via single-step spin-coating of a stoichiometric solution of bismuth tri iodide and methylammonium iodide in DMF. The morphology of MBI depended strongly on the compact and mesoporous layer on the fluorine-doped tin oxide (FTO) substrates (Figure 14.1a, c & e. MBI growth on planar substrates is not continuous which further facilitates the nonuniform growth, however, in the case of mesoporous layers, the nucleation and growth are more uniform. In the case of brookite, the interparticle necking impedes the MBI percolation in the pores and hampers the nucleation and uniform growth of MBI. Planar and brookite MBI samples formed the nonuniform perovskite films, as evident from the cross-sectional scanning electron microscope (SEM) images (Figure 14.1b and d), which lead to the increased shunting pathway because of contact between the spiro-OMeTAD and TiO_2 compact layer (brookite–mesoporous). Additionally, the perovskite loading is very poor in the case of thicker brookite mesoporous layer. In the case of anatase MBI device, being rich in pores, the film coverage is good and the interfaces of various layers are smooth compared to planar and brookite. The continuous interfaces in anatase mesoporous TiO_2–perovskite reduces junction resistance as well as charge recombination and produces higher V_{oc}, as well as better J_{sc}. Highest PCE of 0.2% was obtained with a device with anatase TiO_2 scaffold with good stability for 10 weeks in ambient condition. On the basis of these results obtained for the influence of TiO_2 electron collectors, the most important issue to enhance the Bi-based device cell performance was found to be structuring of continuous interfaces at collector-perovskite heterojunction and suppression of recombination at the interfaces including the grain–grain inter structure of perovskite crystals. With spectral sensitivity for photon collection (<620 nm), MBI perovskite has a potential to yield J_{sc} as high as 10 mAcm^{-2}. The corresponding efficiency level with V_{oc} of 1 V is around ~8%. The carrier diffusion length will also limit the thickness of the MBI absorber. Assuming that the MBI perovskite film is 100 nm thick, efficiency is limited to <4%.

Park et al. expanded the research to mixed halide perovskites such as $(CH_3NH_3)_3Bi_2I_{9-x}Cl_x$.[15] Due to the partial substitution of iodide with chloride in $(CH_3NH_3)_3Bi_2I_{9-x}Cl_x$, the bandgap was shifted from 2.1 eV (X=I) to 2.4 eV (X=Cl, I) assuming a direct character of the bandgap in both cases. The photovoltaic performance in a meso-structured device architecture (FTO/c-TiO$_2$/mp-TiO$_2$/perovskite/Spiro-OMeTAD/Ag), however, was significantly lower (0.003%) compared to $(CH_3NH_3)_3Bi_2I_9$ (0.12%), which can be attributed to the low V_{OC} of only 40 mV.[15]

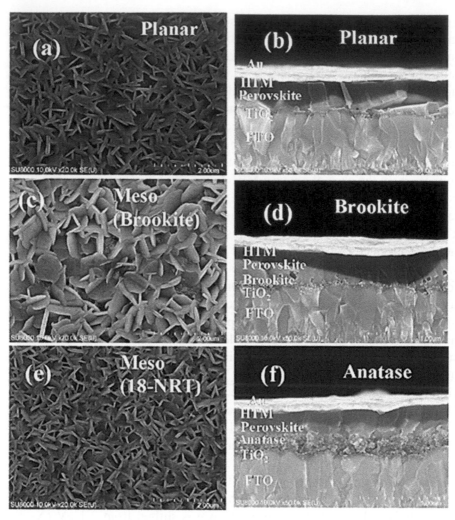

FIGURE 14.1 Scanning electron microscope (SEM) top and a cross-sectional image of MBI perovskite layer deposited on (a, b) TiO_2 compact layer, (c, d) brookite mesoporous, and (e, f) anatase mesoporous.

Source: Reprinted with permission from ref. 13. © 2016 American Chemical Society.

Lehner et al. established a variety of new facile preparation routes for the ternary bismuth iodides $A_3Bi_2I_9$ (A=K, Rb, Cs).[11] By reacting the corresponding binary iodides in a stoichiometric ratio in organic solvents at room temperature or simply ball milling them, high-purity materials were obtained. The optical bandgaps of $A_3Bi_2I_9$ (A=K, Rb, Cs) were derived from Tauc plots of Kubelka–Munk-transformed diffuse reflection Ultraviolet–visible data (Figure 14.2). The absorption edge is located at approximately 2 eV for all three ternary bismuth iodides while the binary exhibits a gap (Eg) of about 1.8 eV. The variation of the $A_3Bi_2I_9$ bandgaps determined in this study only stems from the assumption of direct transitions for A=K, Rb, and an indirect transition for A=Cs.

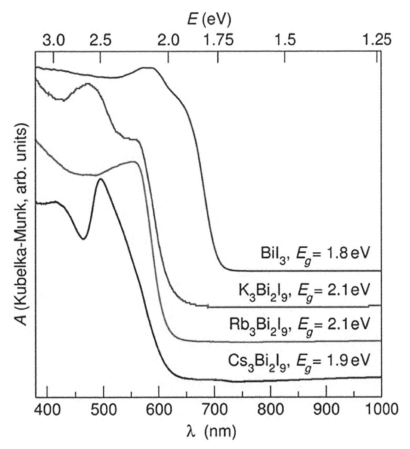

FIGURE 14.2   UV–visible diffuse reflection spectra recorded on powder samples of $A_3Bi_2I_9$ and Bismuth triiodide (BiI_3) transformed into absorbance A.
Source: Reprinted with permission from ref. 11. © 2015 American Chemical Society.

For the indirect gap $Cs_3Bi_2I_9$, a smaller bandgap was found from the Tauc method, while to the eye, the lighter alkali homolog salts appear a slightly darker red. The E$_g$ values are in excellent agreement with the values of 1.98 eV for A=Rb and 1.89 eV for A=Cs that were reported previously.[16]

In the same study, the descending bandgap magnitude of $A_3M_2I_9$ was correlated with the ascending melting points and ascending average atomic numbers. Yet, the anion, structure type, and thus, the connectivity mode and dimensionality of the BiI_6 octahedra have very little influence on E_g. The bandgap of the layered polymorph of $Cs_3Sb_2I_9$ of around 2 eV indicates that the variation of the central metal from Bi to Sb also does not have a strong impact on the magnitude of the bandgap. For the bismuthates, steepness of the edge is greater for the layered structure types compared to the $Cs_3Bi_2I_9$ structure containing isolated anions. A steep edge is desirable, as for related lead halide perovskites, a steep absorption edge has been correlated to little disorder-induced broadening and the absence of deep optically detectable states.

Pure Bismuth triiodide (BiI_3) was also studied for photovoltaic applications with planar TiO_2.[10] BiI_3 belongs to layered heavy metal semiconductors family with interesting anisotropic electronic and optical properties.[10] Lehner et al. also demonstrated the working of planar heterojunction solar cells using BiI_3 as the light absorber with PCE 0.3%, $V_{oc} = 0.22V$, $J_{sc} = 3.85$ mAcm^{-2}, and FF = 0.35.[10] The devices exhibit promising quantum efficiencies (external quantum efficiency 20% in devices with PTAA) with an optical bandgap of 1.8 eV. The relatively low PCE and open circuit voltages were found to arise from the poor match of the low-lying valence band maximum (VBM) of BiI_3 relative to the VBM of the hole transport layers used. Its optical properties and the potential for "defect-tolerant" charge transport properties were tested experimentally by measuring optical absorption and recombination lifetimes by Brandt et al.[17] BiI_3 thin films were synthesized by physical vapor transport (PVT) and solution processing and single crystals were prepared by an electrodynamic gradient vertical Bridgman method.[17]

In Figure 14.3, x-ray diffraction (XRD) spectra for PVT and solution-processed films, as well as their microstructure using SEM, is shown for different regions. It is observed that, as the substrate temperature of PVT BiI_3 increases, the preferred orientation of the BiI_3 layer planes moves from perpendicular to parallel to the substrate, resulting in a different morphology. At the lowest PVT substrate temperature, the film is very thin due to the single zone furnace design, in which growth occurs via a thermal gradient and shows minimal long-range order, and hence, its XRD pattern is dominated by the peaks of the Au substrate. The morphology of films on Au or glass substrates is very similar. The optical bandgap of both the PVT-grown films and solution-processed films was found to be in the range of 1.79–1.80 eV.[17]

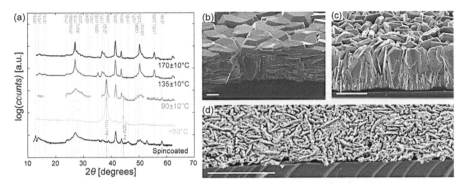

FIGURE 14.3 (a) X-ray diffraction (XRD) spectra as a function of substrate temperature (including gold substrate) and for solution-processed films on glass, showing preferred orientations, and representative micrographs of films deposited (b) near 170°C and (c) near 110°C on glass substrates; (d) Micrograph of solution processed film showing smaller grains. Scale bars represent 5 μm for all micrographs.
Source: Reprinted with permission from ref. 17. © 2015 American Chemical Society.

The carrier recombination lifetime was determined by measuring time-resolved photoluminescence using a time-correlated single photon counting (TCSPC) setup. A 532 nm-wavelength laser was used to excite the sample and a single photon avalanche diode was used to detect photoemission as shown in Figure 14.4. The effective lifetimes were found to be within the range of 180–230 and 190–240 ps for the PVT and solution-processed films, respectively. The single-crystal sample had bi-exponential decay with timescales of 160–260 ps and 1.3–1.5 ns. Given the fact that BiI_3 is reported to be intrinsic (majority carrier type unknown), this recombination lifetime may reflect the sum of electron and hole lifetimes. The mono-exponential decay times of the thin films may be strongly limited by surface recombination, so it can be considered as lower bounds on the bulk Shockley–Read–Hall lifetime. The carrier lifetime of thin films is too short to produce high-performance PV devices. The prospects to improve the lifetime in BiI_3 are suggested by the lifetime difference between single crystal and polycrystalline thin-film materials. This could be achieved by improving phase purity, elemental purity (i.e., controlling contamination), and intragranular structural defect (i.e., dislocation, stacking fault) density will prove essential to increasing the lifetime.

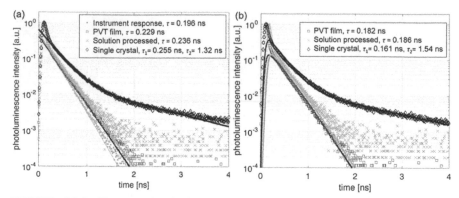

FIGURE 14.4 Time-correlated single photon counting (TCSPC) decay curves for instrument response function, thin-film PVT ($T_{growth} = 110 \pm 10°C$), solution-processed, and single-crystal BiI_3. Fitted raw data is shown in panel a. In panel b, the data has been fit by convolution with the instrument response function. Thin-film samples are fit with mono-exponential decays, while the single-crystal data is fit with a bi-exponential model; time constants are shown in the inset.

Source: Reprinted with permission from ref. 17. © 2015 American Chemical Society.

Recently Johansson et al. reported a layered perovskite structure for $CsBi_3I_{10}$, which was prepared via a solution-based processing method by adjusting the stoichiometric composition of the starting materials CsI and BiI_3.[18] $CsBi_3I_{10}$ features a layered structure similar to BiI_3 alternating with zero-dimensional structures as found in $Cs_3Bi_2I_9$. $CsBi_3I_{10}$ exhibits a bandgap of 1.77 eV similar to BiI_3 and an absorption coefficient in the order of 10^5 cm^{-1}, comparable to lead-based analogs.[18] In comparison to the zero-dimensional $Cs_3Bi_2I_9$ compound (2.03 eV), the layered $CsBi_3I_{10}$ has a lower bandgap, resulting in improved light-harvesting properties. In addition, $CsBi_3I_{10}$ shows improved film formation properties compared to $Cs_3Bi_2I_9$ with more uniform, smoother and pinhole-free layers, which is advantageous for photovoltaic applications. $CsBi_3I_{10}$ was used as absorber material in meso-structured solar cells (FTO/c-TiO_2/mp-TiO_2/perovskite/P3HT/Ag) yielding a PCE of 0.40%, which is significantly higher compared to the $Cs_3Bi_2I_9$ (0.02%) and BiI_3 (0.07%) solar cells obtained in the same device architecture.

KEYWORDS

- bismuth halide perovskites
- optoelectronic applications
- photoluminescence
- heterojunction solar cells
- fluorine-doped tin oxide

REFERENCES

1. Shannon, R.; Shannon, R. D.; Prewitt, C. T. Effective Ionic Radii in Oxides and Fluorides. *Acta Crystallogr. B* **1969,** *25,* 925–46. [Central Research Dept. Experimental Station, EI Du Pont de Nemours and Co., Wilmington, DE and Dept. Earth and Space Sci., SUNY, Stony Brook, NY].

2. Wessel, G. J.; Ijdo, D. J. W. The Crystal Structure of $Cs_3Cr_2Cl_9$. *Acta Crystallogr.* **1957,** *10*(7), 466–468.

3. Chabot, B.; Parthe, E. $Cs_3Sb_2I_9$ and $Cs_3Bi_2I_9$ with the Hexagonal $Cs_3Cr_2Cl_9$ Structure Type. *Acta Crystallogr. Section B* **1978,** *34*(2), 645–648.

4. Eckhardt, K.; et al. Crystallographic Insights into $(CH_3NH_3)_3Bi_2I_9$: A New Lead-Free Hybrid Organic-Inorganic Material as a Potential Absorber for Photovoltaics. *Chem. Commun. (Camb)* **2016,** *52*(14), 3058–3060.

5. Hayden, A.; Evans, H. A.; et al. Mono- and Mixed-Valence Tetrathiafulvalene Semiconductors $(TTF)BiI_4$ and $(TTF)_4BiI_6$ with 1D and 0D Bismuth-Iodide Networks. *Inorg. Chem.* **2017,** *56*(1), 395–401. DOI: 10.1021/acs.inorgchem.6b02287.

6. Öz, S.; et al. Zero-Dimensional $(CH_3NH_3)_3Bi_2I_9$ Perovskite for Optoelectronic Applications. *Sol. Energy Mater. Sol. Cells* **2015.**

7. Lyu, M.; et al. Organic–Inorganic Bismuth (III)-Based Material: A Lead-Free, Air-Stable and Solution-Processable Light-Absorber Beyond Organolead Perovskites. *Nano Res.* **2016,** *9*(3), 692–702.

8. Hoye, R. L. Z.; et al. Methylammonium Bismuth Iodide as a Lead-Free, Stable Hybrid Organic-Inorganic Solar Absorber. *Chem. – A Eur. J.* **2016,** *22*(8), 2605–2610.

9. Okano, T.; Suzuki, Y. Gas-Assisted Coating of Bi-Based $(CH_3NH_3)_3Bi_2I_9$ Active Layer in Perovskite Solar Cells. *Mater. Lett.* **2017,** *191,* 77–79.

10. Lehner, A. J.; Wang, H.; Fabini, D. H.; Liman, C. D.; Hébert, C.-A.; Perry, E. E.; Wang, M.; Bazan, G. C.; Chabinyc, M. L.; Seshadri, R. Electronic Structure and Photovoltaic Application of BiI_3. *Appl. Phys. Lett.* **2015,** *107*(13), 131109. https://doi.org/10.1063/1.4932129.

11. Lehner, A. J.; et al. Crystal and Electronic Structures of Complex Bismuth Iodides $A_3Bi_2I_9$ (A=K, Rb, Cs) Related to Perovskite: Aiding the Rational Design of Photovoltaics. *Chem. Mater.* **2015,** *27*(20), 7137–7148.

12. Pazoki, M.; et al. Bismuth Iodide Perovskite Materials for Solar Cell Applications: Electronic Structure, Optical Transitions, and Directional Charge Transport. *J. Phys. Chem. C* **2016,** *120*(51), 29039–29046.

13. Singh, T.; et al. Effect of Electron Transporting Layer on Bismuth-Based Lead-Free Perovskite $(CH_3NH_3)_3 Bi_2I_9$ for Photovoltaic Applications. *ACS Appl. Mater. Interfaces* **2016**, *8*(23), 14542–14547.

14. Öz, S.; et al. Zero-Dimensional $(CH_3NH_3)_3 Bi_2I_9$ Perovskite for Optoelectronic Applications. *Sol. Energy Mater. Sol. Cells* **2016,** *158*(Part 2), 195–201.

15. Park, B.-W.; et al. Bismuth Based Hybrid Perovskites $A_3Bi_2I_9$ (A: Methylammonium or Cesium) for Solar Cell Application. *Adv. Mater.* **2015**, *27*(43), 6806–6813.

16. Peresh, E. Y.; et al. Influence of the Average Atomic Number of the A_2TeC_6 and $A_3B_2C_9$ (A=K, Rb, Cs, Tl(I); B=Sb, Bi; C=Br, I) Compounds on Their Melting Point and Band Gap. *Inorg. Mater.* **2014**, *50*(1), 101–106.

17. Brandt, R. E.; et al. Investigation of Bismuth Triiodide (BiI_3) for Photovoltaic Applications. *J. Phys. Chem. Lett.* **2015**, *6*(21), 4297–4302.

18. Johansson, M. B.; Zhu, H.; Johansson, E. M. J. Extended Photo-Conversion Spectrum in Low-Toxic Bismuth Halide Perovskite Solar Cells. *J. Phys. Chem. Lett.* **2016**, *7*(17), 3467–3471.

CHAPTER 15

ANTIMONY-BASED PEROVSKITES

Antimony halide perovskites are a potential alternative to lead-based perovskite semiconductors for photovoltaic applications to address the chemical stability and the toxicity issue. Saparov et al. examined $Cs_3Sb_2I_9$ as prospective absorber material in solar cells and found improved stability properties under ambient conditions compared to lead and tin halide perovskite films.[1] $Cs_3Sb_2I_9$ exists in two polymorphs: (1) a zero-dimensional dimer modification (hexagonal) featuring $Sb_2I_9^{3-}$ bi-octahedral units and (2) a two-dimensional layered modification (trigonal).[2] The dimer can be synthesized via solution-based methods using polar solvents, while the layered modification is obtained through solid-state reactions, gas phase reactions (e.g. co-evaporation or sequential deposition of CsI and SbI_3, followed by annealing in SbI_3 vapor) or solution-based methods (e.g. crystallization from methanol or nonpolar solvents). Saparov et al. investigated the layered modification of $Cs_3Sb_2I_9$ as light absorber in perovskite solar cells with the general device architecture of FTO/c-TiO_2/$Cs_3Sb_2I_9$/PTAA/Au. The material exhibited a rather poor photovoltaic performance with an open-circuit voltage (V_{OC}) up to 300 mV, a short-circuit current density (J_{SC}) below 0.1 mAcm^{-2}, and a low overall performance (<1%).[1]

Hebig et al. developed solution-based fabrication and characterization of the lead-free perovskite-related methylammonium antimony iodide $(CH_3NH_3)_3Sb_2I_9$ compound.[3] $(CH_3NH_3)_3Sb_2I_9$ thin films were fabricated by spin-coating SbI_3+CH_3NH_3I as precursors in a mixture of γ- butyrolactone and dimethyl sulfoxide (DMSO)), followed by low-temperature annealing (100–120°C, 30 min) under a nitrogen atmosphere. The crystal structure of $(CH_3NH_3)_3Sb_2I_9$ is shown in Figure 15.1aand the X-ray diffraction pattern (XRD) of the synthesized crystal is shown in Figure 15.1b of the crystalline film confirming the hexagonal structure of the $(CH_3NH_3)_3Sb_2I_9$ but shows a strong preferential growth direction along the c-axis as is evident from the SEM image of Figure 15.1c, these films were deposited by single step deposition. In order to improve the homogeneity of the films and to prevent shunting, solvent engineering techniques were used. This involved a two-step spin-coating (3000/5000 rpm) process with an added drop of toluene as anti-solvent 10 s before the end of step 2. The resulting thin film is shown in Figure 15.1d, which had a very flat and homogeneous surface but much smaller grains and seemed to consist of amorphous material. The XRD measurement of the two-step process with the toluene drop shows no clear peaks, emphasizing the amorphous character of the material. As a pinhole-free absorber layer is essential for the application in photovoltaic devices, two-step spin-coating process was used with toluene dripping to fabricate the thin films for investigation of the optoelectronic properties of $(CH_3NH_3)_3Sb_2I_9$ in this study.

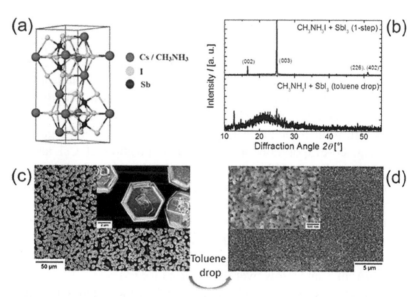

FIGURE 15.1   (a) Crystal structure of $(CH_3NH_3)_3Sb_2I_9$ (space group P63/mmc); (b) X-ray diffraction pattern (XRD) pattern of $(CH_3NH_3)_3Sb_2I_9$ thin films prepared by the one-step spin-coating process (upper panel) and after the toluene drop (lower panel); (c) scanning electron microscope (SEM) image of the $(CH_3NH_3)_3Sb_2I_9$ thin film, showing hexagonal crystals after one-step spin-coating, and (d) SEM image of the thin film with an additional toluene drop during spin-coating with the two-step process.
Source: : Reprinted with permission from ref. 3. © 2016 American Chemical Society.

The absorption properties of the $(CH_3NH_3)_3Sb_2I_9$ thin film were determined by photothermal deflection spectroscopy (PDS) measurements. Figure 15.2a shows the absorption coefficient for Sb-perovskite $((CH_3NH_3)_3Sb_2I_9)$ compared to various Bi-perovskite compounds with different cations

[A = MA⁺, Cs⁺, FA⁺] over a large spectral range. The Bi-based films were prepared in the same way as the Sb-based films and were also fairly amorphous. All the compounds show high absorption coefficients in the range of $\alpha > 105$ cm^{-1}, which is high enough for efficient light absorption in thin film photovoltaic devices. To further study the optical properties of the Sb perovskite, the room-temperature photoluminescence (PL) was measured. Figure 15.2b shows the determined PL spectrum of the Sb perovskite compared to the absorptance of the material. The measured PL signal is relatively weak and broad. The PL peak position is at 1.58 eV, which is shifted by 560 meV to lower energies than its band edge at 2.14 eV. This could imply that the luminescence in Sb-based perovskites originates from radiative recombination involving mid gap states. This implies that the presence of radiative mid-gap transitions suggests that the films suffer from a large amount of non-radiative recombination processes via these mid-gap states.[3]

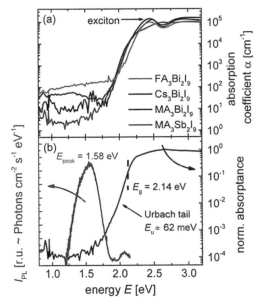

FIGURE 15.2   (a) Comparison of the absorption coefficient of various Bi-based perovskites and $(CH_3NH_3)_3Sb_2I_9$ determined by (PDS) measurements; (b) normalized absorptance of $(CH_3NH_3)_3Sb_2I_9$ (from PDS) with a calculated Urbach tail energy and corresponding room-temperature photoluminescence (PL) spectrum.
Source: Reprinted with permission from ref. 3. © 2016 American Chemical Society.

The thin films of Sb-perovskite as an absorber material for solar cells were tested in a planar heterojunction configuration using the two-step method with the toluene drop to obtain amorphous perovskite films. For comparison, MAPbI$_3$ devices were also fabricated in the inverted structure. For Sb-based solar cells, the following layer stack consisting ITO (120 nm)/PEDOT: PSS (25 nm)/(CH$_3$NH$_3$)$_3$Sb$_2$I$_9$ (300 nm)/PC$_{61}$BM (60 nm)/ZnO-NP (60 nm)/Al (150 nm) was used. Figure 15.3a shows the J–V curve of the Sb perovskite measured under 1 sun conditions, at a scan speed of 0.1 V/s, and showing very little hysteresis. The values for J$_{sc}$ are nearly unaffected by the sweep direction and agree reasonably well with the values obtained from integrating the external quantum efficiency (EQE), as shown in Figure 15.3b. The fill factor (FF) slightly changes from 52 to 55% with the sweep direction, and the V$_{oc}$ also changes from 885 (up) to 896 mV (down). Compared to the bandgap of 2.14 eV, Voc≈0.9 V is relatively low. This difference between the bandgap and V$_{oc}$ can be explained by the high defect density of the obtained material as observed in the PL measurements. Improving the material quality and optimizing the device architecture with suitable contact layers are proposed ways to increase V$_{oc}$ significantly.

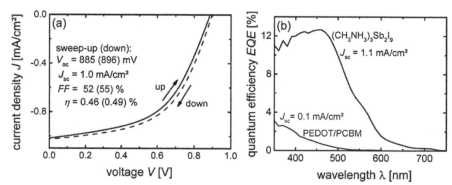

FIGURE 15.3 (a) J–V curve of (CH$_3$NH$_3$)$_3$Sb$_2$I$_9$ solar cell measured with "up" and "down" sweep with a rate of 0.1 V/s; (b) external quantum efficiency (EQE) measurement of the (CH$_3$NH$_3$)$_3$Sb$_2$I$_9$ solar cell compared to the reference device (ITO (120 nm)/PEDOT/PCBM/ZnO-NP/Al).

Source: Reprinted with permission from ref. 3. © 2016 American Chemical Society.

Harikesh et al. recently reported the synthesis of Rb$_3$Sb$_2$I$_9$ in a layered perovskite structure via a low-temperature solution-based route through the reaction of RbI and SbI$_3$.[2] In comparison to the dimer modification of Cs$_3$Sb$_2$I$_9$, the substitution of Cs$^+$ (188 pm) with the smaller Rb$^+$ (172 pm) cation was shown to effectively stabilize the structure in the layered modification. As a consequence, the respective Rb$_3$Sb$_2$I$_9$ perovskite forms a two-dimensional layered structure consisting of corner-sharing BX$_6$ octahedra, which is different to the zero-dimensional dimer modification of $Cs_3Sb_2I_9$ comprising isolated bi-octahedral metal halide units $B_2X_9^{3-}$ (Figure 15.4).

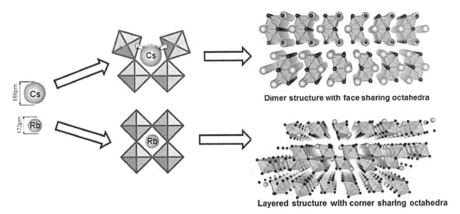

Dimer structure with face sharing octahedra

Layered structure with corner sharing octahedra

FIGURE 15.4 Schematic showing the influence of cation size (A site) on the structure of $A_3Sb_2I_9$.
Source: Reprinted with permission from ref. 2. © 2016 American Chemical Society.

The cross section of the device fabricated in this study is shown in Figure 15.5a, with the following configuration: TiO_2 as an electron transport layer, Poly-TPD as a hole transport layer and gold as the counter electrode. The devices were fabricated using the thin films formed by SbI_3 treatment. This was done to provide an excess of SbI_3 by dissolving it in toluene and dripping it while spin-coating.

The solar cells exhibited a V_{OC} of 0.55 V, a J_{SC} of 2.12 mAcm^{-2}, and an FF of 57% resulting in a PCE of 0.66% as shown in Figure 15.5b. The EQE (Figure 15.5c) plot is in agreement with the absorbance spectrum with the onset of current from around 600 nm and the device exhibited a maximum of around 30% EQE in its absorption range. The internal quantum efficiency (IQE) was calculated using the total reflectance and transmittance $(IQE = EQE/(1 - R - T))$ of the device and exhibited values <32%. This shows that only <32% of the absorbed photons are converted into carriers and collected at electrodes. The device efficiencies were also severely limited by the open circuit voltages of <0.6 V, considering the high bandgap. This difference was attributed to the mid-gap states and was verified by cathodoluminescence study. To get more insight into the recombination behavior of this material, the variation of V_{oc} with the natural logarithm of light intensity was studied as shown in Figure 15.5d. The slope obtained from our analysis yielded a value of 1.95 $K_b T/q$. This indicates that Shockley–Read–Hall-like recombination through defect levels is dominant over bimolecular recombination. Furthermore, the variation of short circuit current with light intensity was also studied as shown. $\alpha < 1$ is generally associated with space charge effects: variation in mobility between carriers or energy barriers in the device. The fitting gave $\alpha = 1$, excluding the possibility of the abovementioned effects. Thus, the only possibility for the reduced V_{oc} is the recombination through defect levels in the material itself or bulk recombination. Exploring alternative device architectures could be a possible strategy to improve the performances along with improving material quality.

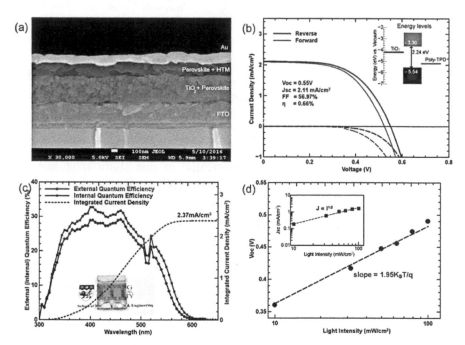

FIGURE 15.5 (a) SEM cross section of the device; (b) I–V curve under forward and reverse scans of the best device with the energy levels of $Rb_3Sb_2I_9$ shown in inset; and (c) the corresponding incident photon-to-current efficiency spectrum with the integrated current density; (d) variation of V_{oc} with natural logarithm of light intensity with the variation of J_{sc} with light intensity plotted in the inset.
Source: Reprinted with permission from ref. 2. © 2016 American Chemical Society.

Antimony halide double perovskite semiconductors with a basic formula $A_2B^I B^{II} X_6$ have been investigated in a theoretical study by Volonakis et al.[4] These materials are based on a heterovalent substitution of Pb^{2+} with an equal number of mono- and trivalent cations to maintain the charge neutrality and form double perovskite structures (elpasolite structure). Volonakis et al. examined halide double perovskites based on monovalent noble metals ($B^I = Cu^+$, Ag^+, Au^+) and trivalent pnictogen cations ($B^{II} = Sb^{3+}$) with Cs^+ as A-site cation and halide (X = Cl, Br, I) as anions. The calculated electronic bandgaps of the examined antimony halide double perovskites are indirect bandgaps and tunable in the visible range, that is, 0.9–2.1 eV (Cs_2CuSbX_6), 1.1–2.6 eV (Cs_2AgSbX_6), and 0.0–1.3 eV (Cs_2AuSbX_6).[4]

New strategies in materials design and bandgap engineering over a wide range by tuning the stoichiometry and compositions, for example, via anion alloying approach to form mixed halide-chalcogenide compounds, enable the development of a remarkable number of novel green lead-free absorber materials. Theoretical calculations predicting promising direct bandgaps and improved optical absorption properties within the visible range compared to lead-based analogs highlight the potential of lead-free perovskite semiconductors for photovoltaics.

KEYWORDS

- open-circuit voltage
- X-ray diffraction pattern
- scanning electron microscope
- photothermal deflection spectroscopy
- external quantum efficiency

REFERENCES

1. Saparov, B.; et al. Thin-Film Preparation and Characterization of $Cs_3Sb_2I_9$: A Lead-Free Layered Perovskite Semiconductor. *Chem. Mater.* **2015**, *27*(16), 5622–5632.
2. Harikesh, P. C.; et al. Rb as an Alternative Cation for Templating Inorganic Lead-Free Perovskites for Solution Processed Photovoltaics. *Chem. Mater.* **2016**, *28*(20), 7496–7504.
3. Hebig, J.-C.; et al. Optoelectronic Properties of $(CH_3NH_3)_3Sb_2I_9$ Thin Films for Photovoltaic Applications. *ACS Energy Lett.* **2016**, *1*(1), 309–314.
4. Volonakis, G.; et al. Lead-Free Halide Double Perovskites via Heterovalent Substitution of Noble Metals. *J. Phys. Chem. Lett.* **2016**, *7*(7), 1254–1259.

PART V
TANDEM PEROVSKITE SOLAR CELLS AND OTHER ASPECTS

ANITA WING YI HO-BAILLIE

CHAPTER 16

PEROVSKITE TANDEM SOLAR CELLS FOR PHOTOVOLTAICS

The bandgap tunability and the superior optical absorption of perovskite thin films make them ideal candidates for tandem solar cells. Figure 16.1 shows the typical configurations of double junction tandem which consists of higher bandgap cells absorbing higher energy (short wavelength) photons in the first instance of illumination. The lower bandgap cell absorbs the filtered light which has lower energy (long wavelength) photons. Multiple junctions are possible with the lower energy photons being "trickled down" the stack. A two-terminal configuration involves direct deposition of a solar cell on top of another. A "buffer" or "window" or "recombination" intermediate layer separates the cells which needs to be transparent and conductive. In this configuration, the cells are connected in series requiring current outputs to be matched. In a four-terminal configuration, cells are independent from each other electrically introducing complexity in wiring and mounting.

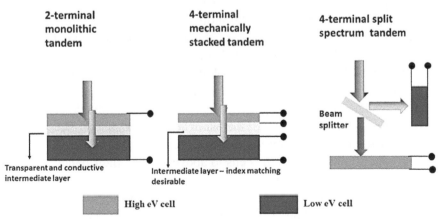

FIGURE 16.1 Schematic of different configurations for a double junction tandem.

The versatility of bandgap tuning as a function of material composition is illustrated in Figure 16.2. While Figure 16.3 shows the ideal bandgap for each sub-cell in a multi-junction tandem for maximum cell performance. This means to realize an efficient double junction cell, the bandgaps of two cells should be around 0.95 eV and just above 1.6 eV, respectively. If the voltage output is close to the sum of the bandgaps, the matched current output is around 21 mA/cm^2 and the fill factor is 80%, 43% energy conversion efficiency is possible. For a triple junction cell, the bandgaps of the three cells should be around 0.9 , 1.4 , and 1.9 eV giving a voltage output of around 4 V. A matched current output of around 14 mA/cm^2 and a fill factor around 80% are required to achieve 45% efficiency. For a four-junction cell, the bandgaps should be around 0.7, 1.1 , 1.5 , and 2 eV. With a voltage output close to the sum of the bandgaps, a matched current output of just above 10 mA/cm^2 and a fill factor above 86%, the cell will be 46% efficient.

In the last 3 years, there has been an increase in research effort on perovskite tandem demonstrations. Tables 16.1–16.4 list the results published to date. Excellent achievements include the demonstration of two-terminal 1 cm^2 Perovskite/Si tandem at 23.6% certified efficiency[1] and two-terminal 1.4 cm^2 Perovskite/Si tandem at 25.2% certified efficiency (measured at maximum power point (MPP).[2] At the end of 2016, two-terminal perovskite/perovskite tandems are emerged. 13.8% efficiency is achieved on a 1 cm^2 device.[3] Smaller areas achieved 17.0–17.8% efficiencies on 0.2 , [3] and 0.02 cm^2, [4] respectively. Although the efficiencies demonstrated are lower than those of a single junction perovskite or Si cell, two-terminal tandem would still be preferred for commercialization due to simplicity. Four-terminal tandems utilizing spectrum splitting[5–7,8] or mechanical stacking[9,10,11,12,13–17] produce efficiencies higher than those of the individual cells, as current matching is no longer a limiting factor. However, the improvement of the cell performance is only 10–15% relative. This is because the higher bandgap cell is under utilized with output voltage far below its bandgap. This is also the case for the lower bandgap cell. In general, whether it be a 1.7 eV perovskite cell, 1.5 eV perovskite cell, 1.2 eV perovskite, or 1.1 eV Si cell, the demonstrated voltage output is generally 60% of the bandgap of the material. This voltage limit brings the efficiency limit down to 26% for a two-terminal tandem. By improving this voltage of the cells to above 70% of their bandgap, 30% tandem efficiency is achievable.

Two-terminal configuration also introduces processing criteria such as processing temperature and solvent compatibility. The annealing temperatures of the top layers should not be too high that it will cause damage to the underlying layers. Likewise, the solvents used for solution processing, the top layers cannot dissolve the underlying layers. As such, physical deposition such as sputtering, atomic layer deposition, chemical vapor deposition, pulsed laser deposition and thermal evaporation are often used for the fabrication of transparent electrodes, interlayers, and even perovskites.[4] Indium tin oxide remains a popular choice for transparent conductive electrode although it is absent in the perovskite/perovskite tandem as reported in[4] which uses a small molecular weight, sublimable organic semiconductors as carrier selective transport layers. However, the lateral conductivity of these layers is unclear and transparent electrodes will probably be required as cell size increases from the 0.02 cm^2 demonstrated in that work.

Barrier layers are used at times to protect the underlying layers from the damage caused by the processing of upper layers. Such barrier layers can introduce parasitic optical absorptions. Cell design with appropriate isolation preventing shunting across the layers also needs to be considered. To minimize this reflection loss, index matching for the air/top cell interface and top cell/bottom cell interface is essential. The former can be realized by depositing anti-reflection coating, which can be a multi-layer stack, to allow a gradual change in refractive index from the transparent electrode to air. The use of carrier transport layer with a refractive index closely matched to the perovskite will also reduce the reflection losses. In the case of Si/Perovskite tandem where multiple bounces maximize the long wavelength absorption in the indirect bandgap Si material, a rear reflector in conjunction with textured surface will be desired. Depositing thin film perovskite solar cell on textured Si cell is one option. Alternatively, textured foil can be placed on the complete device on the illuminating side, which has shown to be effective.[18-19]

To fully appreciate the interference effects from the multiple layers in the tandem stack, their parasitic absorptions and reflection losses, optical analysis using transfer matrix method[20-22] is often used. It is an effective tool in determining best materials and optimum thicknesses for minimum optical losses and even distribution of optical absorptions between the sub-cells, especially when current matching is critical in a two-terminal design.

In summary, for perovskite tandem to be commercially viable, it is imperative that (i) the performance of the single junction cell continues to improve to capitalize on the voltage gain from tandem, (ii) large area processing techniques are to be developed to allow uniform perovskite thin films to be fabricated on commercial size Si or thin film devices, (iii) the lifetime of perovskite solar cells match those of incumbent photovoltaic technologies for the extra cost of fabricating multi-junctions to be recouped thereby, effectively reducing the levelized cost of energy.

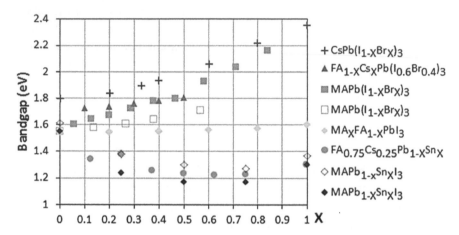

FIGURE 16.2 Evolution of bandgap as a function of X in $MAPb_{1-x}Sn_xI_3$; $FA_{0.75}Cs_{0.25}Pb_{1-x}Sn_xI_3$; $MA_xFA_{1-x}PbI_3$; $MAPb(I_{1-x}Br_x)_3$; $FA_{1-x}Cs_x(PbI_{0.6}Br_{0.4})_3$; and $CsPb(I_{1-x}Br_x)_3$.
Source: Reproduced from refs 23, 24, 25, 26, 27, 28, 29, 30.

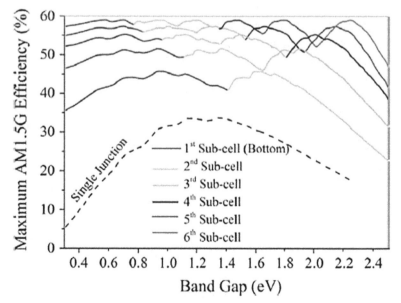

FIGURE 16.3 Maximum efficiency for a bandgap value as part of a multi-junction stack for cases up to and including six bandgaps.
Source: Reprinted with permission from ref 31. © 2016 Elsevier.

TABLE 16.1 Summary of Demonstrated Two-Terminal Perovskite Tandem Cells.

Lower bandgap	Higher bandgap	V_{OC} (V)	J_{SC} (mA/ cm^2)	FF (%)	Eff. (%)	Area (cm^2)	References
Glass/Mo/ Cu$_2$ZnSn (S, Se) $_4$/CdS/ITO	PEDOT: PSS/ CH$_3$NH$_3$PbI$_3$/ PCBM/Al	1.353	5.6	60.4	4.6	0.45	[32]
Glass/Si$_3$N$_4$/Mo/ CIGS/CdS/ITO	PEDOT: PSS/ CH$_3$NH$_3$PbI$_3$Br$_{3-x}$/ PCBM/BCP (5 nm)/ Ca (10–15 nm)	1.45	12.7	56.6	10.9	0.4	[33]
Polymer Cell (ITO/ PEDOT: PSS/ PbSeDTEG8: PCBM/PFN/ TiO2)	Perovskite Cell (PEDOT: PSS PH500/PEDOT: PSS 4083/ CH$_3$NH$_3$PbI$_3$/ PCBM/PFN/Al)	1.52	10.05	67	10.2	0.100	[34]
Metal/ n++BSF/n-Si/ p+Emitter/n++Si T-J	ALD c-TiO$_2$/ mp-TiO$_2$/ CH$_3$NH$_3$PbI$_3$/Spiro-OMeTAD/AgNW/ LiF	1.58	11.5	75	13.7	1	[35]
Ag/AZO/a-Si: H (n)/a-Si: H (i)/c-Si (n)/a-Si: H (i)/a-Si (p)/ ITO	ALD SnO$_2$/ MA$_x$FA$_{1-x}$PbI$_Y$Br$_{3-Y}$/ Spiro-OMeTAD/ MoO$_X$/ITO/LiF	1.8	13	78	18.0	0.1225	[36]
Ag/ITO/a-Si (n)/a-Si (i)/c-Si (n)/a-Si (i)/a-Si (p)/IZO	PCBM/PEIE/ CH$_3$NH$_3$PbI$_3$/Spiro-OMeTAD/MoO$_X$/ ITO/IO: H/ARF	1.701 1.690	16.1 15.9	70.1 77.6	19.2 21.2	1.22 0.17	[37]
Ag/ITO/a-Si (n)/a-Si (i)/c-Si (n)/a-Si (i)/a-Si (p)/IZO	SnO$_2$/PEIE/PCBM/ CH$_3$NH$_3$PbI$_3$/Spiro-OMeTAD/MoO$_X$/ IO: H (10 nm)/ITO (100 nm)/Au	1.718	16.4	71.6	20.5$^+$	1.43	[9]
Ag/SiNP/ITO/a-Si (p+)/a-Si (i)/c-Si (n)/a-Si (i)/a-Si (n+)/ITO	NiO/FA$_{0.83}$Cs$_{0.17}$Pb (I$_{0.83}$Br$_{0.17}$)$_3$/LiF/ PC$_{60}$BM/SnO2/ITO/ LiF/Ag	1.651	18.09	79.0	23.6*	0.99	[1]
Ag/ITO/a-Si (p)/a-Si (i)/c-Si (n)/a-Si (i)/a-Si (n)/nc-Si: H (n+)/ nc-Si: H (p+)	Spiro-TTB/Cs$_x$FA$_{1-x}$Pb (I, Br)$_3$/LiF/ C$_{60}$/SnO$_2$/IZO/ MgF$_2$/Ag	Eff. measured at MPP			25.2*	1.42	[2]

Source: + Denotes stabilized results, * denotes certified results.

TABLE 16.2 Summary of Demonstrated Perovskite/Perovskite Tandem Cells. Most of them are Two-Terminal Except the Last Entry.

Lower bandgap	Higher bandgap	V_{OC} (V)	J_{SC} (mA/cm^2)	FF (%)	Eff. (%)	Area (cm^2)	References
ITO Glass/PEDOT: PSS/$CH_3NH_3PbI_3$/ PCBM	P3HT/ $CH_3NH_3PbBr_3$/c-TiO_2/FTO Glass	1.95	8.4	66	10.8	0.096	[38]
	PTAA/ $CH_3NH_3PbBr_3$/c-TiO_2/FTO Glass	2.25	8.3	56	10.4		
glass/FTO/c-TiO_2/ mp-TiO_2/$CH_3NH_3PbI_3$/ spiro-OMeTAD/	PEDOT: PSS/ PEI/PCBM: PEI/ $CH_3NH_3PbI_3$/ spiro-OMeTAD/ hc-PEDOT: PSS)	1.89	6.61	56	7	0.04	[39]
Ag/BCP/C_{60}/ $FA_{0.75}Cs_{0.25}Sn_{0.5}Pb_{0.5}I_{30}$/ PEDOT: PSS/ITO	PCBM/SnO_2/ $FA_{0.83}Cs_{0.17}Pb$ $(I_{0.5}Br_{0.5})_3$/NiO/ ITO/Glass	1.66	14.5	70	17.0^+	0.2	[3]
		1.76	13.5	56	13.8^+	1	
Au/TaTm: F_6-TCNNQ/ TaTm/$CH_3NH_3PbI_3$/ C_{60}/C_{60}: Phlm	TaTm: F_6-TCCNQ/ TaTm/IPH/ $FA_{0.85}Cs_{0.15}Pb$ $(I_{0.3}Br_{0.7})_3$/IPH/ TiO_2/ITO/Glass	2.29	9.92	78	17.8	0.0264	[4]
ITO Glass/ PEDOT: PSS/ $FA_{0.75}Cs_{0.25}Sn_{0.5}Pb_{0.5}I_3$/ C_{60}/BCP/Ag (80–120 nm)	ITO Glass/NiO/ $FA_{0.83}Cs_{0.17}Pb$ $(I_{0.83}Br_{0.17})_3$/ PCBM (10 nm)/ SnO_2 (4 nm)/ $Zn_xSn_yO_z$ (2 nm)/ ITO/Al (100 nm)	four-Terminal; stacked			20.3^+	0.2	[3]
ITO Glass/PEDOT: PSS/$(FASnI_3)_{0.6}$ $(MAPbI_3)_{0.4}$/C_{60}/BCP/ Ag	FTO Glass/ SnO_2/C_{60}-SAM/ $FA_{0.8}Cs_{0.2}Pb$ $(I_{0.7}Br_{0.3})_3$/spiro-OMeTAD/MoO_x/ sputtered ITO	four-Terminal; stacked			23	0.1	[40]

Source: + Denotes stabilized results.

TABLE 16.3 Summary of Demonstrated Four-Terminal Perovskite/Si Tandem Cells.

			Eff. (%)	Area (cm^2)	References
Ag/ITO/a-Si: H (n)/a-Si: H (i)/c-Si (n)/a-Si: H (i)/a-Si: H (p)/ITO/Ag-grid	FTO Glass/c-TiO$_2$/ mp-TiO$_2$/CH$_3$NH$_3$PbI$_3$/ Spiro-OMeTAD/MoO$_x$/ ITO	Stacked	13.4	1, 0.25	[41]
Al BSF c-Si	LiF/FTO Glass/c-TiO$_2$/ mp-TiO$_2$/CH$_3$NH$_3$PbI$_3$/ Spiro-OMeTAD/AgNW/ LiF	Stacked	17.9*	0.3906	[10]
HIT c-Si	FTO glass/c-TiO$_2$/ CH$_3$NH$_3$PbI$_3$/spiro- OMeTAD/CVD graphene/ Au grid	Stacked	13.2	0.16	[42]
17% Si	Glass/ITO (150 nm)/ PEDOT: PSS (40 nm)/ CH$_3$NH$_3$PbI$_3$ (275 nm)/ PC$_{60}$BM (40 nm)/Al: ZnO np (50 nm)/ITO (500 nm)/ Ag (250 nm)/MgF2 (150 nm)	Stacked	18.0	0.39	[43]
PERL (19.6%) Si	MgF2 (100 nm) ITO glass/c-TiO$_2$/mp-TiO$_2$/ CH$_3$NH$_3$PbI$_3$/Spiro- OMeTAD/MoO$_x$ (10 nm)/ ITO (65)/Au grid (30–35umwide)/MgF2 (175 nm)	Stacked	20.1	0.25	[12]
Al BSF (19.1%) c-Si	FTO glass/TiO$_2$/ CH$_3$NH$_3$PbI$_3$ (350 nm)/ Spiro-OMeTAD (150 nm)/ MoO$_x$ (15 nm)/Au (12 nm)/MoO$_x$ (60 nm)	Stacked	23.6	N. A.	[13]
HIT c-Si	ITO Glass/PTAA/ CH$_3$NH$_3$PbI$_3$/PCBM/C60 (20 nm)/BCP (8 nm)/Cu (1 nm)/Au (7 nm)	Stacked	23.0	0.075	[14]
HIT c-Si	ITO Glass/SnO$_2$/PEIE/ PCBM/CH$_3$NH$_3$PbI$_3$/ Spiro-OMeTAD/MoO$_x$/ IO: H (10 nm)/ITO (100 nm)/Au grid	Stacked	23.0	1	[9]

TABLE 16.3   *(Continued)*

			Eff. (%)	Area (cm^2)	References
	ITO Glass/SnO$_2$/PEIE/ PCBM/CH$_3$NH$_3$PbI$_3$/ Spiro-OMeTAD/MoO$_X$/ IO: H (10 nm)/ITO (100 nm)/Au grid		25.2	0.25	
IBC (24%) Si	FTO Glass/In-TiO$_2$ (50 nm)/mp-TiO$_2$ (80 nm)/ Cs$_{0.05}$ (FA$_{0.83}$MA$_{0.17}$)$_{0.95}$Pb (I$_{0.83}$Br$_{0.17}$)$_3$ (450 nm)/ Spiro-OMeTAD (180 nm)/ MoO$_X$ (13 nm)/ITO (40 nm)/Au grid	Stacked	24.5	0.16	[15]
IBC (24%) Si	Textured cover/ITO quartz/c-TiO$_2$/mp-TiO$_2$/ Rb$_{0.05}$ (FA$_{0.75}$MA$_{0.15}$Cs$_{0.1}$)$_{0.95}$PbI$_2$Br/PTAA/MoO$_X$/ ITO/Au grid	Stacked	26.4	0.1764	[46]
PERL (22.4%)	Fused Silica/ITO/CUSCN/ MAPbI$_3$/PC$_{60}$BM/ZnO: Al np/AgNW	Stacked	26.7	0.104	[44]
HIT c-Si with MgF2	FTO Glass/c-TiO$_2$/ mp-TiO$_2$/CH$_3$NH$_3$PbI$_{3-x}$Cl$_x$/Spiro-OMeTAD/Au (500 nm filter)	Spectrum splitting	28	0.04	[5]
PERL Si	FTO Glass/c-TiO$_2$/ mp-TiO$_2$/CH$_3$NH$_3$PbBr$_3$/ Spiro-OMeTAD/Au/ (500 nm filter)	Spectrum splitting	23.4	1, 0.045	[6]
Screen Printed Si			18.8	1.62, 0.045	
DX3 DSSC	FTO Glass/c-TiO$_2$/ mp-TiO$_2$/(FAPbI$_3$)$_{0.85}$ (MAPbBr$_3$)$_{0.15}$/PTAA/Au/ (771 nm filter)	Spectrum splitting	21.5^	0.1414	[7]
IBC Si	Quartz/ITO (95 nm)/c-TiO$_2$ (50 nm)/mp-TiO$_2$ (180 nm)/(FAPbI$_3$)$_{0.85}$ (MAPbBr$_3$)$_{0.15}$ (180+160 nm)/Spiro-OMeTAD (200 nm)/Au (150 nm)/(no filter)	Spectrum splitting by reflection	23.1	0.09	[8]

Source: ^ Denotes independently verified, * denotes certified results.

TABLE 16.4 Summary of Other Demonstrated Four-Terminal Perovskite Tandem Cells.

			Eff. (%)	Area (cm²)	References
CIGS	LiF/FTO Glass/c-TiO₂/mp-TiO₂/ CH₃NH₃PbI₃/Spiro-OMeTAD/ AgNW/LiF	Stacked	17.3*	0.3896	[10]
CIGS	ITO Glass/c-TiO₂/CH₃NH₃PbI₃₋ₓClₓ/ Spiro-OMeTAD/MoOₓ (10 nm)/Au (1 nm)/Ag (10 nm)/MoOₓ (40 nm)	Stacked	15.5	0.108	[11]
CIGS	FTO glass/ZnO/PCBM/CH₃NH₃PbI₃/ Spiro-OMeTAD/MoOₓ (35 nm)/ I₂O₃: H (170 nm)/Ni/Al/Ni grids (50 nm/2,000 nm/50 nm)	Stacked	20.5	0.517	[45]
CIGS module (7 strips)	Perovskite module (4 strips)	Stacked	17.8	3.67	[16]
CIGS	1 mm Glass/In2O3: H/PTAA/ CH₃NH₃PbI₃/PCBM/ZnO np/ZnO: Al/Ni (50 nm)/Al (4um)/	Stacked	22.1	0.213	[17]
CIS	1 mm Glass/In2O3: H/PTAA/ CH₃NH₃PbI₃/PCBM/ZnO np/ZnO: Al/Ni (50 nm)/Al (4um)/	Stacked	20.9	0.213	[17]

Source: * Denotes certified results.

KEYWORDS

- tandem solar cells
- 4-terminal tandem
- 2-terminal tandem
- mechanical stack tandem
- monolithic tandem
- perovskite tandem

REFERENCES

1. Bush, K. A.; Palmstrom, A. F.; Yu, Z. J.; Boccard, M.; Cheacharoen, R.; Mailoa, J. P.; McMeekin, D. P.; Hoye. R. L. Z.; Bailie, C. D.; Leijtens, T.; Peters, I. M.; Minichetti, M. C.; Rolston, N.; Prasanna, R.; Sofia, S.; Harwood, D.; Ma, W.; Moghadam, F.; Snaith, H. J.; Buonassisi, T.; Holman, Z. C.; Bent, S. F.; McGehee, M. D. 23.6% Efficient

Monolithic Perovskite/Silicon Tandem Solar Cells with Improved Stability. *Nat. Energy* **2017,** *2,* 17009. (Received Sept 2, 2016; accepted Jan 16, 2017; Published Feb 17, 2017)

2. Niesen, B.; Werner, J.; Sahli, F.; Kamino, B.; Bräuninger, M.; Fiala, P.; Yang TC-J; Walter, A.; Moon S-J; Barraud, L.; Paviet-Salomon, B.; Allebé, C.; Monnard, R.; Dupré, O.; Boccard M; Despeisse, M.; Jeangros, Q.; Nicolay, S.; Ballif, C. "High-Efficiency Perovskite/Silicon Tandem Solar Cells", EN08: Low-Cost Tandem Photovoltaic Cells Symposium, 2018 MRS Spring Meeting and Exhibit, Apr 2–6, 2018, Phoenix, Arizona.

3. Eperon, G. E.; Leijtens, T.; Bush, K. A.; Prasanna, R.; Green, T.; Wang, J, T.-W.; McMeekin, D. P.; Volonakis, G.; Milot, R. L.; May, R.; Palmstrom, A.; Slotcavage, D. J.; Belisle, R. A.; Patel, J. B.; Parrott, E. S.; Sutton, R. J.; Ma, W.; Moghadam, F.; Conings,, B.; Babayigit, A.; Boyen, H.-G.; Bent, S.; Giustino, F.; Herz, L. M.; Johnston, M. B.; McGehee, M. D.; Snaith, H. J. Perovskite-Perovskite Tandem Photovoltaics with Optimized Bandgaps. *Science* **2016,** DOI: 10.1126/science.aaf9717. (Received April 27, 2016; resubmitted Aug 15, 2016; Accepted Oct 4, 2016; Published online Oct 20, 2016)

4. Forgács, D.; Gil-Escrig, L.; Pérez-Del-Rey, D.; Momblona, C.; Werner, J.; Niesen, B.; Ballif, C.; Sessolo, M.; Bolink, H. J. Efficient Monolithic Perovskite/Perovskite Tandem Solar Cells. *Adv. Energy Mater.* **2017,** *7,* 1602121. (Received: Sept 22, 2016; Revised: Oct 31, 2016; Published online: Dec 14, 2016)

5. Uzu, H.; Ichikawa, M.; Hino, M.; Nakano, K.; Meguro, T.; Hernández, J. L.; Kim, H.-S.; Park, N.-G.; Yamamoto, K. High Efficiency Solar Cells Combining a Perovskite and a Silicon Heterojunction Solar Cells Via an Optical Splitting System. *Appl. Phys. Lett.* **2015,** *106,* 013506. (Received Oct 5, 2014; accepted Dec 12, 2014; published online Jan 9, 2015)

6. Sheng, R.; Ho-Baillie, A. W. Y.; Huang, S.; Keevers, M.; Hao, X.; Jiang, L.; Cheng, Y.-B.; Green, M. A. Four-Terminal Tandem Solar Cells using CH3NH3PbBr3 by Spectrum Splitting. *J. Phys. Chem. Lett.* **2015,** *6*(19), pp 3931–3934 (Received: July 26, 2015; Accepted: Sept 14, 2015; Published: Sept 14, 2015)

7. Kinoshita, T.; Nonomura, K.; Jeon, N. J.; Giordano, F.; Abate, A.; Uchida, S.; Kubo, T.; Seok, S. I.; Nazeeruddin, M. K.; Hagfeldt, A.; Gratzel, M.; Segawa, H. Spectral Splitting Photovoltaics Using Perovskite and Wideband Dye-Sensitized Solar Cells. *Nat. Commun.* **2015,** *6,* 8834. (Received Aug 14, 2015; Accepted Oct 8, 2015; Published Nov 5, 2015)

8. Duong, T.; Grant, D.; Rahman, S.; Blakers, A.; Weber, K. J.; Catchpole, K. R.; White, T. P. Filterless Spectral Splitting Perovskite–Silicon Tandem System With >23% Calculated Efficiency. *IEEE J. Photovoltaics* **2016,** *6*(6), 1432–1439. (Received May 17, 2016; Revised July 17, 2016 and August 2, 2016; Accepted August 8, 2016; Published August 29, 2016).

9. Werner, J.; Barraud, L.; Walter, A.; Bräuninger, M.; Sahli, F.; Sacchetto, D.; Tétreault, N.; Paviet-Salomon, B.; Moon, S. J.; Allebé, C.; Despeisse, M.; Nicolay, S.; De Wolf, S.; Niesen, B.; Ballif, C. Efficient Near-Infrared-Transparent Perovskite Solar Cells Enabling Direct Comparison of 4-Terminal and Monolithic Perovskite/Silicon Tandem Cells. *ACS Energy Lett.* **2016,** *1,* 474–480. (Received: July 8, 2016; Accepted: July 30, 2016; Published: July 30, 2016)

10. Bailie, C. D.; Christoforo, M. G.; Mailoa, J. P.; Bowring, A. R.; Unger, E. L.; Nguyen, W. H.; Burschka, J.; Pellet, N.; Lee, J. Z.; Gratzel, M.; Noufi, R.; Buonassisi, T.; Salleo,

A.; McGehee, M. D. Semi-Transparent Perovskite Solar Cells for Tandems with Silicon and CIGS. *Energy Environ. Sci.* **2015**, *8*, 956–963. (Received Oct 21, 2014; Accepted Dec 22, 2014)

11. Yang, Y. (M.); Chen, Q.; Hsieh, Y.-T.; Song, T.-B.; De Marco, N.; Zhou, H.; Yang, Y. Multilayer Transparent Top Electrode for Solution Processed Perovskite/Cu(In,Ga)(Se,S)2 Four Terminal Tandem Solar Cells. *ACS Nano* **2015**, *9* (7), pp 7714–7721. (Received May 27, 2015; Accepted June 21, 2015; Published June 22, 2015)

12. Duong, T.; Lal, N.; Grant, D.; Jacobs, D.; Zheng, P.; Rahman, S.; Shen, H.; Stocks, M.; Blakers, A.; Weber, K.; White, T. P.; Catchpole, K. R. Semitransparent Perovskite Solar Cell With Sputtered Front and Rear Electrodes for a Four-Terminal Tandem. *IEEE J. Photovoltaics* **2016**, *VOL. 6, NO. 3, MAY 2016*, 679–687 (Received Dec 14, 2015; Revised Jan 8, 2016; Accepted Jan 17, 2016. Published: Feb 08, 2016)

13. Ren, Z.; Zhou, J.; Ng, A.; Shen, Q.; Shen, H.; Surya, C. Record High Performance of Perovskite/Crystalline Silicon Four-Terminal Tandem Solar Cells. In *Proc. of IEEE 43rd Photovoltaic Specialists Conference* (PVSC), June 5–10 2016, DOI: 10.1109/PVSC.2016.7749719

14. Chen, B.; Bai, Y.; Yu, Z.; Li, T.; Zheng, X.; Dong, Q.; Shen, L.; Boccard, M.; Gruverman, A.; Holman, Z.; Huang, J. Efficient Semitransparent Perovskite Solar Cells for 23.0%-Efficiency Perovskite/Silicon Four-Terminal Tandem Cells. *Adv. Energy Mater.* **2019**, *6*, 1601128. (Received: May 29, 2016; Revised: June 23, 2016; Published online: July 19, 2016)

15. Peng, J.; Duong, T.; Zhou, X.; Shen, H.; Wu, Y.; Mulmudi, H. K.; Wan, Y.; Zhong, D.; Li, J.; Tsuzuki, T.; Weber, K. J.; Catchpole, K. R.; White, T. P. Efficient Indium-Doped TiO$_x$ Electron Transport Layers for High-Performance Perovskite Solar Cells and Perovskite-Silicon Tandems. *Adv. Energy Mater.* **2017**, *7*, 1601768. (Received: Aug 10, 2016; Revised: Sept 16, 2016; Published online: Nov 4, 2016)

16. Schinarakis, K. Record for Perovskite/CIGS Tandem Solar Module, Press Release, http://www.kit.edu/kit/english/pi_2016_133_record-for-perovskite-cigs-tandem-solar-module.php (assessed February, 2017).

17. Fu, F.; Feurer, T.; Weiss, T. P.; Pisoni, S.; Avancini, E.; Andres, C.; Buecheler, S.; Tiwari, A. N. High-Efficiency Inverted Semi-Transparent Planar Perovskite Solar Cells in Substrate Configuration. *Nat. Energy* **2017**, *2*, 16190 (<NS>Received Aug 8, 2016; accepted Nov 14, 2016; published Dec 19, 2016</NS>)

18. Dudem, B.; et al. Ch$_3$NH$_3$PBI$_3$ Planar Perovskite Solar Cells with Antireflection and Self-Cleaning Function Layers. *J. Mater. Chem. A* **2016**, *4*, 7573–7579.

19. Li, K.; et al. Versatile Biomimetic Haze Films for Efficiency Enhancement of Photovoltaic Devices. *J. Mater. Chem. A* **2017**, *5*, 969–974.

20. Jiang, Y. J.; Almansouri, I.; Huang, S. J.; Young, T.; Li Y; Peng, Y.; Hou, Q. C.; Spiccia, L.; Bach, U.; Cheng, Y. B.; Green, M. A.; Ho-Baillie, A. Optical Analysis of Perovskite/Silicon Tandem Solar Cells. *J. Mater. Chem. C* **2016**, *4*, 5679.

21. Grant, D.; Weber, K.; Stocks, M.; White, T. P. Optical Optimization of Perovskite-Silicon Reflective Tandem Solar Cell, OSA Technical Digest (online) (Optical Society of America), paper PTh3B.3 **2015**.

22. Albrecht, S.; Saliba, M.; Correa-Baena, J. P.; Jäger, K.; Korte, L.; Hagfeldt, A.; Grätzel, M.; Rech, B. Towards Optical Optimization of Planar Monolithic Perovskite/Silicon-Heterojunction Tandem Solar Cells. *J. Opt.* **2016**, *18*, 064012.

23. Yang, Z.; et al., Stable Low-Bandgap Pb–Sn Binary Perovskites for Tandem Solar Cells. *Adv. Mater.* **2016,** *28,* 8990–8997.

24. Hao, F.; et al., Anomalous Band Gap Behavior in Mixed Sn and Pb Perovskites Enables Broadening of Absorption Spectrum in Solar Cells. *J. Am. Chem. Soc.* **2014,** *136,* 8094–8099.

25. Eperon, G.; et al., Perovskite-Perovskite Tandem Photovoltaics with Optimized Bandgaps. *Science* **2016,** DOI: 10.1126/science.aaf9717.

26. Pellet, N.; et al., Mixed-Organic-Cation Perovskite Photovoltaics for Enhanced Solar-Light Harvesting. *Angew. Chem.* **2014,** *126,* 3215–3221.

27. Noh, J. H.; Im, S. H.; Heo, J. H.; Mandal, T. N.; Seok, S. I. Chemical Management for Colorful, Efficient, and Stable Inorganic–Organic Hybrid Nanostructured Solar Cells. *Nano Lett.* **2013,** *13,* 1764–1769.

28. Bi, C.; et al., Low-Temperature Fabrication of Efficient Wide-Bandgap Organolead Trihalide Perovskite Solar Cells. *Adv. Energy Mater.* **2015,** *5,* 1401616.

29. McMeekin, D.; et al., A Mixed-Cation Lead Mixed-Halide Perovskite Absorber for Tandem Solar Cells. *Science* **2016,** *351,* 151. DOI: 10.1126/science.aad5845.

30. Beal, R.; et al. *J. Phys. Chem. Lett.* **2016,** *7,* 746–751.

31. Bremner, S. P.; Yi, C.; Almansouri, I.; Ho-Baillie, A.; Green, M. A. Optimum Band Gap Combinations to Make Best Use of New Photovoltaic Materials. *Sol. Energy* **2016,** *135,* 750–757.

32. Todorov, T.; Gershon, T.; Gunawan, O.; Sturdevant, C.; Guha, S. Perovskite-Kesterite Monolithic Tandem Solar Cells with High Open-Circuit Voltage. *Appl. Phys. Lett.* **2014,** *105,* 173902. (Received Sept 5, 2014; accepted Oct 3, 2014; published online Oct 28, 2014)

33. Todorov, T.; Gershon, T.; Gunawan, O.; Lee, Y. S.; Sturdevant, C.; Chang, L. Y.; Guha, S. Monolithic Perovskite-CIGS Tandem Solar Cells via In Situ Band Gap Engineering. *Adv. Energy Mater.* **2015,** *5,* 1500799. (Received: April 22, 2015; Revised: June 19, 2015; Published online: Sept 30, 2015)

34. Chen, C.-C.; Bae, S.-H.; Chang, W.-H.; Hong, Z.; Li, G.; Chen, Q.; Zhou, H.; Yang, Y. Perovskite/Polymer Monolithic Hybrid Tandem Solar Cells Utilizing a Low-Temperature, Full Solution Process. *Mater. Horiz.* **2015,** *2,* 203–211. (Received Dec 9, 2014; Accepted Jan 5, 2015)

35. Mailoa, J. P.; Bailie, C. D.; Johlin, E. C.; Hoke, E. T.; Akey, A. J.; Nguyen, W. H.; McGehee, M. D.; Buonassisi, T. A 2-Terminal Perovskite/Silicon Multijunction Solar Cell Enabled by a Silicon Tunnel Junction. *Appl. Phys. Lett.* **2015,** *106,* 121105. (Received Feb 4, 2015; accepted Feb 23, 2015; published online March 24, 2015)

36. Albrecht, S.; Saliba, M.; Baena, J. P. C.; Lang, F.; Kegelmann, L.; Mews, M.; Steier, L.; Abate, A.; Rappich, J.; Korte, L.; Schlatmann, R.; Nazeeruddin, M. K.; Hagfeldt, A.; Gratzel, M.; Rech, B. Monolithic Perovskite/Silicon-Heterojunction Tandem Solar Cells Processed at Low Temperature. *Energy Environ. Sci.* **2016,** *9,* 81–88. (Received Sept 26, 2015; Accepted Oct 23, 2015)

37. Werner, J.; Weng, C. H.; Walter, A.; Fesquet, L.; Seif, J. P.; De Wolf, S.; Niesen, B.; Ballif, C. Efficient Monolithic Perovskite/Silicon Tandem Solar Cell with Cell Area >1 cm(2). *J. Phys. Chem. Lett.* **2016,** *7,* 161–166. (Received: Dec 3, 2015; Accepted: Dec 19, 2015; Published: Dec 19, 2015)

38. Heo, J. H.; Im, S. H. $CH_3NH_3PbBr_3$ –$CH_3NH_3PbI_3$ Perovskite–Perovskite Tandem Solar Cells with Exceeding 2.2 V Open Circuit Voltage. *Adv. Mater.* **2015,** DOI: 10.1002/

adma.201501629. (Received: April 6, 2015; Revised: Sept 6, 2015; Published online: Oct 27, 2015)

39. Jiang, F.; Liu, T.; Luo, B.; Tong, J.; Qin, F.; Xiong, S.; Li, Z.; Zhou, Y. A Two-Terminal Perovskite/Perovskite Tandem Solar Cell. *J. Mater. Chem. A.* **2016,** *4,* 1208–1213. (Received Oct 29, 2015; Accepted Dec 15, 2015)

40. Zhao, D.; Wang, C.; Song, Z.; Yu, Y.; Chen, C.; Zhao, X.; Zhu, K.; Yan, Y. Four-Terminal All-Perovskite Tandem Solar Cells Achieving Power Conversion Efficiencies Exceeding 23%. *ACS Energy Lett.* **2018,** *3* (2), 305–306

41. Löper, P.; Moon, S.-J.; de Nicolas, S. M.; Niesen, B.; Ledinsky, M.; Nicolay, S.; Bailat, J.; Yum, J.–H.; De Wolf, S.; Ballif, C. Organic–Inorganic Halide Perovskite/Crystalline Silicon Four-Terminal Tandem Solar Cells. *Phys. Chem. Chem. Phys.* **2015,** *17,* 1619–1629. (Received Aug 23, 2014; Accepted Nov 14, 2014)

42. Lang, F.; Glub, M. A.; Albrecht, S.; Rappich, J.; Korte, L.; Rech, B.; Nickel, N. H. Perovskite Solar Cells with Large-Area CVD-Graphene for Tandem Solar Cells. *J. Phys. Chem. Lett.* **2015,** *6,* 2745–2750. (Received: June 4, 2015; Accepted: June 25, 2015; Published: June 25, 2015)

43. Bush, K. A.; Bailie, C. D.; Chen, Y.; Bowring, A. R.; Wang, W.; Ma, W.; Leijtens, T.; Moghadam, F.; McGehee, M. D. Thermal and Environmental Stability of Semi-Transparent Perovskite Solar Cells for Tandems Enabled by a Solution-Processed Nanoparticle Buffer Layer and Sputtered ITO Electrode. *Adv. Mater.* **2016,** *28,* 3937–3943. (Received: October 26, 2015; Revised: December 21, 2015. Published online: February 16, 2016)

44. Quiroz, C. O. R.; Shen, Y.; Salvador, M.; Forberich, K.; Schrenker, N.; Spyropulos, G. D.; Huemueller, T.; Wilkinson, B.; Kirchartz, T.; Spiecker, E.; Verlinden, P. J.; Zhang, X.; Green, M. A.; Ho-Baillie, A.;Brabec, C. J. Balancing Electrical and Optical Losses for Efficient Si-Perovskite 4-Terminal Solar Cells with Solution Processed Percolation Electrodes. *J. Mater. Chem. A* **2018,** *6,* 3583–3592.

45. Fan, F.; Feurer, T.; Jäger, T.; Avancini, E.; Bissig, B.; Yoon, S.; Buecheler, S.; Tiwari, A. N. Low-Temperature-Processed Efficient Semi-Transparent Planar Perovskite Solar Cells for Bifacial and Tandem Applications. *Nat. Commun.* **2015,** *6,* 8932. (Received: July 31, 2015; Accepted: Oct 19, 2015; Published online: Nov 18, 2015)

46. Duong, T.; Wu, Y.; Shen, H.; Peng, J.; Fu, X.; Jacobs, D.; Wang, E.-C.; Kho, T. C.; Fong, K. C.; Stocks, M.; Franklin, E.; Blakers, A.; Zin, N.; McIntosh, K.; Li, W.; Cheng, Y.-B.; White, T. P.; Weber, K.; Catchpole, K. Rubidium multi-cation perovskite with optimized bandgap for perovskite-silicon tandem with over 26% efficiency. *Adv. Energy Mater.* **2017,** *7,* 1700228. (Accepted)

CHAPTER 17

LIFE CYCLE ASSESSMENTS

Life cycle assessment or life cycle analysis (LCA) is an important tool for assessing technology based on its environmental impacts associated with all stages of production (from raw material extraction and processing, manufacture), distribution, use (including maintenance and repair), and end-of-life (disposal or recycling). LCA is useful as it quantifies the key environmental impacts in each process and material into categories such as impacts on global warming potential, human health (toxicity including carcinogenic and noncarcinogenic effects), respiratory system by inorganics, freshwater and marine ecosystems (including ecotoxicity and eutrophication, acidification), malodorous air, ozone depletion, photochemical oxidation (summer smog), land use and human health by ionization radiation, land-use, terrestrial ecotoxicity, non-renewable resource depletion, and energy payback time (EPBT) which is most relevant to energy generation technology. The results are important as they identify "hot spots" which guide research on designs, use of these materials and manufacturing processes that not only gives best energy performance but with the lowest environmental impacts.

As research on perovskite solar cells grow exponentially since 2012, LCA on this technology started to appear in 2015 evaluating the environmental impacts of various cell designs, materials, and processing sequences see Table 17.1. The system boundary in these studies is typically cradle to gate due to the lack of information on cell or module deployment. The functional unit is typically 1 kWhr. Ecoinvent is the most commonly used database in these studies. With regards to life cycle inventories, some of the data are adapted from existing ones for common materials or from previous studies, while some are determined from the material usage from laboratory scale syntheses and processes. These data are likely to produce conservative (high) LCA impacts until more efficient, better utilization industrial scale processes are established. However, as many of these LCA are "first of its kind," it is not uncommon to have oversights in these early studies especially, when there is no "standard" process sequence for perovskite solar cell

TABLE 17.1 Summary of Life Cycle Analysis on Perovskite Solar Cells.

Cell structure	System boundary	Cell Eff. (%)	Lifetime (years)	Process	Solar insolation (kWh/m²)	Performance ratio (%)	Energy payback time (EPBT) (years)	Global warming potential (GWP) (g CO₂-eq/kWh)
[1] FTO glass/c-TiO₂/MAPbI₃ (Cl)/spiro-OMeTAD/ Ag	Cradle to gate	15.4	1.00	Solution except perovskite and metal by thermal evaporation	1700	80	17.32	5480
[1]			15.00				1.15	366
[2]			5.00				CBD	1100
[1] ITO glass/ PEDOT: PSS/ MAPbI₃ (Cl)/ PCBM/Al	Cradle to gate	11.5	1.00	Solution except metal by thermal evaporation	1700	80	16.54	5240
[1]			15.00				1.10	350
[2]			5.00				CBD	1050
[3] FTO glass/c-TiO₂/mp-TiO₂/ MAPbI₃/spiro-OMeTAD/ Au	Cradle to gate	9.1	2.00	Solution except metal by thermal evaporation	1960	80	0.27	82.5
[3]		9.1	2.00		1700	75	0.31	93
[2]		9.1	5.00		1700	75	0.27	32
[3] ITO glass/ ZnO/MAPbI₃/ spiro-OMeTAD/ Ag	Cradle to gate	11	2.00	Solution except metal by thermal evaporation	1960	80	0.19	60.1
[3]		11	2.00		1700	75	0.23	67.3
[2]		11	5.00		1700	75	0.19	24
[4] FTO glass/c-TiO₂/ MAPbI₃ (Cl)/ spiro-OMeTAD/ Ag	Cradle to gate	15.4	1.00	Solution except perovskite by dual source evaporation	1700	80	NR	5415

TABLE 17.1 (Continued)

	Cell structure	System boundary	Cell Eff. (%)	Lifetime (years)	Process	Solar insolation (kWh/m²)	Performance ratio (%)	Energy payback time (EPBT) (years)	Global warming potential (GWP) (g CO₂-eq/kWh)
[4]	ITO glass/ PEDOT: PSS/ MAPbI₃ (Cl)/ PCBM/Al	Cradle to gate	11.5	1.00	Solution except metal by thermal evaporation	1700	80	NR	5198
[4]	FTO glass/c-TiO₂/mp-TiO₂/	Cradle to gate	6.4	1.00	Solution except c-TiO₂ by	1700	80	NR	10,621
[2]	MASnI₃₋ₓBrₓ/ spiro-OMeTAD/ Au	Cradle to gate	6.4	5.00	ALD and metal by thermal evaporation	1700	80	NR	1880
[4]	ITO PET/ FEDOT: PSS/ MAPbI₃ (Cl)/	Landfill	9.2	1.00	Solution except metal by thermal evaporation	1700 / 1700	80 / 80	NR	0.08
[4]	FCBM/Al	Incineration/lead recovery			Solution except metal by thermal evaporation			NR	<0.02
[5]	FTO glass/	Cradle to gate	6.5	1.00	Solution	1000	75	55.4	5867
[5]	TiO₂ nanotube/	gate	6.5	1.00		1700	75	32.57	3448
[5]	MAPbI₃/liquid		10	1.00		1700	75	21.14	2222
[5]	electrolyte/Pt		15	1.00		1700	75	14.2	1476
[2]			6.5	5.00		1700	75	32.57	414
[2]	FTO glass/SnO₂/ MAPbI₃/carbon	Cradle to gate	15	5.00	Solution (perovskite by spraying)	1700	75	1.05	99

TABLE 17.1 *(Continued)*

	Cell structure	System boundary	Cell Eff. (%)	Lifetime (years)	Process	Solar insolation (kWh/m^2)	Performance ratio (%)	Energy payback time (EPBT) (years)	Global warming potential (GWP) (g CO$_2$-eq/kWh)
[2]	FTO glass/SnO$_2$/MAPbI$_3$/CuSCN/MoO$_x$/Al	Cradle to gate	15	5.00	Solution (perovskite by spraying) except MoO$_x$ and metal by thermal evaporation	1700	75	1.3	147
[2]	FTO glass/SnO$_2$/MAPbI$_3$/CuSCN/MoO$_x$/Al	Cradle to gate	15	5.00	Solution except perovskite, MoO$_x$ and metal by thermal evaporation	1700	75	1.54	205
[6]	HIT Si cell	Cradle to grave	20.00	20	N/A	1700	75	2	161
[6]	p-n Si cell	Cradle to grave	16.00	20	N/A	1700	75	1.6	48
[6]	FTO/c-TiO$_2$/MAPbI$_3$/Spiro-OMeTAD/MoO$_3$/ITO/Ag grid	Cradle to grave	17.00	1	Solution except metal by thermal evaporation	1700	75	1.1	6639
[6]	FTO/c-TiO$_2$/MAPbI$_3$/Spiro-OMeTAD/MoO$_3$/ITO/Au grid	Cradle to grave	17.00	1	Solution except metal by thermal evaporation	1700	75	1.1	6668

TABLE 17.1 (Continued)

Cell structure	System boundary	Cell Eff. (%)	Lifetime (years)	Process	Solar insolation (kWh/m²)	Performance ratio (%)	Energy payback time (EPBT) (years)	Global warming potential (GWP) (g CO₂-eq/kWh)
[6] ITO/PEDOT: PSS/MAPbI₃/PCBM/ZnO/ITO/Al grid	Cradle to grave	17.00	1	Solution except metal by thermal evaporation	1700	75	0.9	4327
[6] HIT Si/c-TiO₂/MAPbI₃/Spiro-OMeTAD/MoO₃/ITO/Ag grid	Cradle to grave	27.00	1	Solution except metal by thermal evaporation	1700	75	1.5	5930
[6]			1 (P/Si)+19 (Si)				1.7	393
[6]			5 (P/Si)				1.5	1186
[6]			5 (P/Si)+15 (Si)				1.7	368
[6] HIT Si/c-TiO2/MAPbI₃/Spiro-OMeTAD/MoO₃/ITO/Au grid	Cradle to grave	27.00	1	Solution except metal by thermal evaporation	1700	75	1.5	5932
[6]			1 (P/Si)+19 (Si)				1.7	394
[6]			5 (P/Si)				1.5	1187
[6]			5 (P/Si)+15 (Si)				1.7	368
[6] p-n Si/ITO/PEDOT: PSS/MAPbI₃/PCBM/ZnO/ITO/Al grid	Cradle to grave	24.00	1	Solution except metal by thermal evaporation	1700	75	1.3	3729
[6]			1 (P/Si)+19 (Si)				1.6	273
[6]			5 (P/Si)				1.3	746
[6]			5 (P/Si)+15 (Si)				1.6	249

Note: C3D—Values cannot be determined. NR—Values not reported.

TABLE 17.2 Summary of EPBT and GWP of Other Photovoltaic Technologies for Comparison.

	EPBT (years)			GWP (g CO_2-eq/kWh)		
	[3]	[5]	[2]	[3]	[5]	[2]
mono-crystalline silicon (mono-Si)	2.38	1.7–2.7	4.10	37.6	29 to 45	23
multi-crystalline silicon (multi-Si)	1.90	1.5–2.6	3.10	29.1	23 to 44	20
ribbon-Si[3]; a-Si[5,2]	1.41	1.8–5.5	2.30	21.7	18 to 50	12
Cadmium telluride (CdTe)	0.63	0.75–2.10	1.00	16.3	14 to 35	12
Copper indium gallium diselenide (CIGS)	N/A	N/A	1.70	N/A	N/A	17
Organic photovoltaics (OPV)[3] or dye-sensitized solar cells (DSSC)[5]	0.35	0.6–12.6	N/A	78.2	19 to 120	N/A

fabrication. The wide range of values reported in these studies is a reflection of this. Therefore, sensitivity analysis and uncertainty analysis can be helpful in determining critical inputs that require more rigorous data collection.

The first[1] of the perovskite LCA compares two perovskite solar cells with different architectures (standard polarity vs. inverted polarity) and different deposition methods (solution process vs. dual source thermal evaporation) for the perovskite light-absorbing layer (chlorine incorporated $MAPBI_3$). A sensitivity analysis is carried out to show the effect of cell lifetime (1–15 years) on total environmental impacts. Conclusions in that report include the higher toxicity impact from MAI compared to PbI_2 due to solvent used and higher total environmental impact from FTO glass, PCBM, spiro-OMeTAD, and Ag compared to indium tin oxide (ITO) glass, TiO_2, PEDOT: PSS, and Al, respectively. The conclusion that vapor deposited (dual-source thermal evaporation) perovskite uses lower energy than solution processed perovskite is debatable as it does not appear to take into account the very low thermal evaporation rate and therefore, it takes long process time. In addition, it appears that the energy required for maintaining low chamber pressure for thermal evaporation has not been considered. On the other hand, the energy required to maintain the glovebox environment for the perovskite solution process has been considered.

The EPBT reported in this work, changes with the device lifetime, which is atypical. Although the report concludes that there is no compelling reason to dismiss lead (Pb) containing perovskites, the scope of the analysis is limited to cradle to gate without considering the impact from landfill or recycling at the end of life or the possibility of Pb leakage during the production or operation.

The second LCA work[3] compares the perovskite cells that use TiO_2 versus ZnO as the hole blocking layer fabricated on FTO glass versus ITO glass, respectively and use different metals (Au for TiO_2-based vs. Ag for ZnO-based cells). A sensitivity analysis is carried out looking at the effect of performance ratio (which is a ratio of actual power output to nominal output of the solar module), primary energy consumption (used in the cradle to gate stages), module efficiency and insolation (the amount of sun's energy on the module in $kWh/m^2/year$ which is location dependent) on EPBT and global warming potential (GWP). GWP is a relative measure of heat trapped by greenhouse gas emissions in the atmosphere. It is often expressed as the equivalent mass of carbon dioxide that will trap the same amount of heat by the greenhouse gasses (g CO_2-eq/kWh). GWP is dependent on the module lifetime, while EPBT remains constant. Results from sensitivity analysis show that EPBT and CO_2 are most sensitive to changes in performance ratio and primary energy consumption. ITO substrate is found to consume about 2.5 times the primary energy embedded in the FTO substrates. This is due to the use of precious metal indium during manufacturing resulting in large energy consumption. The high mass ratio (95%) of the ITO glass over the module also means that the ITO glass dominates the impact categories of acidification, eutrophication, human toxicity, ionizing radiation, land use, malodorous air, depletion of abiotic resources, and terrestrial ecotoxicity. The report also highlights that the undesirable impacts (eutrophication, fresh water aquatic ecotoxicity, fresh water sediment ecotoxicity, human toxicity, land use, marine aquatic ecotoxicity, marine sediment ecotoxicity, depletion of abiotic resources, stratospheric ozone depletion, terrestrial ecotoxicity, acidification, and malodorous air) from the use of gold. This is because "the production of gold from ores not only requires a large amount of energy, but also leads to the release of toxic mine drainage into lakes and rivers. The drainage contains nitrates, sulfides, arsenic, antimony, and mercury, which can cause acidification and eutrophication, and are extremely harmful to aquatic organisms. Therefore, the use of gold is environmentally expensive, and the replacement of gold results in a substantial reduction in most environmental impacts." The intensive energy use and low utilization by thermal evaporation (for metal electrode deposition in this work) is highlighted. The impacts of solvents are also reported. The emission or evaporation of ethanol, acetone, isopropanol, and chlorobenzene contributes to photochemical oxidation (summer smog). Chloroform (used for ZnO precursor) contributes to stratospheric ozone depletion impact.

Again, as acknowledged in the work, the impacts from usage, transport and disposal of the solar modules are not considered. In this work, within the cradle to gate system boundary, EPBT of perovskite solar cells is the lowest compared to other photovoltaic technologies, see Table 17.1 and 17.2. However, the lifetime of perovskite solar module has to be at least 5 years for its GWP impact to be compatible with those of the incumbent photovoltaic technologies.

The findings of this work highlight the importance of producing perovskite solar module with stable performance (to maintain high-performance ratio over its lifetime) using processes and materials that consume (i) minimum primary energy (e.g., elimination of gold and other precious metals; elimination of thermal evaporation for the deposition of electrodes) and (ii) minimum amount of organic solvents. In cases, when organic solvents cannot be eliminated, recycling should be considered.

Serrano-Lujan et al.[4] then reported LCA on three kinds of perovskite solar cells. The first two cells have identical "standard polarity" and "inverted" structures as the devices analyzed by Espinosa et al.[1] The third one is tin (Sn)-based cell with standard polarity structure except it uses gold instead of silver as the rear electrode. In addition, the Sn-based cell has an extra step of depositing meso-porous TiO_2 for the infiltration of Sn-based perovskite. A cradle to gate analysis (system boundary 1) is carried out for these three cells with an assumed lifetime of one year. The Pb-containing cell is further analyzed but for the end of life scenarios only. The Pb-containing cell for system boundary two also has a modified cell structure. Low temperature processes are used to process this standard polarity cell on ITO polyethylene terephthalate (PET) substrate. For the disposal phase, two scenarios are considered. The first scenario involves landfill assuming the worst case scenario where 70% of Pb will be leached from the cells to the soil during the first year. The second scenario involves incineration assuming the best-case scenario where 98% of Pb will be recovered from the ashes. This work suggests that Sn-based perovskites cells do not seem to have advantages over Pb-containing ones based on the higher environmental impacts from the former. This is because of the inferior design and performance assumed for the Sn-based cell, which uses gold, and with a cell performance less than half of the Pb-containing counterpart. In reality, the Sn-based cells are likely to be less stable than the Pb-containing ones, too. The advantage of Pb recovery from incineration versus landfill can be seen. However, this is swamped by the impacts (e.g., freshwater ecotoxicity, human toxicity, respiratory inorganics, freshwater eutrophication, and marine eutrophication) from the waste treatment of the PET substrate. The work also argues that Sn is 6.3 times more expensive than Pb, fresh ecotoxicity is higher from emitted Sn than emitted Pb, and the production of Sn has larger impact than Pb in terms of terrestrial ecotoxicity and GWP. Therefore, given the lower lifetime and performance of Sn-based perovskite solar cells and only very small amount of metal present in the perovskite layer, Pb-containing perovskite is still a rational choice especially, when a recycling end of life scenario is in place.

Zhang et al.[5] analyzed the environmental impacts of a less efficient liquid perovskite cell using TiO_2 nanotube based on dye-synthesized solar cell from cradle to gate without considering nanowaste and nano-emissions as acknowledged in the paper. A sensitivity analysis shows the effect of solar insolation, conversion efficiency, and lifetime on the environmental impacts. Results show that a module lifetime of 5 years is a minimum to reduce the impacts by an order of magnitude. The majority of the environmental impacts come from the raw material extraction for TiO_2 nanotube and perovskite due to the solvents used. Methyl-ammonium iodide (MAI) has higher impact than PbI_2 because of the amount of solvent used. This is consistent with finding by Espinosa et al.[1] The EPBT's calculated in this work are in the same order of magnitude as those calculated by Espinosa et al.[1] but are higher than those by Gong et al.[3] Perovskite cell performance plays a key role in bringing the EPBT down to match those of incumbent photovoltaic technologies.

Celik et al.[2] analyzed impacts from cells fabricated by spraying or thermal evaporation process. Optimistic efficiency (15%) and lifetime (5 years) are assumed for these cells with relatively new planar structures. Cells that use carbon as the rear electrode do not have hole transport layers. Cells that use inorganic hole transport layer CuSCN has the MoO_x/Al stack on the rear. Devices with such structure at the nominated efficiency are yet to be demonstrated. Nevertheless, results show that solution processed perovskites have lower impacts (except for marine eutrophication due to solvents used) than thermally evaporated perovskites due to lower energy consumption. Similar conclusion is drawn on the low impact of Pb due to the small amount present in the perovskite layer. Similar to most of the perovskite solar cell LCA, this analysis is limited to cradle to gate without considering the scenarios of having to dispose Pb or recycle Pb. Nevertheless, the calculated EPBT of 1–1.5 years is comparable to those of incumbent technologies. GWP will improve and match those of incumbents as lifetime of the perovskite module improves. This assumes a relatively simple cell structure with inexpensive electrode materials and respectable conversion efficiency.

Lunardi et al.[6] conducted a LCA on three perovskite/silicon (Si) tandem cell structures. The first two structures have a standard polarity perovskite cell on a hetero-junction (HIT) cell using silver (Ag) or gold (Au) grid for the top electrode. The third structure has an inverted perovskite cell on a commonly available p-n junction (n-type emitter on the "top") Si cell using aluminum (Al) grid for the top electrode. The impacts assessed include global warming, human toxicity, freshwater eutrophication and ecotoxicity and abiotic depletion potential impacts. Two scenarios are assumed in terms of the lifetime of the tandem device. The first scenario assumes a lifetime of one year for the top perovskite cell, which then turns opaque making it impossible for the bottom Si cell to continue energy generation. The second scenario assumes a lifetime of one year for the top perovskite cell which "expires gracefully" becoming transparent for the bottom Si cell to continue energy generation for the rest (19 years) of its lifetime. The EPBT of these tandem cells in different scenarios (as the amount of energy generated will be different) are also calculated and compared with the standalone Si and perovskite single junction cells. EPBT of the perovskite cells calculated are similar to those by Celik et al.,[2] A sensitivity analysis to the lifetime of the tandem for scenario two is carried out for all of the impacts considered.

It is found that transparent perovskite cells with ITO and a metal grid have lower impacts than opaque perovskite solar cells that have full metal contact due to lower metal usage and therefore, the associated environmental impacts. This advantage is reflected in the perovskite/Si tandem that utilizes a top transparent perovskite cell. Aluminum is a better material choice than Ag and Au for electrode as it has the lowest environmental impacts in all categories. Spiro-OMeTAD has a higher impact (higher energy consumption) than PEDOT: PSS which is consistent with previous findings in.[1, 3, 5] p-n junction Si cell has lower impact than the HIT cell due to the absence of amorphous Si and ITO depositions in the former. Therefore, an inverted perovskite Si cell on a p-n junction (with n-type emitter on the "top" which is a common configuration of commercial cells) using Al grid without the use of Spiro-OMeTAD is an environmental friendly pathway towards perovskite/Si tandem. Perhaps not surprisingly, the lifetime of the perovskite cells has to match that of Si bottom cell for the tandem to be viable alternative to Si cell alone.

KEYWORDS

- life cycle analysis
- environmental impact
- energy payback time
- perovskite cradle to grave
- perovskite toxicity

REFERENCES

1. Espinosa, N.; et al. Solution and Vapour Deposited Lead Perovskite Solar Cells: Ecotoxicity from a Life Cycle Assessment Perspective. *Sol. Energy Mater. Sol. Cells* **2015,** *137,* 303–310.
2. Celik, L.; et al. Life Cycle Assessment (LCA) of Perovskite PV Cells Projected from Lab to Fab. *Sol. Energy Mater. Sol. Cells* **2016,** *156,* 157–169. DOI:10.1016/j.solmat.2016.04.037.
3. Gong, J.; et al. Perovskite Photovoltaics: Life-Cycle Assessment Of Energy and Environmental Impacts. *Energy Envion. Sci.* **2015,** *8,* 1953-1968. DOI: 10.1039/c5ee00615e.
4. Serrano-Lujan, L.; et al. Tin- and Lead-Based Perovskite Solar Cells Under Scrutiny: An Environmental Perspective. *Adv. Energy Mater.* **2015,** *5,* 1501119.
5. Zhang, J.; et al. Life Cycle Assessment of Titania Perovskite Solar Cell Technology for Sustainable Design and Manufacturing. *ChemSusChem.* **2015,** *8,* 3882–3891.
6. Lunardi, M. M.; Ho-Baillie, A. W. Y.; Alvarez-Gaitan, J. P.; Moore, S.; Corkish, R. A Life Cycle Assessment of Perovskite/Silicon Tandem Solar Cells. *Prog. Photovoltaics.* **2017,** *25,* 679-695. DOI: 10.1002/pip.2877.

CHAPTER 18

COMMERCIAL PROSPECTS AND MANUFACTURING COSTS

The strong absorption of perovskites allowing submicron device thicknesses lends itself to wider photovoltaic applications other than the traditional flat panel rigid installations. Flexible, device integrated, building integrated, vehicle intergraded devices are some of the applications worthy to be explored. The bandgap tunability of perovskite by varying its compositions opens up opportunities for chromaticity control apart from tandem applications as discussed in previous section. These strengths of perovskite solar cells are not necessarily found in other thin film semiconductor material. Highly transparent or semi-transparent or semi-opaque modules can be used as energy generating glazing or facades for buildings.

Companies have started to emerge with the aim of commercializing perovskite solar cells or tandem solar cells containing perovskites. They include Greatcell in Australia; Oxford in the UK; Iris PV and Solar-Tectic in the US; Xiamen Weihua Solar and Microquanta Semiconductor in China; and Saule in Poland.

One of the key questions related to commercializing perovskite solar cells is the cost of manufacturing. Answers to this question is important for commercial decisions on the level of investments and for setting performance and quality (e.g., lifetime) targets for the technology to be a viable product. Most importantly, cost analysis allows high cost components to be identified during the process development.

A typical method of calculating manufacturing cost of a technology is a bottom-up "Cost of Ownership" (CoO) approach.[1-3] A process sequence, location of factory, size of production in terms of yearly output, yield and throughput are defined. Inputs include material costs, equipment cost, cost of utilities, labor cost, building cost, maintenance cost, and overheads. Outputs of the cost analysis include the cost of each process step and a breakdown into the main cost components. For the input data, it is common to source information for each material and equipment item from publications, cost reports, price lists, equipment suppliers, or solar cell manufacturers. From the latter two sources, the data will be the most accurate scaled to industrial production. However, the data are likely to be commercially confidential.

In addition, for an emerging technology at its early stages of development, there can be significant uncertainties in the process and costs that impact many of the cost inputs, and some cost inputs can be an order of magnitude more uncertain than others. At the times, industrial scale data are simply not available when the technology is only entering pilot scale production or at lab-scale development.

Perovskite solar cell is an example of early stage photovoltaic technology. Perovskite solar cells have been described as a lower cost alternative to other solar cell technologies because they can be fabricated using solution processes with small material usage and low temperature requirement. However, detailed cost analysis is limited.[4]

Cai et al.[5] analyzed two perovskite on glass modules with structures (i) FTO Glass/(TiO$_2$/ZrO$_2$/carbon)/perovskite and (ii) ITO Glass/PEDOT: PSS/perovskite/PCBM/Ca/Al. The manufacturing costs of these two process sequences are estimated to be $30 and 41/m^2. Module efficiency at levelized cost of electricity (LCOE) 12% and a lifetime of 15 years would yield a LCOE at 3.5–4.9 US cents/kWh making it a competitive energy generation source. At the time of reporting, the cell architectures analyzed[6-7] have only been demonstrated on small areas. For their cost analysis, Cai et al. have assumed that these structures can be manufactured in large modules of high-efficiency serially connected cells using equipment and processes adapted from thin film silicon and dye sensitized solar cell manufacturing lines.

Chang et al.[8] recognize the uncertainties in the cost input data for perovskite cell technology and therefore, developed a costing method that builds on the commonly used CoO approach but factors the uncertainties into the calculations through a Monte Carlo analysis. In their analysis, three values are selected for each cost parameter which are the "nominal," "low", and "high" values representing the uncertainty range. The impact of the uncertainty in each parameter is assessed using a Monte Carlo analysis by generating a few thousand scenarios. For each scenario, the value of each cost parameter is generated randomly according to its two half normal distribution, and then the cost calculations are completed using these generated values. The distribution of the cost outputs from the scenarios can then be analyzed to understand the uncertainty of these cost estimates. The advantage of this method is that it provides a very quick assessment of the process sequence allowing for uncertainties without the long and involved process of collecting "sufficiently accurate" data, which at times are unavailable.

Using this method, the manufacturing cost estimate is delivered very rapidly with a list of cost drivers that highlight the opportunities for the greatest cost improvement. The uncertainty analysis identifies key sources of cost uncertainty highlighting areas that require better data sourcing. Chang et al. applied this method to analysis three perovskite cell structures (Figure 18.1) based on demonstrated process sequences (Table 18.1) to produce functional laboratory-scale modules of series-connected cells.

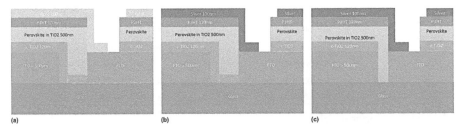

(a)   (b)   (c)

FIGURE 18.1   Solar cell structures analyzed by Chang et al. for sequences A, B, and C. *Source:* Reprinted with permission from ref 8. © 2017 John Wiley and Sons.

TABLE 18.1   Process sequences A, B, and C, and the two limit sequences L1 and L2 analyzed by Chang et al.

A	B	C	L1	L2	Process description	Equipment	Major materials
X	X	X	X		FTO glass receipt		FTO glass
X	X				FTO patterning.	Laser	
X	X	X	X		Glass cleaning	In-line glass Washer	acetone, ethanol, DI water
X	X				Sacrificial metal mask	Screen printer	Silver paste
X	X	X			TiO$_2$ spray pyrolysis	Spray pyrolysis tool	TAA, acetylacetone, ethanol
X	X				Chemical lift off	Chemical etch bath	HCI, DI water
			X	X	Pattern FTO and TiO$_2$	Laser	
X	X	X			Scaffold print	Screen printer	TiO$_2$ ink, terpineol, ethylcellulose
X	X	X			Perovskite Pbl$_2$ coat	Blade coater	Pbl$_{2P}$ DMF
X	X	X	X		Laser pattern Pbl$_2$	Laser	
X	X	X			Dip coat MAI	Dip coater	MAI, isopropanol

TABLE 18.1 *(Continued)*

A	B	C	L1	L2	Process description	Equipment	Major materials
X	X	X			Coat HTM	Blade coat	P3FIT, chloro-benzene, LFTFSI, tert-butylpyridine
X	X	X	X		Laser pattern HTM	Laser	
X					Evaporation of gold contact	Evaporator	Gold, masks
	X	X			Evaporation of silver contact	Evaporator	Silver, masks
X	X	X	X		Ribbon and encapsulate	Laminator and cure.	Ribbon, solder, glass, EVA, edge seal
X	X	X	X		J-box, test and package	Module tester	Junction box, packaging

HTM, hole-transport material.

Note: Sequence L1 is the limiting case when perovskite cell is free. Sequence L2 is a "free" module, and so there are no processes included in that sequence.

Source: Reprinted with permission from ref 8. © 2017 John Wiley and Sons.

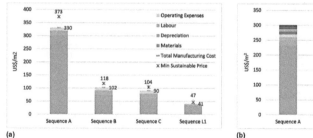
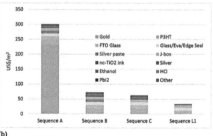

FIGURE 18.2 (a) Manufacturing cost breakdown and minimum sustainable price and (b) material cost breakdown for sequences A, B, C, and L1 in USD/m² determined by Chang et al. The values are calculated using nominal input values.

Source: Reprinted with permission from ref 8. © 2017 John Wiley and Sons.

The minimum sustainable price (MSP) and the manufacturing cost breakdown, material cost breakdown for all of the perovskite solar cell sequences analyzed by Chang et al is shown in Figure 18.2a, b. The MSP is the required selling price that provides a return to the manufacturer equal to the weighted average cost of capital by taking cash flow and net present value into consideration.[2] The MSP, which is an estimated cost to the customer, rather than the manufacturing cost, is used to calculate the LCOE by Chang et al. The values are calculated using nominal input values. Material costs contribute to the vast majority of the module costs. In particular, the cost of gold alone is larger than the cost and MSP of crystalline-Si modules (cost = USD 136/m², ³ or MSP=USD 112/m², ⁹ to USD 164/m², ³) or cadmium telluride modules (cost=USD 86/m² or MSP = USD 101/m², ¹⁰). Therefore, it is imperative that gold is eliminated in the commercial production of perovskite solar cells.

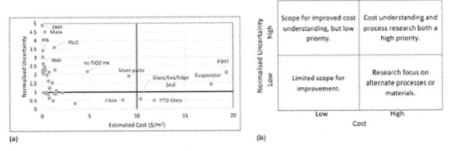

(a)   (b)

FIGURE 18.3   (a) The normalized uncertainties as a function of median cost for the main cost components of sequence B, divided into four regions by solid black lines. (b) The categorization of different research priorities for each region.
Source: Reprinted with permission from ref 8. © 2017 John Wiley and Sons.

The improvement in manufacturing cost for sequence B is apparent making it comparable to incumbent technologies. Further improvement is seen in sequence C that removes the sacrificial metal mask process for patterning the c-TiO₂ layer. The manufacturing cost of limiting sequence L1, where the cell is "free" highlights the significant costs of super-state glass, encapsulation, and junction box. This shows the importance of improving module efficiency. Not only does it drive down the cost in USD/W and LCOE, it also attracts a premium in module price when $/W is also a metric.

Figure 18.3a plots the "normalized uncertainty" versus median estimated cost for some of the cost components. The "normalized uncertainty" is the difference between 90th and 10th percentile cost divided by the median cost. The plot can be divided into four regions which are very useful for identifying high-cost components and cost components with great uncertainties, Figure 18.3b. This means that cost components in the bottom left region are of low priority as the materials or the processes are relatively cheap or their consumptions are insignificant. On the other hand, cost components in the bottom right region are the key cost drivers that should be eliminated or replaced in new cell designs or alternative processes identifying research priorities. In the case of perovskite solar cells, substrate cost and moduling cost dominate the cost structure, which is commonly found in other thin film technologies. Cost components on the top right region have great uncertainties and will be the focus of further cost analysis activities, including better sourcing of data and better understanding of the process. In the study by Chang et al. the whole-transport material P3HT is in this category. The advantage of Chang's method is the rapid identification of these cost components for addressing the key cost drivers and uncertainties. This in turn improves the cost analysis process and guides the research into low-cost alternatives for the early stage technology.

A LCOE by Chang et al. shows that for a perovskite module at a cost of around USD 120/m^2 to be competitive in 2015 with incumbent photovoltaic technologies at a LCOE of USD 0.19/kWh, the module power conversion efficiency at 18% and a lifetime of 20 years are required. Further analysis shows that to meet the SunShot LCOE target of USD 0.09/kWh in 2020, the manufacturing (including moduling and balance of system) cost needs to be reduced to around USD 50/m^2 if the same module power conversion efficiency (18%) and the same lifetime (20 years) are maintained.

Following this work, Chang et al. extended their analysis to roll-to-roll perovskite modules.[11] The major cost drivers identified are P3HT and PCBM, the use of evaporation for the rear metal deposition, and the transparent ITO coating. Improvements in these areas would lead to a manufacturing cost at USD37/m^2 and the technology would be competitive with existing flexible photovoltaic modules on a $/W and power to weight basis if a geometric fill factor of 68 is assumed and the module is 10% efficient with a 3 year lifetime. For the R2R perovskite module to be competitive with Si and CdTe flat plate technologies, a power conversion efficiency of 15% with a 15 year lifetime would be required.

In conclusion, new innovations are required to overcome commercial barriers such as (i) small device size, which can be addressed relatively easily as scalable processes are constantly being developed and (ii) the lack of longevity with the current devices. The latter is more challenging. The presence of lead albeit small but in readily soluble form can still makes it difficult for this new technology to enter into the market. Nevertheless, the outlook of perovskite solar cells remains positive. The versatility of the material, the ease of fabrication, the photovoltaic enabling attributes of the materials make it very easy for researchers, even those new to the field, to establish a baseline process for reasonable efficient perovskite solar cell. This is the main reason for the phenomenal growth in research activities in this field. Not only can researchers experiment with new ideas, they can experiment with old ideas, too that may have been previously too hard to execute. Building on the rapid progress in the last few years, and together with the growing number of research activities, pathways to overcome these challenges and barriers will be discovered.

KEYWORDS

- **perovskite techno economical analysis**
- **perovskite cost analysis**
- **levelized cost of electricity**
- **minimum sustainable price**
- **cost of ownership**

REFERENCES

1. Kelly, W. W. Factory Commander. http://www.wwk.com/products.html (accessed: June 3, 2016).
2. Powell, D. M.; Winkler, M. T.; Goodrich, A.; Buonassisi, T. Modeling the Cost and Minimum Sustainable Price of Crystalline Silicon Photovoltaic Manufacturing in the United States. *IEEE J. Photovoltaics* **2013**, *3*(2), 662–668.
3. Powell, D. M. c-Si PV Cost Model. https://pv.scripts.mit.edu/wp-content/uploads/2013/10/c-Si-Solar-Cost-Modelv20.xlsx (accessed: June 3, 2016).
4. Asif, A. A.; Singh, R.; Alapatt, G. F. Technical and Economic Assessment of Perovskite Solar Cells for Large Scale Manufacturing. *J. Renewable Sustainable Energy* **2015**, *7*(4), 043120.
5. Cai, M.; Wu, Y.; Chen, H.; Yang, X.; Qiang, Y.; Han, L. Cost-Performance Analysis of Perovskite Solar Modules. *Adv. Sci.* **2016**, *4*, 1600269.
6. Chen, J.; Rong, Y.; Mei, A.; Xiong, Y.; Liu, T.; Sheng, Y.; Jiang, P.; Hong, L.; Guan, Y.; Zhu, X.; Hou, X.; Duan, M.; Zhao, J.; Li, X.; Han, H. Hole-Conductor-Free Fully Printable Mesoscopic Solar Cell with Mixed-Anion Perovskite $CH_3NH_3PbI_{(3-x)}(BF_4)_x$. *Adv. Energy Mater.* **2015**, *6*, 1502009.
7. Wu, C. G.; Chiang, C. H.; Tseng, Z. L.; Nazeeruddin, M. K.; Hagfeldt, A.; Grätzel, M. High Efficiency Stable Inverted Perovskite Solar Cells Without Current Hysteresis. *Energy Environ. Sci.* **2015**, *8*(9), 2725–2733.
8. Chang, N. L.; Ho-Baillie, A. W. Y.; Basore, P. A.; Young, Evans, R.; Egan, R. J. A Manufacturing Cost Estimation Method with Uncertainty Analysis and its Application

to Perovskite on Glass Photovoltaic Modules. *Prog. Photovoltaics Res. Appl.* **2017,** *25,* 390–405. DOI: 10.1002/pip.2871.

9. Jones-Albertus, R.; Feldman, D.; Fu, R.; Horowitz, K.; Woodhouse, M. Technology Advances Needed for Photovoltaics to Achieve Widespread Grid Price Parity. *Prog. Photovoltaics Res. Appl.* **2016,** *24,* 1272–1283.

10. Redlinger, M.; Lokanc, M.; Eggert, R.; Woodhouse, M.; Goodrich, A. The Present, Mid-Term, and Long-Term Supply Curves for Tellurium; and Updates in the Results from NREL's CdTe PV Module Manufacturing Cost Model (Presentation), 2008, http://www.nrel.gov/docs/fy16osti/64507.pdf (accessed:April 15, 2016).

11. Chang, N. L.; Ho-Baillie, A. W. Y.; Vak, D.; Gao, M.; Green, M. A.; Egan, R. J. Manufacturing Cost and Market Potential Analysis of Demonstrated Roll-to- Roll Perovskite Photovoltaic Cell Processes. *Sol. Energy Mater. Sol. Cells* **2018,** *174,* 314–324.

INDEX